Tourism and Gentrification in Contemporary Metropolises

Tourism gentrification is a critical shaping force of socio-economic and contemporary urban landscapes. This book aims to be the first substantive text on this subject, explaining the multiple and complex relationships between tourism and gentrification and their outcomes and manifestations in contemporary metropolises. This is achieved by drawing on in-depth case analyses addressing the different issues at stake.

Part I deals with the manifestations of tourism gentrification and the ways it affects urban landscapes through heritagization and urban regeneration strategies. Part II looks at the correlations between tourism gentrification and culture. Finally, the last two parts aim to identify and examine forms and expressions of tourism gentrification, distinguishing among the actors, beneficiaries, and victims of the phenomenon while looking at its implications for intra-metropolitan territories and metropolitan governance.

The book approaches these issues in an innovative way, by looking at a variety of metropolises in a diverse range of countries and by dealing with the different relations and management issues generated by gentrification in relation to tourism. Through interdisciplinary approaches, this groundbreaking text sheds light on the role tourism plays in contemporary metropolises, furthering knowledge of urban tourism. For these reasons, it will be of particular interest to scholars and students of tourism, urban studies, geography, anthropology and sociology.

Maria Gravari-Barbas is an architect and geographer. She is Professor and Director of the Institute for Research and High Studies of Tourism (IREST) and a researcher at the Interdisciplinary Team for Tourism Research (EIREST), Paris 1 Panthéon-Sorbonne University. Her research focuses on the contemporary production of urban spaces in relation to tourism, architecture, heritage and festive phenomena.

Sandra Guinand is an urban-planner and geographer. She is a FNS Fellow, visiting researcher at the Institut für Geographie und Regionalforschung, Vienna University, associate researcher at EIREST, Paris 1 Panthéon-Sorbonne University, and Centre Jacques Berque. Her research interests focus on urban regeneration projects, socio-economical transformations of urban landscape, with a specific focus on heritage processes and public–private partnerships.

Contemporary Geographies of Leisure, Tourism and Mobility

Series Editor: C. Michael Hall, Professor at the Department of Management, College of Business and Economics, University of Canterbury, Christchurch, New Zealand

The aim of this series is to explore and communicate the intersections and relationships between leisure, tourism and human mobility within the social sciences.

It will incorporate both traditional and new perspectives on leisure and tourism from contemporary geography, e.g. notions of identity, representation and culture, while also providing for perspectives from cognate areas such as anthropology, cultural studies, gastronomy and food studies, marketing, policy studies and political economy, regional and urban planning, and sociology, within the development of an integrated field of leisure and tourism studies.

Also, increasingly, tourism and leisure are regarded as steps in a continuum of human mobility. Inclusion of mobility in the series offers the prospect to examine the relationship between tourism and migration, the sojourner, educational travel, and second home and retirement travel phenomena.

For a full list of titles in this series, please visit www.routledge.com/series/SE0522
The series comprises two strands:

Contemporary Geographies of Leisure, Tourism and Mobility aims to address the needs of students and academics, and the titles will be published in hardback and paperback. Titles include:

10. **Tourism and Climate Change**
    Impacts, adaptation and
    mitigation
    *C. Michael Hall, Stefan Gössling
    and Daniel Scott*

11. **Tourism and Citizenship**
    *Raoul V. Bianchi and Marcus L.
    Stephenson*

Routledge Studies in Contemporary Geographies of Leisure, Tourism and Mobility is a forum for innovative new research intended for research students and academics, and the titles will be available in hardback only. Titles include:

67. **Co-Creation in Tourist
    Experiences**
    *Edited by Nina K. Prebensen,
    Joseph S. Chen and Muzaffer
    Uysal*

68. **International Tourism
    Development and the Gulf
    Cooperation Council States:
    Challenges and Opportunities**
    *Edited by Marcus L. Stephenson
    and Ala Al-Hamarneh*

Tourism and Gentrification in Contemporary Metropolises

International Perspectives

**Edited by Maria Gravari-Barbas
and Sandra Guinand**

LONDON AND NEW YORK

First published 2017
by Routledge

2 Park Square, Milton Park, Abingdon, Oxfordshire OX14 4RN
52 Vanderbilt Avenue, New York, NY 10017

Routledge is an imprint of the Taylor & Francis Group, an informa business

First issued in paperback 2019

British Library Cataloguing-in-Publication Data
A catalogue record for this book is available from the British Library

Library of Congress Cataloguing-in-Publication Data
A catalog record has been requested for this book

ISBN: 978-1-138-64278-2 (hbk)
ISBN: 978-0-367-36898-2 (pbk)

Typeset in Times New Roman
by Swales & Willis Ltd, Exeter, Devon, UK

Contents

Lists of figures   vii
List of tables   x
List of contributors   xii

1 Introduction: addressing tourism-gentrification processes
  in contemporary metropolises   1
  MARIA GRAVARI-BARBAS AND SANDRA GUINAND

PART I
Tourism gentrification, heritagisation and urban regeneration   23

2 A new colonisation of a Caribbean city: urban regeneration
  policies as a strategy for tourism development and
  gentrification in Santo Domingo's Colonial City   25
  JESÚS M. GONZÁLEZ-PÉREZ

3 Airbnb and tourism gentrification: critical insights from
  the exploratory analysis of the 'Airbnb syndrome'
  in Reykjavík   52
  ANNE-CÉCILE MERMET

4 Tourism gentrification in the cities of Latin America: the
  socio-economic trajectory of Cartagena de Indias, Colombia   75
  SAIRI T. PIÑEROS

PART II
Tourism gentrification and the urban cultural
and creative turn   105

5 The Barrio Chino as last frontier: the penetration of
  everyday tourism in the dodgy heart of the Raval   107
  ALAN QUAGLIERI DOMÍNGUEZ AND ALESSANDRO SCARNATO

6 Rome: a cultural capital with a poor working-class heritage: strategies of touristification and artification 134

SARAH LILIA BAUDRY

7 Grunge authenticity: the tenement as upscale tourist destination 153

ELISSA SAMPSON

PART III
Who are the tourism gentrifiers? 179

8 The sharing economy and its role in metropolitan tourism 181

NATALIE STORS AND ANDREAS KAGERMEIER

9 Post-tourism on the waterfront: bringing back locals and residents at the Seaport 207

SANDRA GUINAND

10 Playing for/with time: tourism and heritage in Greece and Thailand 233

MICHAEL HERZFELD

PART IV
Forms and expressions of tourism gentrification: a critical analysis 253

11 Tourism and urban changes: lessons from Lisbon 255

TERESA BARATA-SALGUEIRO, LUIS MENDES AND PEDRO GUIMARÃES

12 The rent gap re-examined: tourism gentrification in the context of rapid urbanisation in China 276

ZENG-XIAN LIANG

13 Super-gentrification and hyper-tourismification in Le Marais, Paris 299

MARIA GRAVARI-BARBAS

Index 329

Figures

1.1   Talinn, Rotermanni Kvartal   2
1.2   Helsinki beach   6
1.3   New York High Line   9
1.4   Mercat de Santa Catarina, Barcelona   14
2.1   Gran Santo Domingo: Ozama region   28
2.2   Location of tourist establishments in 2014 within the current
      Colonial City defined by the MPCC   38
2.3   Location of Airbnb's holiday rentals (15 October 2015) within
      the social and urban structure of the Colonial City as
      defined by the MPCC   39
2.4   Summary of the strategic plan proposal for the Colonial City   41
2.5   The risks of tourism 'dumbing down' in the Colonial City   42
2.6   The location of low-income households and residents living
      in rented properties on the social and urban structural map
      of the Colonial City   44
2.7   Location of empty homes (over 20 per cent of homes on
      a block) on the social and urban structural map of
      the Colonial City   45
2.8   Construction of gentrified spaces: property revaluation
      and new residential condominiums   46
2.9   Ground level land use on El Conde and Las Damas
      streets in 2014   47
3.1   Evolution of the number of foreign visitors in Iceland,
      1949–2014   58
3.2   The evolution of real estate values in the 'Reykjavík West'
      districts, 1990–2014   59
3.3   The evolution of real estate values (kr/m^2) between 2009
      and 2014 and the number of Airbnb listings   61
3.4   The three main features of the short-term rental market
      in Reykjavík   62

viii *Figures*

3.5   Location of Airbnb listings in the Reykjavík urban area   63
3.6   Location of hotels in the Reykjavík urban area   64
3.7   Comparison between the actual monthly average values for
      long-term rentals and the theoretical monthly average values
      for short-term rentals   66
4.1   Cartagena UNESCO World Heritage Site   77
4.2   Accommodation growth statistics, 1990–2014   87
4.3   Restaurant growth statistics, 1990–2014   91
4.4   Superposition of land use data and the location of accommodation
      and tourist services   94
4.5   Local restaurant, 2012   96
4.6   International cafe, 2014   97
5.1   The system of cultural facilities of Raval   114
5.2   The impressive Filmoteca de Catalunya building and a popular
      residential block in the Illa Robador Area   116
5.3   Flea market in South Raval   118
5.4   South Raval resident population by nationality, comparison
      between 1996 and 2005   120
5.5   Experiencing 'cool Barcelona'. Tourist flat in the South Raval   124
5.6   Spatial correlation between the number of Airbnb listings
      and foreign residents from EU-15 countries   125
6.1   Photograph of MURo Museum   142
6.2   Photograph of MAAM Museum   143
6.3   Overview of street-art urban spots in Rome   145
7.1   W 99 Orchard Tenement Museum laundry   160
7.2   Two Bridges Neighborhood Map   165
7.3   LMDC map   167
7.4   103 Orchard   171
7.5   2014 media alert   172
8.1   Airbnb in the urban neighbourhoods of Berlin   186
8.2   Development of the indexed accommodation capacity in Berlin and
      arrivals and overnight stays of tourists between 1992 and 2014   187
8.3   Personality profile of respondents by level of participation   192
8.4   Motivations of sharing-economy accommodation
      users – differentiated by Airbnb and Couchsurfing users   193
8.5   Interaction between hosts and guests   196
9.1   Location of South Street Seaport in Lower Manhattan   211
9.2   A view of Harbor Place, with the WTC and the National Aquarium   214
9.3   View of the Seaport with the Museum Gallery from Water Street   215
9.4   South Street Seaport Marketplace's signage and new
      Fulton market building   216
9.5   Concerts in front of the Marketplace building on Pier 17   217

9.6 Seaport new branding 220
9.7 Column presenting perspectives of the envisioned project 221
9.8 South Street Seaport becomes Seaport Culture District
for a season 223
9.9 Dining, entertainment, fashion, culture . . . 225
9.10 Pictures of Hudson Valley farmers against the future
building for the Fulton Stall Market 226
11.1 Demographic dynamics in central Lisbon 261
11.2 Number of hotels in Lisbon, between 1989 and 2015 264
11.3 Short-rental apartments in Lisbon 265
11.4 *Baixa-Chiado*: its functional retail structure and the
changes witnessed between 2007 and 2015 268
12.1 Map of Shenzhen OCT 282
12.2 Current view of Guanghua Street and Guangqiao Street 283
12.3 Current view of Huiwen and Lihai Garden and Guihua Garden 285
12.4 Current view of Water Bank unit of Portofino Garden
(since 1998) 287
12.5 The evolution of the rent gap in China 294
13.1a Paris arrondissements and the Le Marais area 301
13.1b The Le Marais area 301
13.2 The Smokestack of the *Société des Cendres*, one of the
last vestiges of Le Marais' industrial past 304
13.3 Median prices per square metre (apartments) in the 3rd and 4th
arrondissements (Le Marais) and evolution, 2008–2013 309
13.4 A clothing boutique installed in a former bakery with
protected facades 312
13.5 The Saint-Paul Hammam, currently occupied by COS 313
13.6 The *Société des Cendres*, currently occupied by Uniqlo 314
13.7 A Sunday in rue Francs Bourgeois, Le Marais 316
13.8 Urban landscape layers in Le Marais 322

Tables

2.1	World Heritage Cities in the Caribbean region in 2015	26
2.2	The main urban regeneration plans in the Colonial City of Santo Domingo carried out within the framework of development cooperation	34
2.3	Priority actions set out in the strategic and regulatory plans	35
2.4	Indicators according to dominant use at street block level	43
3.1	Statistical insight of the Airbnb database	57
3.2	An assessment of multi-unit hosts in Reykjavík	65
4.1	Buildings restored and their use, 1960–1983	82
4.2	Store types by neighbourhood, 1990–2014	89
4.3	Land use by neighbourhood	93
5.1	Evolution of resident population by nationality	121
11.1	Retail change between 2007 and 2015 in downtown Lisbon	267
12.1	Temporal analysis of tourism gentrification in the Shenzhen OCT community	281
12.2	Temporal analysis of tourism gentrification in the Beijing OCT community	288
12.3	The evolution of the site of the OCT community in 1992 and 1999	289
12.4	The status of Beijing OCT community in 2015	292
13.1	Number of residents in Le Marais	305
13.2	Comparison of the population evolutions in Le Marais, 1st and 2nd arrondissements of Paris and the rest of the Paris intra-muros districts	306
13.3	Le Marais' attractiveness. Quotations from interviews with estate agencies, 2010	307
13.4	Second homes in Le Marais. Excerpts from interviews with local associations, 2010	308
13.5	Prices per square metre for apartments sold in Le Marais during the period August 2015 to July 2016	309

13.6	Median prices of apartments in Paris per square metre	310
13.7	The commercial structure of Le Marais	315
13.8	Number of visitors at the most important attractions of Le Marais or in the Marais vicinity	317
13.9	Different representations of Le Marais in the Paris guides	318
13.10	Number of comments about Le Marais and Le Marais attractions on the TripAdvisor website	319
13.11	Selected comments from TripAdvisor	319

Contributors

Sarah Lilia Baudry holds a Master's degree in Urban Planning (Panthéon-Sorbonne University) and in History (Ecole des Hautes Etudes en Sciences Sociales). She is currently doing her PhD at the University Paris-Diderot, UMR Geographie Cités Institute under the supervision of Dominique Rivière. She focuses on public spaces in Rome and related urban policies and urban practices.

Teresa Barata-Salgueiro studied at the University of Chicago and holds a PhD in Geography from Lisbon University. She is a Professor of Geography at the Centro de Estudos Geograficos, Lisbon University. Her research interests are urban spaces, real estate activities and social exclusion. She has particularly studied the cities of Lisbon and Porto.

Jesús M. González is Associate Professor at the University of the Balearic Islands (Spain). He has been a visiting researcher at twelve European and American universities. His main research focuses on the urbanization processes and tourism in Caribbean countries. Dr González is currently Chairman of the Urban Geography Group of the Association of Spanish Geographers.

Maria Gravari-Barbas is an architect and geographer. She is Professor and Director of the Institute for Research and High Studies of Tourism (IREST) and a researcher at the Interdisciplinary Team for Tourism Research (EIREST), Paris 1 Panthéon-Sorbonne University. Her research focuses on the contemporary production of urban spaces in relation to tourism, architecture, heritage and festive phenomena.

Pedro Guimarães holds a PhD in Geography. He is currently post-doctoral researcher at IGOT, Lisbon University. His research interests are related to tourism gentrification and urban regeneration issues.

Sandra Guinand is an urban-planner and geographer. She is a FNS Fellow, visiting researcher at the Institut für Geographie und Regionalforschung, Vienna University, associate researcher at EIREST, Paris 1 Panthéon-Sorbonne University, and Centre Jacques Berque. Her research interests focus on urban regeneration projects, socio-economical transformations of urban landscape, with a specific focus on heritage processes and public–private partnerships.

Michael Herzfeld is Ernest E. Monrad Professor of the Social Sciences in the Department of Anthropology at Harvard University. He is the author of numerous volumes and a member of the editorial boards of several journals. His research in Greece, Italy, and Thailand has addressed, inter alia, the social and political impact of historic conservation and gentrification, the discourses and practices of crypto-colonialism, social poetics, the dynamics of nationalism and bureaucracy, and the ethnography of knowledge among artisans and intellectuals.

Andreas Kagermeier is Professor of Leisure and Tourism Geography at Trier University (Germany). His main research interests are destination management and new trends in tourism product development, urban tourism and leisure mobility. His recent activities focus on the role of experience orientation in tourism product development as well as methodological aspects of measuring visitor perception in leisure and tourism.

Luis Mendes is a geographer. He is Guest Lecturer at the School of Education of Lisbon and a PhD Candidate in Geography at IGOT, Lisbon University. His research focuses on gentrification and urban regeneration issues.

Anne-Cécile Mermet received a PhD in urban geography from the Université Paris 1 Panthéon-Sorbonne in 2012. She is currently post-doctoral researcher at the Geography Institute, Neuchâtel University, Switzerland. Her research focuses on the link between heritage, retail and gentrification in city centres.

Sairi Tatiana Piñeros is a PhD student in Geography at IREST-EIREST, Paris 1 Panthéon-Sorbonne University. Her research interests include tourism practices, geographical and tourism imaginaries, perception of space, urban changes and digital footprints.

Alan Quaglieri Domínguez is a PhD candidate and associate researcher at GRATET, Research Group on Territorial Analysis and Tourism Studies at Universitat Rovira I Virgili, Barcelona. His research focuses on post-tourism geography, the creative city and global cultural consumption.

Elissa Sampson is an urban geographer. She is a Visiting Scholar and Lecturer at Cornell University. Her research interests focus on how the past is actively used to create new geographies of migration, memory and heritage.

Alessandro Scarnato holds a PhD from the Universitat Politècnica de Catalunya. He is currently post-doctoral researcher at UPC – Escola d'Arquitectura de Barcelona. His research interests include the production and aestheticization of landscape and tourism related issues.

Natalie Stors is Junior Lecturer and PhD candidate at the Chair of Leisure and Tourism Geography at Trier University (Germany). In her Master thesis she analysed the development and implication of the touristification process, taking Copenhagen as an example. Her PhD thesis deals with various aspects of the sharing economy in tourism, including a case study of Berlin.

Zeng-Xian Liang is Association Professor at the School of Tourism Management at Sun Yat-Sen University, Guangzhou, China. He is particularly interested in the interaction between urban tourism and social space reconstruction in modern China. His research focuses on theme parks, urban tourism and tourism related real estates.

1 Introduction

Addressing tourism-gentrification processes in contemporary metropolises

Maria Gravari-Barbas and
Sandra Guinand

Contemporary metropolises – socio-economic and morphological changes in the urban landscape

Since the 1980s, cities have been experiencing tremendous change. As the world's population continues to grow more urban (54 per cent of the world's population is currently living in urban areas) (Nations-Unies, 2014) urban territories have been expanding, leading to the metropolization phenomenon, "human and material wealth concentration phenomenon in the biggest cities"[1] (Ascher, 1995), characterized by new centralities as well as urban sprawl (Halbert, 2010). Meanwhile, the flow of globalization and the structural shift in the economy and new technologies have contributed to accelerate the transformation of cities (Friedmann, 1986). These moves have left behind vacant industrial sites, while contributing to inner-city reinvestments by bringing back new residents, commercial and socio-cultural activities (Laska *et al.*, 1982). But it has also changed the way people move around, look at and interact with each other and the world.

A wide range of literature has been analysing the changes that affect contemporary metropolises through the concept of gentrification. When Ruth Glass first coined the term in 1964, she most likely did not anticipate the numerous variations the concept would hold less than forty years later. Associated with the changes in the conduct of urban affairs and the simultaneous modification of the urban landscape, gentrification has been extensively studied, mostly in connection with urban regeneration policies (Lees *et al.*, 2008; Smith, 2006). First, as urban regeneration was implemented as a means to improve inner cities through housing policies (Cameron, 1992; Couch, 2003), voices were quick to denounce its failure to primarily allocate public resources to the affluent instead of those most in need. Since then, and exacerbated by the increasing competition for resource allocation between territories, urban regeneration has become a global strategy pursued more or less aggressively by contemporary metropolises (Smith, 2002). The neoliberal turn has indeed opened new territories to capital accumulation and real estate speculation, pushing further the new urban frontier (Smith, 1996) but has also considerably altered actors' governance around urban agendas (Mosedale, 2014). With failing urban economies and shrinking public budgets (at all levels), public authorities have been more and more reliant on private funders, diversifying their economies and revenues.

As such, capitalizing on land by recapturing former industrial or vacant sites and transforming them into mixed-used neighbourhoods with luxurious condominiums targeting the upper social class has been a trend developed from London to Beijing, especially on waterfronts (Figure 1.1). This dramatic transformation of former blue-collar landscapes into high-end facilities has been investigated and designated by scholars as "new build gentrification" (Davidson and Lees, 2005, 2010; He, 2010; Rérat *et al.*, 2010). Yet, these gentrification patterns are also the result of structural change in mobility and residential preferences, which have mainly been looked at under the "back to the city" movement (Bidou-Zachariassen, 2003) and the geography of "induced-displacement" (Van Criekingen, 2008).

Although academic literature analyses the extent of gentrification in tourist cities (Bures and Cain, 2008) little attention has been paid to tourism and tourist mobility as main factors of gentrification in metropolitan areas (Gotham, 2005). Tourism gentrification is difficult to grasp, as it is affected by the changing patterns of tourism flow; it is however a critical shaping force of socio-economic and contemporary urban landscapes.

Figure 1.1 Talinn, Rotermanni Kvartal

Source: Sandra Guinand, 2014

This edited volume brings together different perspectives and geographies on the tourism gentrification phenomenon. It intends to explain the subtle balance that exists between the multiple and complex relationships of tourism and gentrification, and their outcomes and manifestations in contemporary metropolises. The contributions by the various authors in this book draw on in-depth case analyses addressing the different issues at stake raised by tourism gentrification. They examine the changes induced in the urban landscape and the impacts for urban design, planning and tourism. As tourism gentrification is a phenomenon imbedded in the broader context of globalization and forms of (neo)liberalism (Harvey, 2007), the book also sheds light on the new actors, prescribers, beneficiaries or victims of the phenomenon. It shows how it can be implemented or resisted as well as how local actors cope or surrender. By bringing light on the features of the phenomenon, the constellations of actors behind it, its governance, its outcomes on urban space and consequences for metropolitan territories, this book aims at bringing a solid basis for further, more specific, research on tourism gentrification. This introductory chapter will position tourism as a gentrification process and/or result thereof, and drawing on examples from the contributions, it will identify its key components and attempt to replace it in the larger debate on gentrification.

Tourism as a gentrifying process

When did tourism become a gentrifying process? This question may be difficult to answer as tourism has from its very beginning had an impact on the places being visited, by the implementation of new infrastructures or services offered, but also by the new and fresh perspectives the tourist "gaze" (Urry, 2002) has brought on places and landscapes.

Tourism, urban development and gentrification have always been strongly imbedded. Kevin Gotham was one of the first researchers to formulate the expression in his article on the Vieux Carré French Quarter of New Orleans (Gotham, 2005). For the author, tourism gentrification can be defined as "the transformation of a middle-class neighbourhood into a relatively affluent and exclusive enclave marked by a proliferation of corporate entertainment and tourism venues" (ibid., 1102). Gotham argues that the concept brings out the dual processes of globalization and localization imbedded in urban redevelopment, since tourism is characterized by international global actors (hotel chains, car rentals, tour operators, etc.), linked to the service industry (communications, finance, etc.) while at the same time investing at the local level by developing local culture, products and places for consumption that will appeal to visitors. For Gotham, this nexus between the global and the local in tourism (Milne and Ateljevic, 2001) cannot be disconnected. However, if the author insists on the forces of corporate tourism industries influencing space development and consumption, we will argue in this book that local actors, local inhabitants, as well as the tourists themselves, also contribute to the gentrification phenomena. Furthermore, as Herzfeld demonstrates in this volume (Ch.10), local actors do act, cope and structure their environment in order to take advantage of the tourism economy. If "tourism gentrification provides the conceptual link between production-side and demand-side explanations of gentrification while avoiding one-sided conceptions" (Gotham, 2005: 1103), this book shows that actors are intermingled in this process.

The relationship between tourism and gentrification is indeed complex and diverse. Tourists are attracted by gentrified and gentrifying neighbourhoods. The SoHo cast-iron buildings, inhabited since the 60s by artists and urban bohemians (Zukin, 1989), became, concomitantly with their gentrification, a prime urban destination for strolling, shopping or gourmet-fooding. Tourism often follows urban gentrifiers (Bridge, 2007; Schlichtman and Patch, 2014) and invests, in return on the 'rediscovered, 'trendy' urban areas that are experiencing the gentrification phenomena. This can also be understood by both the physical and the symbolic changes that these areas experienced after their gentrification: heritagization, improvement of the urban infrastructure and public spaces, with simultaneous creation of shops that cater to new residents and are also very attractive to tourists (farmers' markets and gourmet shops, designer retail shops, specialized bookstores, art galleries, etc.). These phenomena were, for instance, observed in the Le Marais area of Paris, where tourism developed in this neighbourhood after its heritagization and the general up-grading of its image and accessibility (Maria Gravari-Barbas, this volume, Ch. 13).

In some other cases however, tourism comes first. Some projects are planned and designed from their inception to cater to the visitors' economy. It is during a second stage that these touristified urban areas are occupied by new individual owners, renters and consumers as well as other institutional and collective social actors (real estate agents, developers, mortgage lenders, etc.) (Hamnett, 1991) who are attracted by the services and the atmosphere created and promoted in order to attract local, regional or international visitors. This can be the case with derelict but centrally located places (former industrial, port or warehouse areas) that have experienced capital and human disinvestment since the second part of the twentieth century. These areas, often transformed into "tourism playgrounds" (Judd, 2003), can attract (sometimes with the help of public intervention) residential gentrification. This was, for instance, the case in Baltimore, where the private/public-led Inner Harbor redevelopment generated new-build gentrification phenomena in the adjacent areas.

Beyond this dual tourism and residential gentrification relationship, tourism also induces commercial gentrification. Tourism investment of a neighbourhood often contributes to a change in commerce and businesses' landscape, which obviously goes far beyond the stereotypical image of the souvenir postcard, T-shirt shops and guided tours. The introduction of tourism in an urban area indeed implies the bringing of bigger, international and more sophisticated markets, which will also impact on the quality of restaurants, bars and shops as they are replaced by chains or trendy and upscale establishments.

Finally, tourism gentrification can also be seen as an internal phenomenon: the continuous upgrading of former middle-class, popular or marginal tourism areas into exclusive and high-end tourism ventures (Guinand, this volume, Ch. 9). For instance, Chapuis *et al.* (2015) analysed the upgrading transformation of the

Red-Light district (De Wallen) in Amsterdam. This area, an identified tourism destination related to 'window brothels' experienced a tourism gentrification through a renovation plan aiming at changing the image of tourism and its practices (from mass tourism to qualitative tourism).

Global tourists as international gentrifying elites

Research confirms that boundaries between tourists and locals have become increasingly blurred (Bock, 2015; Condevaux *et al.*, 2016) due, among other reasons, to the fact that tourists are increasingly blending in with the local population. International tourists are seeking authentic and local experiences, "exploring ordinary but lively and diverse neighborhoods and visiting cafes, bars and markets that were previously almost exclusively frequented by locals" (Bock, 2015). This quest for 'authenticity' of local life, as opposed to the (past) tourist hot-spots and designated attractions, combined with the ability of transnational elites to (technically, socially and economically) live their life in selected places around the world and an "elective affinity" between tourists and the upper-class (Chapuis *et al.*, 2015), have considerably impacted not only on the city centres but also on the peripheries. This phenomenon is also connected with the tourism experiential turn (Gravari-Barbas, 2015) made possible by the incredible advancements of technology and the rise of social media and interfaces. Tourist behaviours have considerably changed during recent years. The off-the-beaten-path tourism preferences (Gravari-Barbas and Delaplace, 2015) as well as the desire to live 'like a local', bring tourism to former working and more remote neighbourhoods which in turn become desirable and touristically marketed. These dynamics change the image, the patterns and the social composition in contemporary metropolises. They contribute in new and different ways to the gentrification phenomenon.

Contemporary mobilities, 'poly-topical' living and gentrification

The growing trend of second-home ownership (or multiple-home ownership), for leisure and leisure-related investment purposes, has seen the rise of a transnational class (Sklair, 2000) able to be 'at home' and to live 'like a local' in different contexts around the world.

Stock (2012) discusses the "poly-topical" living phenomenon, in which "different places can be familiar and identity-related, and not only a place of residence".[2] Facilitated by contemporary means and communication/technology, information and transportation, "geographical individualization" (Stock, 2012) occurs where the choice of possible places to visit and live becomes higher; people become more distant from their place of residence; and individual spaces for living ("individual spatial trajectories") are more autonomous from one another (ibid). People become "geographically plural": they are simultaneously involved and "invested" in different places. According to Stock this means that (a) individuals become temporary residents of one or many places; (b) individuals have the ability to transform foreign places into familiar ones; (c) depending on intentionalities, one place could be the referent of multiple significations' constructions; (d) individuals are able to construct different geographical identity referents; (e) individuals are able to free themselves from local conditions, which goes hand in hand with the distancing from the place of residence: this "detachment" from local conditions also means to be able to manage and deal with other spatial scales; to inhabit a place at the local level, but also at regional and national ones, etc. (ibid.) (Figure 1.2).

Researchers have suggested a relation between the transnational elite and the gentrifying class. Seminal papers (Friedmann and Wolff, 1982; Ley, 1980: 243) mainly identify professional transnational mobility (Sassen, 1991) due to the shift of productive capacities from an industrial to a post-industrial focus, in global cities. Recent works, however, have emphasized personal experiences and desires as the drivers of decision-making. Rofe (2003) presents an "esoteric" theoretical path taken by King (1993, in Rofe, 2003) for examining the transnational elite and their lifestyles. Drawing on King's work, Rofe introduces the "urban myth" of transnational gentrifying elites who *genuinely* believe they live in a world culture" (1993: 152, in Rofe, 2003). As Rofe suggests, according to this perspective, the "intersection between landscapes's gentrification and the transnational elites becomes explicit" (Rofe, 2003: 2521) as the built and

Figure 1.2 Helsinki beach
Source: Sandra Guinand, 2014

social environment (the "consumptionscapes") come to be the template for the gentrifying elites' identity. The factors behind transnational gentrifying elite are obviously not intrinsically touristic; they are more lifestyle choices-oriented. We can however stress the proximity and even the permeability between lifestyle and tourism choices (the elective affinity), as these trends are expressed in selected cities around the world (Gravari-Barbas, this volume, Ch. 13).

Airbnb and the metropolises – towards new phenomena of tourism gentrification

The so-called "explorer-tourists" searching for contact and interaction with the local population (Stors and Kagermeier, this volume, Ch. 8) are particularly attracted by the possibilities offered by social and business operators to "share" the living quarters of the "locals". This is probably the most powerful myth of contemporary tourism, the total "tourism *graal*": it even goes beyond MacCannell's backstage (1976) to transpose roles as one becomes, for one or two days, a local. The locality then becomes the ultimate value for the global visitors.

The Airbnb phenomenon needs to be understood in this sense: far from being part of the sharing economy (Hamari *et al.*, 2016) it exemplifies the desire of tourists and more generally of transnational classes to go beyond traditional 'commercial' accommodations.

This aspiration is not without consequences as Airbnb has become a 'gentrifiying machine' in most major urban and metropolitan destinations as growing concerns have showed in San Francisco, New York and Lisbon. The conversion of housing into accommodation for visitors results in a process of major social changes as has been observed in Berlin (Novy, 2017) or Barcelona (Lopez-Gay and Cocola Gant, 2016; Russo and Arias-Sans, 2017; Quaglieri and Scarnato, this volume, Ch. 5). Researchers have indicated that apartment conversions into Airbnbs cause an out-migration of residents, a shortage in housing, and price increases, which also exclude other residents from the possibility of moving into the area (Cocola Gant, 2016). Airbnb gentrification can even be related to the "super-gentrification" phenomenon (Lees 2003) as it occurred in the Marais (Gravari-Barbas, this volume, Ch. 13). As Cocola Gant (2016:7) has shown, the Airbnb phenomenon corresponds to "a snowball process [. . .]. It leads to a form of collective displacement never seen in classical gentrification, that is to say, to a substitution of residential life by tourism".

Tourism as an upscale culture consumption

Since its introduction by Veblen (1953), the concept of "conspicuous consumption" has been used by tourism research in order to describe and analyse contemporary tourism practices. Philips and Back (2011) define four expressions of conspicuous consumption in destinations: interpersonal mediation, status demonstration, materialistic hedonism, and communication of belonging. These expressions can interestingly be interpreted in relation with gentrification. Indeed, tourist desire to be part of the local and to share everyday experiences (sleep in a 'real' apartment, shop at the local market, visit 'ordinary' places and go off-the-beaten-track) is not incompatible with the desire to visit the most exclusive cultural places, the state-of-the-art temporary exhibitions or the "starchitectural" museums (Gravari-Barbas and Renard-Delautre, 2015). Tourism gentrification is, therefore, not only caused by the phenomenon of a tourism 'invasion' in former non-touristic neighbourhoods, but also by the global upscaling of metropolitan destinations through exclusive and sophisticated places and services (gourmet shops, Michelin-star restaurants, exclusive designer boutiques, state-of-the-art galleries, high-end museums, etc.). These (in large part newly created) places, which cater, if not exclusively, then primarily to the transnational visiting class, also contribute to dramatically change social and urban patterns. They are expressed by phenomena of commercial gentrification, as experienced in Santo Domingo (González-Pérez, this volume, Ch. 2) which are subsequently impacting residential gentrification, artification (Baudry, this volume, Ch. 6) and aestheticization (Featherstone, 1991) (Guinand, this volume, Ch. 9) expressed by new relationships between formerly unrelated domains: gastronomy, fashion, art, architecture, design, etc. closely interact with each other and offer total and exclusive experiences (Lipovetsky and Serroy, 2013).

Harnessing the global – local tourism strategies and gentrification

Fierce competition between territories and shrinking public budgets have forced cities to look for strategies in order to foster economic growth and development. Tourism more than ever plays an important role in the economy by being generally associated with city rebirth (renaissance and beautification), revitalization or urban regeneration. For instance, the Inner Harbor Festival Market proposed by James Rouse in the late 1970s, once completed, put Baltimore at the forefront of media attention and pinned it on the map of tourist destinations. This urban project geared toward consumption and entertainment on the waterfront also became an economic model for redevelopment and new urban facilities. For many years, James Rouse and Martin Millspaugh, his business partner, travelled the world inspiring city authorities in search of new ideas to put on their urban agenda. Today much of the real estate development occurring on the north-eastern shore of Baltimore is still very much linked to Rouse's project. Since then other economic models have emerged: Bilbao and its iconic architecture, Liverpool and its reuse of warehouses, culture-led reinvention of city centres through heritagization processes or alternative social scenes as in Berlin, etc. All these models strengthen tourism as a policy pursued by urban agendas in order to harness global flows and create a marketing window for international attention.

The growing share of tourism industry in urban projects is compelling as it touches wide areas of urban and social life: infrastructure, equipment, housing, food industry, etc. Tourists or visitors do, indeed, push for and require responses and adjustments regarding amenities, urban design and place-making. Interestingly, some of the given responses can be very creative and imaginative.

As for the creative class precepts (Florida, 2002) cities have been eager to harness the tourism economy to the conduct of urban affairs. However, gentrification is usually the end result of such implemented policies: a desired and expected outcome or an uncontrolled consequence that came with the improvement of the neighbourhood. For instance, the barrio Chino, in the Raval area of Barcelona, was not targeted by city authorities as the nexus for gentrification policies. But its popularity for visitors and the improvements undertaken by the city pre-pared the stage for gentrification (Quaglieri and Scarnato, this volume, Ch. 5). Tourism raises mixed and ambiguous prospects as well as tensions and conflicts (Colomb and Novy, 2017). As a multidimensional process bringing urban qual-ity and amenities, it can be perceived as a good thing for the neighbourhood improvement by some residents, while, at the same time, considerably altering and modifying its social and economic environment. This is for example the case with the South Street Seaport Historic District in New York (Guinand, this volume, Ch. 9) where the new services provided and the amenities offered clearly target one segment of the population and visitors. Thus, capitalizing on tourist mobility and visitors can take different forms and strategies leading to diverse outcomes in the Metropolis.

Tourism, heritage restoration and neighbourhood transformation

Heritage and cultural sites are important features of tourism and what tourists are looking for (Figure 1.3). In the US, for instance, Boston almost became a tourist city by accident (Ehrlich and Dreier, 1999) since it was mostly its flair as

Figure 1.3 New York High Line

Source: Sandra Guinand, 2014

a 'walkable city' offering historic sites and well-preserved heritage in the down-town area with a pleasant and safe urban environment that attracted visitors in the 1980s. The restoration of Quincy Market and Faneuil Hall as places for retail and entertainment added to the destination's popularity. Although these places have evolved since their first renovation, they are still part of the city's attractiveness to this day.

Cities have been rediscovering their historical past and heritage features mainly through tourists or visitors. For instance, in Marseille, in the south of France, the growing number of cruise-ship visitors inspired city officials to bring monu-ments and industrial heritage at the forefront (Guinand, 2015). The city has been undergoing major transformation through the *Euroméditerranée* redevelopment project, notably with the newly constructed buildings of MUCEM and *Villa de la Méditerranée* on the former Pier J4 next to the Saint-Jean Fort, a major identity symbol of the city. Located between the new *Euroméditerranée* sector and the *Vieux port*, visitors have also rediscovered *Le Panier*, one of the oldest neighbour-hoods of Marseille and until recently home of Corsican and Italian immigrants. But the growing interest of visitors has put pressure on this once 'popular' area, where artists and artisans have settled, attracted by the low rent and real estate prices. Today, some of the original residents have left the area as it has become too much of a 'museum district' with numerous tourists striding its streets and the rise in Airbnb rentals (Rescan, 2015).

The growing interest shown in heritage preservation in public policies or by individuals in creating an enjoyable cityscape to wander around and spend time in, appears to be today a common feature of urban development as many of the contributions in this book indicate. What has been overlooked, however, are the different processes and characteristics behind this phenomenon. For instance, the situation in Cartagena, Columbia, (Piñeros, this volume, Ch. 4) shows how tourism was fuelled by second-home ownership of the intellectual bourgeoisie who suddenly reinvested in the cultural and historical values of old houses in the city centre. Interestingly, the 'museumification' process that resulted in a tremendous change in the social and economic function of the city core was predominantly led by the private sector with the agreement, but little help, of public authorities.

In the case of the Barrio Chino in the Raval, Quaglieri and Scarnato (this volume, Ch. 5) show how intense international migration from EU-member coun-tries has been the catalyst for a mundane 'touristification of everyday life' which anchors itself in the 'authentic' landscape and social dynamics of the neighbour-hood. Accompanied by extensive public interventions and investments for the 1992 Olympic Games, the quest of this last frontier has led to major changes in the neighbourhood's real estate market and socio-cultural features.

The New York Tenement Museum, presented by Sampson in this book (Ch. 7), is another interesting example of tourism-gentrification in which herit-age is a salient feature. In this case, what is particularly striking is the importance of the immaterial dimension that transcends the buildings and legitimizes their reason to remain. For instance, it is the stories and memories of the immigrants as valuable traces of history that are being promoted and willingly experienced by visitors. But as the social and cultural profiles of the visitors change over time, so do the historical settings in the visited 'witness' apartments. Moreover, the Tenement Museum plays an ambiguous and contradictory role in the neighbourhood, taking advantage of its worldwide success in order to increase its cash flow while at the same time inducing displacing ordinary people and stories in the few remaining fringes of New York's Lower East Side.

Heritage, with its material and immaterial dimensions as a component of the experiential and commodification of culture turn, is thus a strong feature of tourism gentrification. And as the museumification process moves forward, visitors are keen to (re)discover new territories and cultural landscapes, pushing the new frontier for gentrification further.

Luxury leisure developments and their attributes

As heritage plays an important role as an identity symbol and imaginary to which the individual can relate, so do architectural forms linked to "starchitects" in the context of (neo)liberalism (Harvey, 2007). These ambitious, creative forms are another way to assert the city's presence on the international scene. Starchitecture represents a brand that the city is selling, or a means to sell itself. As Judd and Fainstein put it:

> Cities are just like any other consumer product. They have adopted image advertising, a development that can hardly escape any traveller who opens an airline magazine and reads its formulaic articles on the alleged culinary and cultural delights of Dallas, Frankfurt or Auckland. (1999: 4)

Architecture and the urban landscape are part of this cultural differentiation process of an urban marketing strategy. It also acts as a window for capitalism, representing attributes a contemporary metropolis is supposed to possess if it wants to remain competitive (Guinand, 2015; Valença, 2016). These new urban forms give a measure of the modernity and progress and how 'entrepreneurial' a city is. For example, the new One World Tower in New York or the Petronas Twin Towers in Kuala Lumpur are world images that most of us have seen through various media. But these signs in the urban landscape do not stand alone as symbolic capital; they are usually part of a wider strategy of urban redevelopment aiming at attracting financial capital and its elites (Harvey, 2001). The well-known Guggenheim case in Bilbao for instance stemmed out of a planned strategy to turn a once worker and industrial city into an international, dynamic and creative tertiary centre that people would like to visit or settle in. As Plaza and Haarich (2015) have shown, Bilbao succeeded as a brand with its Guggenheim museum but also inscribed the city name in the global network. However, these strategies geared towards the international scene and its actors sometimes fail to bring in and respond to local realities. In Bilbao, as tourists and conference delegates sharply increased in the years following the museum opening (Plöger,

2007) the socio-economic disparities were not addressed as real-estate in the city core rose, sharply increasing issues around gentrification (Vicario and Martinez-Monje, 2003).

The transformation or the upgrading of the skyline marks a turn, a new start, in a city's economy. It also celebrates the intensification of the aesthetic capitalism phenomenon (Lipovetsky and Serroy, 2013) that has reached each of the objects of our daily life. Although more and more prevalent, this marketing strategy nevertheless aims at making one feel that he or she is enjoying the ultimate experience. The redevelopment project of the Seaport in New York (Guinand, this volume, Ch. 9) and the strategies deployed by the developer follow these lines. The story-telling process, the staged images and the temporary events deployed at the Seaport all target the visitor's emotions and senses in order to capture their attention. Going to the Seaport should be an experience *per se* for the locals and the tourists. But the luxurious aesthetic paradigm used to transform the mundane routine of this piece of land is powerfully oriented to attract local wealthy elites as well as transnational ones.

For some cities, tourism gentrification is indeed a deliberate strategy. This is clearly the case in the OCT district in Beijing and Shanghai, as described by Liang (this volume, Ch. 12). These newly upscale neighbourhoods which were once industrial or factory sites have been significantly gentrified in a very short period of time, after the urban interventions of the OCT development agency. These neighbourhoods host events, high-end facilities and bold architecture, like the new opera house in the Beijing OCT, and have become major national tourist attractions. These new facilities are also products and responses for the new rich Chinese that the state-led capitalist regime has favoured. As a transnational capitalist class, made-up of transnational corporate executives, globalising bureaucrats, politicians and professionals, and consumerist elites, have emerged on a global scale (Sklair, 2000), public authorities, as in the case of China, or private developers, are keen to fulfil their needs or propose new products. As Lees (2003) has shown in the case of super-gentrification in Brooklyn, new capitalists have no difficulty paying full price for the purchase of houses, condos and other facilities. As they usually work for international corporations with branches or offices around the globe, these people are extremely mobile, constantly travelling, and prone to stay in luxurious hotels or purchase a second-residential unit in whatever place suits them. However, as the different case studies show in this book, responses and transformation (as this is a dialectical process) in urban forms, amenities and facilities vary tremendously from one place to another; as Bridge (2007) points out there is no such thing as one global gentrifier and the concept of transnational elites is not a homogeneous one, since aesthetic trajectories and localisms of cosmopolitan knowledge prevail.

From downtown tourism-oriented activities to post-tourism settings

The competitive advantages of a city, as we have seen, are also defined by the tourism economy of how many 'visitors' or 'entries' a city gets. Cities must, therefore, be attractive, if not fun (as James Rouse used to say) (Demarest, 1981) and festive. Therefore, major entertainment districts have become a recurrent trend in metropolises such as the former warehouses area around King Street West in Toronto which welcomes bars, nightclubs, movie theatres, performing arts scenes as well as sports clubs and family entertainments (Darchen, 2013). In the United States, festival marketplaces in the late 1970s–80s set the movement by recapturing land, people and economic activities in downtown areas. These waterfront redevelopment projects were offering small retail shops, restaurants, entertainment venues, events and the opportunity to stroll down the waterfront among historic buildings in a safe and clean environment. Critically described as theme parks (Sorkin, 1992; Boyer, 1992), these places nevertheless managed to refocus the middle-class suburbanites' attention back to the city as well as attract visitors. Extensive research has shown the role of the entertainment industry in the redevelopment of downtown areas, either by their direct implication in the transformation of urban forms and commercial functions as in the evident case of Time Square in New York or Potsdamer Platz in Berlin (Sassen, 1991; Gotham, 2005) or by having a strong influence as a model for the design of urban places as in the case of Inner Harbor, Baltimore (Gravari-Barbas, 1998).

Today, these precepts are still very much prevalent as the Chinese cases developed by Liang testify (this volume, Ch. 12), where housing developments are planned hand-in-hand with amusement parks. However, entertainment has taken a new turn as the cultural economy and the commodification of urban landscape and experience have become more predominant (Zukin, 1995). City authorities' policies and private interests have been very keen on turning city centres into "theme parks" (Sorkin, 1992; Mitchel and Staeheli, 2006). Most central urban areas around the globe have become safe and clean spaces either through restrictive and conservative policies, such as the one implemented by Rudolph Giuliani in the 1990s in New York, or through private initiatives such as business improvement districts (Schaller and Modan 2005; Dubresson, 2015), conservancies or other type of private interests (profit and non-profit). The already highly gentrified historic areas, like the Marais in Paris, have a lot to share with open-air museums or staged scenes; they differentiate themselves from more 'ordinary' and 'plain' areas but still play on the 'authentic' register as unique and exclusive places. These highly visited sites have experienced a dramatic transformation in their function as in the case of Cartagena, Columbia, or the nature and type of businesses and shops as in the case of Lisbon or Santo Domingo (in this volume). Tourism gentrification implies displacement of people, but also of commercial activities. These newly created spaces transcend the idea of a tourist bubble (Judd, 1999) to become, as Sklair states: "a transnational social space of capitalist globalisation, characterised by ubiquitous chain stores and restaurants, taking advantage of elements of indigenous or traditional cultural traits which in turn offer the visitors of the place the opportunity to indulge or pretend to a bourgeois lifestyle" (2009: 532).

Scholars have shown that tourism plays as a cultural intermediary (Zukin, 1991; Ley, 2003; Gotham, 2005; Colomb and Novy, 2017) in the gentrification process since it disseminates images, discourses, imaginaries about places being 'authentic', 'aesthetic', 'exotic', 'nostalgic', etc. which in return attract new visitors and customers and boost development growth. But as the race for competition between territories is still running, and with it the quest for distinction and uniqueness, the lines between tourists, visitors and residents are being blurred (Bock, 2015; Condevaux *et al.*, 2016). Visitors are blending with locals as the ordinary becomes exceptional and the exceptional ordinary (Figure 1.4). The quest for 'uniqueness' and 'local' are to be found in the margins while becoming at the same time a city-brand. Thus, competitive cities (the actors behind them) play on these different registers, notably through social media, while embracing at the same time the vocabulary of globalization and entrepreneurship. This post-tourism phenomenon (Urry 2002; Rojek, 1997) goes hand in hand with gentrification syndromes: as visitors are more and more looking for new and 'off the beaten track' places, cities try to capture the visitor's (local or foreign) imagination, offering an array of activities and places to explore. But as new opportunities and territories are being explored and brought into the network of places to be or visit, they risk the transformation and loss of their uniqueness.

Figure 1.4 Mercat de Santa Catarina, Barcelona

Source: Sandra Guinand, 2007

Setting the research agenda on tourism gentrification issues

The chapters presented in this book are very diverse. The most emblematic case studies in recent bibliography such as Barcelona, Paris or Berlin are compared with more recently researched European cases such as Rome or Lisbon, as well as less well known cities such as Reykjavík or Rethymnon (Greece). The scope is clearly international and cases from the Caribbean (the colonial cities of Cartagena or Santo Domingo), the United States (New York) and Asia (Bangkok) shed light on the different, and even antinomic, ways gentrification phenomena interact with tourism.

The different case studies analysed in this book clearly show the interaction between tourism and gentrification which appear as reciprocally supporting phenomena. Authors offer different perspectives in showing that gentrification phenomena are related to different actors, stakeholders and contexts. In some cases they are resulting from the global economy trends against which the locals (local decision makers, local residents, etc.) try to resist. In others, however, they are embraced and encouraged by the locals and are the expression of local decision-making. We intend to present here the main issues raised in the volume as a means to shape future research in tourism gentrification.

The heritagization syndrome

The case of Santo Domingo's Colonial city illustrates the problematic relationship between heritage policies and gentrification where 'heritage success is social failure'. The deliberate use of tourism by local authorities as the main focus for urban regeneration led to a 'schismatic city'. The conclusions are not different in the case of the Colonial city of Cartagena de Indias, Colombia, where the heritagization of this now UNESCO-listed colonial centre played a major role in its gentrification process. The same conclusions are shared for Le Marais in Paris, where gentrification was the pre-announced effect of the heritagization process. Similarly the case of Lisbon exemplifies how tourism fuelled urban change and heritage redevelopment. In all these cities, authors showed the dynamic evolution of gentrification related to tourism.

The Airbnb syndrome

According to Quaglieri and Scarnato's estimation (this book) from data provided by the platform *Inside Airbnb*, in Barcelona, despite the emphasis on the invention of new models of development and decongestion of tourism pressures, Airbnb presents the same spatial patterns as more conventional hospitality sectors: it targets the most attractive tourism-oriented areas of the city. The authors show that the Airbnb tourism rentals act in a predatory way, contributing to the gentrification of central neighbourhoods (the Raval for this study).

The case of Reykjavík (a newcomer city in the European tourism landscape) (De Freytas-Tamura, 2016) offers another perspective on this phenomenon where the impact of the so-called 'sharing economy' on the social structure of the city is significant (Mermet, this volume, Ch. 3). In the context of the major economic crisis the country faced at the beginning of the year 2000, Airbnb rentals played a crucial role in the permanent or temporary eviction of residents.

Airbnb's rapid development pushed local governments to come up with new regulation policies. In Paris, Madrid and Berlin, local governments, particularly concerned by the Airbnb impact on neighbourhoods, took decisive measures to control it, without any significant effects until now.

Looking at the Berlin case, Kagermeier and Stors (this volume, Ch. 8) present an alternative view of the Airbnb syndrome to the prevalent one. They estimate that the rules implemented in the city in order to control tourism rentals underline the difficulty of fully understanding and grasping the implications of the Airbnb phenomenon, since the policies tend to overestimate the platform's influence on the real estate market.

These two different perspectives are not only explained by the situation of the two cities on the tourism map and by their 'tourism anteriority', but also by inherent difficulties in current research to operate a cross-cut comparative study between the cases observed in different cities.

The artification syndrome

The case of peripheral neighbourhoods in Rome illustrates the ways tourism commodifies former industrial vestiges. Factories, wastelands and warehouses have become the desirable symbols of a new order of development based no longer on production but on aesthetic consumption. Interestingly enough, this trend not only concerns 'shrinking cities' like Berlin or Detroit, cities that entered the tourism market through aestheticization and the desirability for alternative places; it also concerns places such as Rome, characterized by what Judd and Fainstein (1999) call "place luck": places that have been part of the first tourism development in Europe since the eighteenth century because of the beauty of their monuments and historical sites. As the case of Rome shows, artification and aestheticization are necessary conditions for tourism gentrification outside the historical centre.

The museumfication syndrome

Tourism gentrification is related to cultural phenomena as expressed by cultural urban events or cultural infrastructure. Cultural venues located in former derelict areas played a major role in the transformation of these places into desirable destinations and, eventually, as attractive places to live. The case of the Tenements Museum shows, however, how ironic cultural transformation can be. How can a museum and national landmark dedicated to immigration (the Tenement Museum), whose discourse celebrates the role immigrants played in the national history, serve as a means for further gentrification of the Lower East Side? Tourism gentrification can transform the stigmata of immigration into exclusive aesthetic arguments that contribute to the integration of this poor cosmopolitan area into the globalized fluxes of the transnational mobility of capital and people.

However, this rather dark portrait of cities depicted by most authors of this volume, which announces the future exclusive use of central urban cores (historical centres, ethnic neighbourhoods and former old industrial areas) by the global capital and transnational elites, is challenged by an alternative analysis proposed in the cases of Rethemnos and Bangkok (Herzfeld, this volume, Ch. 10): What if gentrification does not come from external forces but from within? What if, far from destroying the lives of residents, gentrification is perceived as the means to offer the possibility of surviving as a community with improved circumstances? In his challenging analysis of Rethymnon and Bangkok, Herzfeld stresses the positive effects of 'self-gentrification' made possible thanks to tourism development. However, he insists on the fact that this 'self-salvation' depends on factors that keep residents firmly in place – which in Barcelona, Berlin, Paris or New York is far from being the case . . .

Tourism gentrification: a challenge for the 21st-century city

The chapters in this book focus on the dual interconnection between tourism and gentrification: tourism is understood both as a result of gentrification *and* as a pre-condition for gentrification. They also show the dynamic relation of these phenomena in time: far from being a stable relation, early gentrification can encourage tourism in its beginnings, while it can fiercely resist it at a subsequent stage.

Far from being consensual and globally shared or accepted, tourism gentrification phenomena can be conflicting. Tourism (and tourists) appears today as a premium challenge for social or medium-range housing. In tourism-gentrified areas, apartments are offered not only to the local and national upper class (as used to happen in the first cases of gentrified areas), or to the transnational elite's second-home owners, but also to the masses of tourists. This is a major shift: tourism gentrification not only involves the rich or the super-rich, but also the tourism middle-class, which by the means of the 'shared economy' in which it participates, globally becomes one of the major stake-holders of the historical centres and other tourism-gentrified areas.

Notes

1 Authors' own translation.
2 Authors' own translation.

References

Ascher, F. (1995) *Métapolis ou l'avenir des villes*, Paris: Odile Jacob.

Bidou-Zachariasen, C. (ed.) (2003) *Retours en ville*, Paris: Descartes & Cie.

Bock, K. (2015) 'The Changing Nature of City Tourism and its Possible Implications for the Future of Cities', *European Journal of Futures Research*, vol. 3, no. 20, available at: http://link.springer.com/article/10.1007/s40309-015-0078-5/fulltext.html. [accessed 31 Jan 2017]

Boyer, M. C. (1992) 'Cities for Sale: Merchandising History at South Street Seaport', in Sorkin, M. (ed.), *Variations on a Theme Park*, New York: Noonday Press, pp. 181–204.

Bridge, G. (2007) 'A Global Gentrifier Class?' *Environment and Planning A*, vol. 39, no. 1, pp. 32–46.

Bures, R., Cain, C. (2008) *Dimensions of Gentrification in a Tourist City*, available at http://paa2008.princeton.edu/papers/81623. [accessed 31 Jan 2017]

Cameron, S. (1992) 'Housing, Gentrification and Urban Regeneration Policies', *Urban Studies*, vol. 29, no. 1, pp. 3–14.

Chapuis A., Gravari-Barbas, M. and Jacquot S. (2015) 'Tourism/Gentrification: Sex, Gender and Crossed Resistances', paper presented at the AAG, available at: www.academia.edu/12256317/Tourism_Gentrification_sex_gender_and_crossed_resistances. [accessed 31 Jan 2017]

Cocola Gant, A. (2016) 'Holiday Rentals: The New Gentrification Battlefront', *Sociological Research Online*, vol. 21, no. 3, available at http://www.socresonline.org.uk/21/3/10.html. [accessed 31 Jan 2017]

Colomb C. and Novy J. (2017) *Protest and Resistance in the Tourist City*, London and New York: Routledge.

Condevaux A., Djament-Tran G. and Gravari-Barbas M. (2016) 'Before and After Tourism(s). The Trajectories of Tourist Destinations and the Role of Actors Involved in "Off-The-Beaten-Track Tourism: A Literature Review', *Via@*, vol. 1, no. 9, available at http://viatourismreview.com/2016/10/avantetaprestourisme-analysebiblio/ [accessed 31 Jan 2017]

Couch, C. (2003) *City of Change and Challenge: Urban Planning and Regeneration in Liverpool*, Aldershot; Burlington: Ashgate.

Darchen, S. (2013) 'The Creative City and the Redevelopment of the Toronto Entertainment District: A BIA-Led Regeneration Process', *International Planning Studies*, vol. 18, no. 2, pp. 188–203.

Davidson M. and Lees L. (2005) 'New-build "Gentrification" and London's Riverside Renaissance', *Environment and Planning A*, vol. 37, no. 7, pp. 1165–1190.

Davidson M. and Lees L. (2010) 'New-Build Gentrification: Its Histories, Trajectories, and Critical Geographies', *Population, Space and Place*, vol. 16, no. 5, pp. 395–412.

De Freytas-Tamura, K. (2016) 'Secret to Iceland's Tourism Boom? A Financial Crash and a Volcanic Eruption', *New York Times*, 16 November 2016, available at: www.nytimes.com/2016/11/17/world/europe/reykjavik-iccland-tourism.html?_r=0. [accessed 31 Jan 2017]

Demarest, M. (1981) 'Living: He Digs Downtown', *Time*, 24 August 1981, available at http://content.time.com/time/magazine/article/0,9171,949385,00.html. [accessed 31 Jan 2017]

Dubresson, A. (2015) 'Partenariats public-privé au Cap', *Séminaire La ville à l'ère globale : transformations urbaines, réceptions et défis*, 7 April, 2015, Centre Jacques Berque, Rabat, Maroc.

Edensor, T. (2007) 'Mundane Mobilities, Performances and Spaces of Tourism', *Social & Cultural Geography*, vol. 8, no. 2, pp. 199–215.

Ehrlich, B. and Dreier, P. (1999) 'The New Boston Discovers the Old: Tourism and the Struggle for a Liveable City', in Fainstein S. and Judd, D. (ed.), *The Tourist City*, New Haven and London: Yale University Press, pp. 155–178.

Featherstone, M. (1991) 'Postmodernism and the Aestheticization of Everyday Life', in Lash, S., Friedman, J. (ed.), *Modernity and Identity*, Oxford: Blackwell.

Florida, R. (2002) *The Rise of the Creative Class (and How It's Transforming Work, Leisure, Community and Everyday Life)*, New York: Basic Books.

Florida, R. (2012) 'Cities and the Creative Class', *City & Community*, vol. 2, no. 1, pp. 3–19.

Friedmann, J. (1986) 'The World City Hypothesis', *Development and Change*, vol. 17, no. 1, pp. 69–83.

Friedmann, J. and Wolff, G. (1982) 'World City Formation: An Agenda for Research and Action, *International Journal for Urban and Regional Research*, vol. 6, no. 3, pp. 309–344.

Gotham, K. (2005) 'Tourism Gentrification: The Case of New Orleans' Vieux Carré (French Quarter)', *Urban Studies*, vol. 42, no. 7, pp. 1099–1121.

Gravari-Barbas, M. (1998) 'La "festival market place" ou le tourisme sur le front d'eau. Un modèle urbain américain à exporter', *Norois*, vol. 178, no. 1, pp. 261–278.

Gravari-Barbas, M. (2015) 'Winescapes : Tourism et artialisation, entre le local et le global', *CULTUR Revista de Cultura e Turismo*, vol. 8, no. 3, available at : http://periodicos. uesc.br/index.php/cultur/article/view/374. [accessed 31 Jan 2017]

Gravari-Barbas, M. and Delaplace, M. (2015) 'Le tourisme urbain "hors des sentiers battus" Coulisses, interstices et nouveaux territoires touristiques urbains', *TEOROS*, no. 34.

Gravari-Barbas, M. and Jacquot, S. (2017) 'No Conflict? Discourses and Management of Tourism-Related Tensions in Paris', in Colomb, C. and Novy, J. (ed.), *Protest and Resistance in the Tourist City*, London: Routledge.

Gravari-Barbas, M. and Renard-Delautre, C. (2015) *STARCHITECTURE(S) Celebrity Architects and Urban Space*, Paris: L'Harmattan,

Guinand, S. (2015) *Régénérer la ville. Patrimoine et politiques d'image à Porto et Marseille*, Rennes: Presses universitaires de Rennes.

Halbert, L. (2010) *L'avantage métropolitain*, Paris, PUF.

Hamari J., Sjöklint M. and Ukkonen A. (2016) 'The Sharing Economy: Why People Participate in Collaborative Consumption', *Journal of the Association for Information Science and Technology*, vol. 67, no. 9, pp. C1–C1, 2045–2306.

Hamnett, C. (1991) 'The Blind Men and the Elephant: The Explanation for Gentrification', *Transactions of the Institute of British Geographers*, no. 16, no. 2, pp. 173–189.

Harvey, D. (2001) 'The Art of Rent. Globalization and the Commodification of Culture', in Harvey, D., *Spaces of Capital: Towards a Critical Geography*, New York: Routledge, pp. 394–411.

Harvey, D. (2007) 'Neoliberalism and the City', *Studies in Social Justice*, vol. 1, no. 1, pp. 2–13.

He, S. (2010) 'New-build Gentrification in Central Shanghai: Demographic Changes and Socioeconomic Implications', *Population, Space and Place*, vol. 16, no.5, pp. 345–361.

Judd, D. R. (1999) 'Constructing the Tourist Bubble', in Judd, D. and Fainstein, S. (ed.), *The Tourist City*, New Haven and London: Yale University Press, pp. 35–53.

Judd, D. R. (2003) *The Infrastructure of Play: Building the Tourist City*, New York: Cleveland State University.

Judd, D. and Fainstein, S. (1999) *The Tourist City*, New Haven and London: Yale University Press.

King, A. D. (1993) 'The Global, the Urban and the World', in King, A. D. (ed.), *Culture, Globalization and the World-System: Contemporary Conditions for the Representation of Identity*, London: Macmillan, pp. 149–154.

Laska S., Seaman J. and McSeveney D. (1982) 'Inner-City Reinvestment: Neighborhood Characteristics and Spatial Patterns Over Time', *Urban Studies*, vol. 19, no. 2, pp. 2155–2165.

Lees L. (2003) 'Super-Gentrification: The Case of Brooklyn Heights, New York City', *Urban Studies*, vol. 40, no. 12, pp. 2487–2509.

Lees, L., Slater, T. and Wyly, E. (2008) *Gentrification*, New York: Routledge.

Ley, D. (1980) 'Liberal Ideology and the Postindustrial City', *Annals of the Association of American Geographers*, no. 70, pp. 238–258.

Ley, D. (2003) 'Artists, aestheticization and the field of gentrification', *Urban Studies*, vol. 40, no. 12, pp. 2527–2544.

Lipovetsky G. and Serroy, J. (2013) *L'esthétisation du monde. Vivre à l'âge du capitalisme artiste*, Paris: Editions Gallimard.

Lopez-Gay, A. and Cocola Gant, A. (2016) 'Cambios demográficos en entornos urbanos bajo presión turística: el caso del barri Gòtic de Barcelona', in Domínguez-Mújica, J. and Díaz-Hernández, R. (ed.), *XV Congreso Nacional de la Población Española*, Fuerteventura: Asociación de Geógrafos Españoles, pp. 399–413.

McCannell, D. (1973) 'Stages Authenticity: Arrangement of Social Space in Tourist Settings', *The American Journal of Sociology*, vol. 79, no. 3, pp. 589–603.

MacCannell, D. (1976) *The Tourist. A New Theory of the Leisure Class*, Berkeley and Los Angeles: University of California Press.

Milne, S., Ateljevic, I. (2001) 'Tourism, Economic Development and the Global Local Nexus: Theory Embracing Complexity', *Tourism Geographies*, vol. 3, no. 4, pp. 369–93.

Mitchel, D. and Staeheli, L.A. (2006) 'Clean and Safe? Property Redevelopment, Public Space, and Homelessness in Downtown San Diego', in Lowe, S. L. and Smith, N. (ed.), *The Politics of Urban Space*, London, New York: Routledge, pp. 143–176.

Mosedale, J. (2014) 'Political Economy of Tourism: Regulation Theory, Institutions and Governance Networks', in Hall, C. M., Lew, A. A. and A. M. Williams (ed.), *The Wiley-Blackwell Companion to Tourism*, Oxford: Wiley-Blackwell, pp. 55–65.

Nations-Unies. (2014) *Rapport sur les perspectives d'urbanisation*, New York: ONU.

Novy, J. (2017) 'The Selling (Out) of Berlin and the *De-* and *Re-politicization* of Urban Tourism in Europe's "Capital of Cool"', in Colomb, C. and Novy J., *Protest and Resistance in the Tourist City*, London: Routledge.

Philips, W-J. and Back, K-J. (2011) 'Conspicuous Consumption Applied to Tourism Destination', *Journal of Travel & Tourism Marketing*, vol. 28, no. 6, pp. 583–597.

Plaza, B. and Haarich, S. (2015) 'The Guggenheim Museum Bilbao: Between Regional Embeddedness and Global Networking', *European Planning Studies*, vol. 23, no. 8, pp. 1456–1475.

Plöger, J. (2007) *Bilbao City Report*, London: CASE.

Rérat P., Söderström, O. and Piguet, E. (2010) 'New Forms of Gentrification: Issues and Debates', *Population, Space and Place*, vol. 16, no. 5, pp. 335–344.

Rescan, M. (2015) 'A Marseille, le centre-ville résiste toujours à la gentrification', *Le Monde,* 16 June 2015, available at www.lemonde.fr/recherche/#SGSeEZTGSQGVHKge.99. [accessed 31 Jan 2017]

Rofe, M. (2003) '"I Want to be Global": Theorising the Gentrifying Class as an Emergent Elite Global Community', *Urban Studies*, vol. 40, no. 12, pp. 2511–2526.

Rojek, C. (1997) 'Indexing, Dragging and the Social Construction of Tourist Sights', in Rojek, C. and Urry, J. (ed.), *Touring Cultures*, London: Routledge.

Russo, A-P. and Arias-Sans, A. (2017) 'The Right to Gaudí. What Can We Learn from the Commoning of Park Güell, Barcelona?', in Colomb, C. and Novy, J. (ed.), *Protest and Resistance in the Tourist City*, London: Routledge.

Sassen, S. (1991) *The Global City: New York, London, Tokyo*, Princeton: Princeton University Press.

Schaller, S. and Modan, G. (2005) 'Contesting Public Space and Citizenship Implications for Neighborhood Business Improvement Districts', *Journal of Planning Education and Research*, vol. 24, no. 4, pp. 394–407.

Schlichtman, J. and Patch, J. (2014) 'Gentrifier? Who, Me? Interrogating the Gentrifier in the Mirror', *International Journal of Urban and Regional Research*, vol. 38, no. 4, pp. 1491–1508.

Sklair, L. (2000) 'The Transnational Capitalist Class and the Discourse of Globalization', *Cambridge Review of International Affairs*, vol. 14, no. 1 pp. 67–85.

Sklair, L. (2009) 'The Emancipatory Potential of Generic Globalization', *Globalizations*, vol. 6, no. 4, pp. 525–539.

Smith, N. (1986) 'Gentrification, the Frontier and the Restructuring of Urban Space', in Smith, N. and Williams, P. (ed.), *Gentrification of the City*, London: Allen & Unwin.

Smith, N. (1996) *The New Urban Frontier: Gentrification and the Revanchist City*, London and New York: Routledge.

Smith, N. (2002) 'New Globalism, New Urbanism: Gentrification as a Global Urban Strategy', *Antipode*, vol. 34, no. 3, pp. 427–450.

Smith, N. (2006) 'Gentrification Generalized: From Local Anomaly to Urban "Regeneration" as Global Urban Strategy', in Fisher, M. and Downey, G. (ed.), *Frontiers of Capital: Ethnographic Reflections on the New Economy*, Duke University Press, pp. 191–206.

Sorkin, M. (ed.) (1992) *Variations on a Theme Park: The New American City and the End of Public Space*, New York: Hill and Wang.

Stock, M. (2012) 'L'hypothèse de l'habiter poly-topique : pratiquer les lieux géographiques dans les sociétés à individus mobiles', *EspacesTemps.net*, available at www.espaces temps.net/articles/hypothese-habiter-polytopique/. [accessed 31 Jan 2017]

Urry, J. (2002) *The Tourist Gaze*. London: Sage Publications.

Van Criekingen, M. (2008) 'Towards a Geography of Displacement. Moving Out of Brussels' Gentrifying Neighbourhoods', *Journal of Housing and the Built Environment*, vol 23, no. 3, pp. 199–213.

Valença, M. (2016) *Arquitectura de Grife na Cidade Contemporanea. Tudo Igual, mas Diferente*, Rio de Janeiro: Mauad X.

Veblen, T. (1953) [1899] *The Theory of the Leisure Class: An Economic Study of Institutions*, The Mentor Edition. New York: The Macmillan Company.

Vicario, L. and Martinez-Monje, P. (2003) 'Another Bilbao Effect? The Generation of a Potentially Gentrifiable Neighbourhood in Bilbao', *Urban Studies*, vol. 40, no. 12, pp. 2383–2400.

Zukin, S. (1989) *Loft Living: Culture and Capital in Urban Change*, New Brunswick: Rutgers University Press.

Zukin, S. (1991) *Landscapes of Power: From Detroit to Disneyworld*, Berkeley, CA: University of California Press.

Zukin, S. (1995) *The Cultures of Cities*, Cambridge and Oxford: Blackwell.

Part I

Tourism gentrification, heritagisation and urban regeneration

2 A new colonisation of a Caribbean city

Urban regeneration policies as a strategy for tourism development and gentrification in Santo Domingo's Colonial City

Jesús M. González-Pérez

Introduction

Santo Domingo's Colonial City is the main historic centre of the Dominican Republic and one of the nine Caribbean historic centres that have been declared World Heritage Sites by UNESCO. A first morphological analysis reveals an on-going trend towards 'favelisation' (informal settlement) in the city's centre and serious urban segregation issues. The segregation shows itself not only in key social and economic aspects, but also as cultural/ethnic exclusion: the Colonial City can be depicted as a 'Hispanic' environment, while popular neighbourhoods are identified as 'African' environments. An important part of the Colonial City is significantly orientated towards tourism.

In this context, urban regeneration is a consequence of the combination of aid towards development and poorly coordinated urban public policies. Currently, the proposals in the Master Plan for the Colonial City (MPCC), which has valuable funding from the Inter-American Development Bank (IDB), focus on tourism and real estate specialisation in the historic centre of Santo Domingo as a method for the recovery of heritage sites, economic revitalisation and their internationalisation.

We have structured this work into four sections. The first looks at the current status of gentrification of cities in the Caribbean, gearing them more towards tourism. The second offers a morphological analysis of Santo Domingo, highlighting internal socio-urban contrasts. Thirdly, we review urban and strategic planning and its impact on 'touristification' and gentrification. Finally, we will conclude by setting out some of the main consequences from gentrifying the Colonial City.

The 'touristification' and gentrification of Caribbean cities

Tourism is the economic mainstay of island nations in the Caribbean. The latest tourism boom (1990s) has included urban tourism in old towns, incorporating hotels and residential real estate. Urban regeneration and spotlighting cultural heritage are policies that directly tap into these new interests.

Since 1990, during the neoliberal boom in Latin America and the Caribbean, historic city centres took a starring role in line with their gradual commercialisation (Gutiérrez, 2008). Tourism, and the inscription of cities on the World Heritage list, triggered the internationalisation process of historic city centres in the Caribbean. Eight of the twelve listed by World Heritage in the region were awarded this status after 1989. The declaration of World Heritage of a certain city itself represents a strategic action of inestimable value, essentially due to the effects it has on tourism and real estate development. From the very day of this declaration, the city tends to become a consumer product: a place to visit and somewhere to invest in (González and Lois, 2010). Nine of these cities are located on islands and four of them on Cuba alone (see Table 2.1). Most of the UNESCO perimeters correspond to the areas of the original colonial city. The most important exception is Trinidad, which includes the Valley de los Ingenios.[1] The historic centre and fortification system are generally included in the boundary.

Table 2.1 World Heritage Cities in the Caribbean region in 2015[2]

City	Entry	Registered sector	Surface area Central Zone (CZ)/ Buffer Zone (BZ)
Havana (Cuba)	1982	Old Havana and its Fortification System	214 ha (CZ)
San Juan, Puerto Rico (US)	1983	La Fortaleza and San Juan National Historic Site	–
Cartagena (Colombia)	1984	Port, Fortresses and Group of Monuments, Cartagena	–
Trinidad (Cuba)	1988	Trinidad and the Valley de los Ingenios	–
Santo Domingo (Dominican Republic)	1990	Colonial City of Santo Domingo	106.0 ha (CZ)
Coro (Venezuela)	1993	Coro and its port (La Vela)	–
Sta. Cruz de Mompox (Colombia)	1995	Historic Centre of Santa Cruz de Mompox	–
Willemstad (Netherlands Antilles, Netherlands)	1997	Historic Area of Willemstad, Inner City and Harbour, Curaçao	86.0 ha (CZ) 87.0 ha (BZ)
Saint George (Bermuda, United Kingdom)	2000	Historic Town of St George and Related Fortifications, Bermuda	257.5 ha (CZ)
Cienfuegos (Cuba)	2005	Urban Historic Centre of Cienfuegos	70.0 ha (CZ) 105.0 ha (BZ)
Camagüey (Cuba)	2008	Historic Centre of Camagüey	54.0 ha (CZ) 276.0 ha (BZ)
Bridgetown (Barbados)	2011	Historic Bridgetown and its Garrison	187.0 ha (CZ) 321.0 ha (BZ)

Source: Original production based on UNESCO and the Organization of World Heritage Cities (http://whc.unesco.org/en/list/)

The efforts to revitalise historic city centres as tourist attractions and commercial hubs by providing services and infrastructures triggers gentrification, the privatisation of public spaces and new forms of surveillance (Dürr and Jaffe, 2012). On the one hand, the difficulty in controlling the processes linked to touristification make historic centres new mass tourism destinations, in the Fordism vein (hotel tourism in the 1950–1970s), combined with post-Fordist (residential tourism from 1980) approaches associated with second homes. On the other hand, renovating heritage is, in many instances, being treated as a business. Forced population displacement due to the new type of resulting segregation is one of the tangible effects. There is also a new risk for Caribbean cities: turning areas into 'added extras' for sun and sand destinations, as places for a day out. This means that most tourist activity is concentrated within an area contained in just a few streets seeing life, tourists returning to their all-inclusive resort for the night, whether this be a hotel in Bávaro, Varadero, Crane Beach, South Shore Park or a cruise ship temporarily moored in one of the cities.

Gentrification in Latin America and the Caribbean was far behind that seen in the United States and Europe, usually associated with different periods in gentrification processes. The main forces behind gentrification arose from 1990 onwards, which explains why most bibliographic references are post-2000. Gentrification analysis in Latin America was first introduced by Jones and Varley (1999), who analysed the conservation processes in the colonial city centre of Puebla (Mexico). Some of the main works on gentrification in Latin America (Lungo and Baires, 2001; Vergara, 2013; Hiernaux and González, 2014; Janoschka, Sequera and Salinas, 2014) barely mention Caribbean examples. There is no specific scientific literature on gentrification in Santo Domingo. Although the gentrification process is a generalised urban strategy rooted in neoliberal urban policies (Smith, 2002), it is a socialist city that has captured the attention of scientific literature: Havana (Scarpaci, 2000a, 2000b, 2002; Scarpaci, Segre and Coyula, 2002; Bailey, 2008; Shaw, 2008; Achtenberg and Currents, 2013; González, 2015). The debate around the Cuban capital looks set to increase in the coming years, after relations between Cuba and the United States began to thaw in December 2014. In Puerto Rico, key texts on gentrification (and resistance to it) include work on Alto del Cabo in Santurce (Arocho, 2007), Santurce (Rodríguez, 2008; Suárez, 2009), Guanaybo (Suárez, 2010) and San Juan (De Jesús, 2011). Other important literature on island cities includes St. John's (Thomas, 1991) and Kingston (Semple, 1999).

Urban structure and morphology of the Colonial City of Santo Domingo

The Colonial City of Santo Domingo is seeking to invigorate its economy and increase its competitiveness in the region through tourism and property investment. In recent years, the changes in urban planning have had a major impact on the city with serious problems in terms of segregation and urban poverty. In this light, the main aim here is to look at social and urban processes and transformation that are mainly linked to gentrification and tourism within the framework of urban regeneration policies in the Colonial City of Santo Domingo. The historic centre has undergone a major economic transition through tourism in recent years. As is common in fourth-wave gentrification, the financialisation of housing and pro-gentrification urban policies stand out (Lees, 2003; Lees *et al.*, 2008; Wilson and Tallon, 2011), and in the case at hand, with a high involvement of public administrations as well as international organisations. The methodology combines theoretical analysis with important fieldwork undertaken in 2013 and 2014, focused around field observation, data collection on the ground and interviews with agents and experts in analysis of the city and planning. This is rounded out with a review of documents related to urban, tourism and strategic plans.

The Colonial City is one of the 65 areas in the Distrito Nacional, which itself has no internal administrative organisation into neighbourhoods or districts (see Figure 2.1). The urban surface area of Santo Domingo increased by 1,056 per cent between 1978 and 2002. The Colonial City is a small area of 106 hectares within this urban sprawl, representing 0.07 per cent of the total land area. Whilst the population of the Distrito Nacional (Ozama or Metropolitan region) increased by 22.26 per cent from 2002 to 2010 (608,116 inhabitants), the colonial area lost 3,656 inhabitants (−30.13 per cent). Consequently, the Colonial City (8,477 inhabitants in 2010) has lost community representation within the metropolitan area, falling from 0.44 per cent in 2002 to 0.23 per cent in 2010.

Figure 2.1 Gran Santo Domingo: Ozama region

Source: Original production. Mapas Gaar S.A.

One of the biggest transformations in the historic centre occurred during the first period of the Joaquín Balaguer government (1966–1978). The city underwent major *Haussmannian* renovation aimed at splitting up working-class areas that had resisted the North American invasion (1965–1966, Operation Power Pack, Dominican Civil War) and making them more accessible. The old streets were lengthened with modern avenues that broke up the urban layout and contributed to the creation of large blocks. The historic centre was fragmented and urban centrality dissipated (Dilla, 2010). In the mid-twentieth century, the historic centre operated as a capital centre, housing political, legislative, educational, religious and commercial activities. Today, only the Catholic Church has maintained its operational and management centres. At the start of the new millennium, residential (42 per cent of property) and commercial (35 per cent) use predominates. Institutions and offices occupy 8 per cent, whilst 11 per cent of property lies empty.

The Colonial City has 2,277 properties over 116 blocks. It borders Avenida Mella to the north, Paseo Presidente Billini (pier) to the south, the Ozama River to the east and Palo Hincado street to the west. The area laid out by Ovando when he founded the city (known as Ciudad Ovandina) ends at Billini street. Modern urban fabric is plentiful in the northern sectors, in areas that join up to the twentieth-century city. The historic centre is a consolidated area, closely unified and clearly identifiable in the urban design of Santo Domingo. Based on the area's main morphological features and urban processes, the historic centre should be understood as a communion of different sectors:

1   The pedestrian street of El Conde is the most important functional and pivotal thoroughfare in the historic centre. Alongside the central section comprising the streets of Isabel la Católica, Las Damas (the first urban street designed by Governor Ovando) and Arzobispo Meriño, this carriageway is the most representative area of the historic centre with abundant heritage sites. The historical and heritage value of the buildings diminishes from Colón Park (Cathedral) heading westwards. Tourism use is overarching and concentrates retail and restaurant outlets geared towards tourists in the historic city. Mixed land use predominates: tertiary-residential, tertiary-commercial and tertiary-hotel.

2   The neighbourhoods of Santa Bárbara, San Antón and San Lázaro are home to 'excluded' communities (inhabited by people of low economic level). Due to the lack of upkeep, most homes are called 'casas bomba' ('bomb homes') or 'casas arrabalizadas' (slum suburban homes). Although located within the city walls, Santa Bárbara was the first neighbourhood to spring up outside the old Ciudad Ovandina. Santa Bárbara was home to workers from the limestone quarries that supplied material for the construction of the colonial buildings and, to a lesser degree, workers from the port of Santo Domingo. Some of the homes in Santa Bárbara were 'invaded', i.e. occupied during the revolutionary uprising in response to the 1962 *coup d'état* which entrenched the population in a part of the Colonial City. Despite the victory of the counter-revolutionary movement, some homes continue to be in the hands of these 'invaders'. At present, printing presses are the area's

identifying mark (42 out of the 84 business units). In turn, most of the population of San Antón comes from rural exodus. Some of the best examples of vernacular architecture are located in this neighbourhood: usually single-storey homes made from wood and painted in bright colours. Chances of joining the speculative property market could drive their gentrification.

3   The streets with the greatest wealth of heritage in the southern half are mainly residential and tourist areas. Internal social differences are important. First, architectural quality is important in the colonial area, between Conde and Billini streets. It has a thriving property market with a palpable threat from second homes and gentrification. Tertiary-residential, tertiary-commercial and institutional uses dominate in this first southern sector. Second, the area between Billini and Arzobispo Portes streets is interesting in terms of urban design, although it is not colonial in origin. Residential use is gradually surpassing tertiary and commercial uses. Third, the blocks between Arzobispo Portes and the pier serve as a buffer zone located inside the city walls. These blocks are smaller, of lower architectural value and home to a working-class population. This predominantly residential use combines with interesting local commerce that adds value to the urban landscape, with a high presence of corner shops, some of which operate as larger markets/bars ('colmadones').[3] Lastly, the most southerly area in the neighbourhood borders Avenida Presidente Billini and a run-down jetty.

4   The northern area of the old walled town is not uniform. The urban structure and colonial architecture are less predominant than in the area south of Conde street. This probably plays a role in the lower level of touristification. Mixed and residential uses predominate. There are notable differences between the colonial sector (between Conde and Mercedes streets) and the more renovated blocks located in the north, bordering Avenida Mella. This avenue was built in the early 1900s. Its route follows the outer northern border of the historic centre. Although there are still some architectural examples from the first half of the twentieth century, the area around Avenida Mella has lost many of its historical features. It is today a popular commercial thoroughfare, bordering Santo Domingo's Chinatown (outside the city walls).

The historic centre of Santo Domingo comprises diverse urban areas and different social structures. The differences between the neighbourhoods in the north and the more touristic, renovated and gentrified areas in Ciudad Ovandina have become more acute in recent years. There are also major pockets of poverty and areas suffering physical and environmental degradation sitting alongside tourist areas. In this sense, the poor suburban properties ('casas arrabalizadas', local parlance for shanty towns) occupy a large part of the historic city. According to the 2010 Population and Housing Census (ONE, 2010), 16.36 per cent of households in the historic centre have a medium-low level of sanitation, 12.03 per cent have a zinc roof, material for the exterior walls is not cement or stone in 185 homes, 196 homes have overcrowding problems, 48 extreme overcrowding problems and 107 households are without any separate bedrooms.

Urban policies: the failure of comprehensive regeneration

The history of urban regeneration in the Colonial City is a reflection of how difficult it is to produce a common project for the city. Act 6232 (1963) set up Urban Planning Offices that are attached to town councils. Santo Domingo City Council (Resolution 61/82) (1982) set up the governing commission for the colonial area aimed at coordinating the institutions that operate there and the activities they carry out. Nonetheless, the government body responsible for regeneration is underused, legislation is contradictory and responsibilities are split between different institutions, leading to a complex share of responsibilities. In this context, the influence exercised by overseas bodies and institutions, including NGOs and supranational organisations such as the World Bank, is important. As analysed below, the history of urban planning in the Colonial City is intimately linked to technical and budgetary support from supranational institutions and agencies for development cooperation.

Planned evictions and expulsions

Planned evictions have been part of city planning strategies since 1950 and are an everyday occurrence in Santo Domingo (Franco, 2010). These evictions affect different areas of the city – from the historic centre to outlying areas that have risen in real estate value thanks to their strategic location or to business interests. Three major eviction phases can be identified in Santo Domingo since the mid-twentieth century (Cela, 1992; Morel and Mejía, 1998; Verma, 2006):

a   The last period of the Trujillo dictatorship.
b   Balaguer Government: 1966–1978.
c   Balaguer Government: 1986–1992:

 - Mass evictions (early 1987–late 1988).
 - Revved-up construction (late 1988–mid-1990).
 - Crisis and adjustment (mid-1990–early 1991).
 - The moment of truth; Fifth Centenary Celebration (1991–1996).

The mass evictions in 1967 were justified as necessary to 'clean up the city'. The reconstruction of sections of the city walls and the work in the Plaza de Armas stand out from 1986 and 1987. The so-called *Santo Domingo Northern Zone Development Plan*, produced by the National Planning Office and the German Gesellschaft für Technische Zusammenarbeit, once again planned major evictions under the pretext of necessary urban regeneration. The aim was to boost commercial and tourist pull in the Colonial City, increasing land values as a basis for a financially viable project (Morel and Mejía, 1998). The economic crisis and the economic restructuring agreements (1990–1991) signed with the IMF (1990–1991) worsened the situation. Economic liberalisation and the need to cut inflation placed a brake on building homes for rehousing. Under the initiative of the Consejo de Unidad Popular (People's Unity Council), around 1,000 families took over the streets in October 1990. The government refused to negotiate with the demonstrators and consequently, in December 1990, the United Nations Human Rights Council condemned the Dominican government's eviction scheme for violating the right to housing (Faxas, 2007).

The peak in eviction occurred under the government of Juan Balaguer (known as 'the Great Evictor') with forced mass evictions between 1991 and 1996 due to celebrations for the Fifth Centenary of the Discovery of America. During these years, Santo Domingo sought to beautify the city by taking advantage of the anniversary and opening up spaces for business and tourism investment. The historic centre was seen as the main enclave. The construction of the Columbus Lighthouse and other work transformed Santo Domingo into an archipelago of new middle-class subdivisions (Verma, 2006). Balaguer also aimed to Haussmannise the traditional cores of urban resistance. His principal target was the huge low-income upper town area of Sábana Perdida, northeast of the city centre (Verma, 2006: 105). Valiente is one of the many neighbourhoods built in the mid-1990s by families evicted from the city centre. COPADEBA (Committee for the Defence of Neighbourhood Rights) and the Ciudad Alternativa (Alternative City) project are just some of the main critical voices and expressions of resistance. The former is linked to the church and deemed a popular neighbourhood movement. Ciudad Alternativa (1989), comprising experts and a neighbourhood organisation, was an alternative proposal to the official urban regeneration strategy.

There are no official data on how many people were affected by displacement. Morel and Mejía (1998) calculate that between 1986 and 1992, 400,000 inhabitants were impacted by forced displacements in Santo Domingo, 40 neighbourhoods were bulldozed and around 180,000 residents and 30,000 families were affected by planned evictions. The displacements almost always occurred from the centre to the outskirts (Ciudad Almirante, Los Frailes, Sabana Perdida, La Victoria, Guaricano, Los Alcarrizos and Americanos). The risk of eviction in the Colonial City is currently due to speculative pressure and higher land prices.

Urban regeneration and development cooperation in Santo Domingo

The history of urban planning in the Colonial City is intimately linked to technical and budgetary support from supranational institutions and agencies for development cooperation. Traditionally, regeneration projects for historic centres were coordinated by public culture or heritage institutions and financed with public funds, with occasional contributions from external funding. In turn, overseas cooperation was limited to funded or co-funded local projects. Nevertheless, in the new planning paradigm, many institutions responsible for regeneration have sought support from overseas development agencies and, less successfully, from private partners. These external collaborators do not limit themselves to co-fund local programmes, but do influence the new city design model and set urban regeneration targets and strategies. Santo Domingo is part of this new trend.

The Spanish government was one of the first to take part in saving the historic centre of Santo Domingo. The first contribution was made at the end of the 1950s. The government of dictator Francisco Franco funded the restoration of the Diego Colón Palace and the church of La Compañía de Jesús (Prieto, 2008). International organisations, however, have mostly contributed to finance the (re)designing of the city. For instance, one of the first contributions came from the Esso Plan (1967), which was funded from oil exploration *royalties*. Touristification and segregation were the basis of the plan, which is an example of external dependence and speculative practices (see Table 2.2).

Rehabilitation funded with development support all look at the historic centre with high touristic development potential. New tourist uses and property speculation have underpinned strategies for renovating the city since the first private funding plans in the 1960s to the most recent co-funded plans from supranational institutions from 2000 onwards. The other values inherent in comprehensive regeneration come second, including all those linked to resettlement policies and access to housing for the poorest.

Urban and strategic planning serving touristification and gentrification

Legal protection of the Colonial City dates back to 1969. UNESCO included it on the list of World Heritage Cities in 1992. The main background to current thinking comes from the Regulatory Plan Draft Bill (1988) and the Study of the Colonial City of Santo Domingo (1999). The focus on monuments, museums and tourism has prevailed amongst rehabilitation activities since the initial interventions in the 1960s. Currently, international development plays a key role in the regeneration of the Colonial City. In this way, the two main ongoing plans are the Revitalisation Project for the Santa Bárbara Neighbourhood, at district level with participation from AECID (the Spanish International Development Agency) and the MPCC, which includes the entire historic centre with the IDB as the main funding body.

We will now focus our analysis on the ongoing Master Plan for the Colonial City (MPCC)[4] which is split into three instruments: the Strategic Plan (2004), the Regulatory Plan (2006) and the Project Profile Catalogue. It was designed by Italian consultants Lombardi and Associati. The strategic plan sets out two principles that are the basis of the MPCC proposals: recovering residential space, including the most rundown areas (Santa Bárbara, San Antón and Avenida Mella), and strengthening the tourism system. In other words, gentrification and touristification are the basis for urban regeneration policies. The Regulatory Plan is mainly guidelines and prescriptions for town and country planning. Architectural objectives prevail, with few references to social or environmental concerns. The priority lines are zoning rules and land use. The use plans, organised into dominant uses and preferential professions, summarise the proposals for land use. On the one hand, specific regulations are differentiated by building type and, on the other, the dominant use sectors, which are assigned specific functional vocations (see Table 2.3).

Table 2.2 The main urban regeneration plans in the Colonial City of Santo Domingo carried out within the framework of development cooperation

Name and year	Features and main objectives
Esso Plan (Standard Oil) Santo Domingo Colonial (1967)	Transform the Colonial City into a tourist-cultural sector. This recovery was deemed incompatible with people's rights to live in the historic centre. Many houses were demolished to liberate the old city walls and transform some streets into exclusive enclaves for museums and cultural accoutrements.
OAS. 'Terms of Reference for a Study of Pre-Investment about the Use of Cultural Heritage in the Colonial Zone of Santo Domingo' (1982)	Promoting cultural tourism use and socioeconomic profitability of the project.
OAS. Cuna de América Plan (1991)	Favouring tourism use, the conservation of cultural heritage and reactivating tourism and economic activity. It included removing the poorest members of society from the historic city through eviction and subsequent resettlement in other parts of the city.
IDB. Revitalization of the Colonial City of Santo Domingo (1999)	Producing a revitalisation programme for the Colonial City.
IDB. Master Plan Colonial City (MPCC) of Santo Domingo (2006)	The main rehabilitation plan for the historic centre. Producing strategic guidelines for socioeconomic development of the historic centre. It is split into three instruments.
AECID. Revitalisation Project for the Santa Bárbara Neighbourhood (2007)	This is structured into four main components with different activities planned based on each: (a) improving environmental and sanitary conditions; (b) recovering public spaces and heritage; (c) improving living conditions in homes; (d) strengthening commercial and service activity.
IDB. Tourism Development Programme – Colonial City of Santo Domingo (2011)	Boosting Dominican tourism sector competitiveness through diversifying attractions, generating higher revenue for the local population and reducing pressure on the coast. This is structured into three components: (a) developing key tourist attractions; (b) local integration into tourism development; (c) strengthening tourism management.

Source: Original production

Table 2.3 Priority actions set out in the strategic and regulatory plans

1	Social housing construction projects in the northern area	Plaza de San Antón: renovation of a row of wooden houses. Santa Bárbara: mixed conservation activities for current buildings with a high heritage value, replacement buildings and new builds, as applicable, including two blocks outside the current edge of the Colonial City aimed at strengthening the functional vocation of this working-class residential area. Avenida Mella: building a car park.
2	High-level urban development in the southern residential area	This is based on two rows of street blocks from the Paseo Presidente Billini towards the centre. Aim: favouring the transition process from an average housing area to an upper middle and high-income residential area. This includes the construction of shopping parades on the Paseo Presidente Billini and in the main squares.
3	Urban development in the port area	Tourist port: located between the Puerta de la Misericordia to the current port area, near the Paseo Presidente Billini, and construction of a service station for these activities on the first block outside the city walls, facing Avenida George Washington. Transforming the current port area into a shopping area, including a car park and a pedestrian bridge connecting to Las Damas street and Ozama fortress. Tree-lined boulevard: along the Ozama River, between the port and Puerta de San Diego, including a pedestrian walkway to connect the future boulevard with the pedestrian street that runs from Conde-Las Damas. Extending and shoring up the car parks near Puerta de San Diego.
4	Re-zoning El Conde street	Pedestrianising the street linking Independencia Park with Conde street. Renovating buildings from the early 20th century with historical/architectural value to house private offices and high-quality homes. Connecting the pedestrian system of El Conexión with Catedral and España squares to promote a touring route around the historic centre for tourism and shopping.
5	Tourism Area Development	Work to bolster cultural attractions as museums, improving the look of the city (removing overhead cables), creating tourism routes (signposting), as well as different actions in the Plaza de España (underground car park and constructing a single-storey building for commercial and tourist activities).
6	Changing and improving the current traffic transit system	Traffic improvements on Avenida Mella, pedestrianising El Conde street up to Independencia Park, reordering some public transport routes, limiting traffic by building partial access roads to cul-de-sacs and new car parks.

Source: Original production based on Lombardi and Associati data (2004, 2006)

The MPCC is committed to spatial contextualisation and prioritises targets related to entrepreneurial cities. Firstly, the document highlights the need to bolster the historic centre within the framework of increasing competitiveness at different scales: internal (compared to other sectors in the metropolitan area) and external (Caribbean region). Secondly, boosting the residential side supposes a strategic target where the presence of owner-residents is valued positively. In this way, and under the guise of revaluating the old town and improving the condition of historic buildings, extending high-quality residential use is being sought after, with the clear shift in residential social make-up this implies. According to the plan, the southern area with its high level of heritage sites has the most potential to become a neighbourhood of this kind. Thirdly, the plan proposes activities that without the relevant corrective measures could lead to social segregation. In this sense, the plan argues for the opportunity to promote middle- and upper-class neighbourhoods with lower income households being located in so-called mixed-use neighbourhoods. There is a risk of a *gated community* being built which, stylistically speaking, matches the morphology of historic centres. In short, renovating the Colonial City includes many principles inherent to neoliberal cities. In its zoning, land-use planning is clearly lacking more social and inclusive initiatives and proposals.

Tourism as a pillar for economic regeneration

The Esso Plan aimed at transforming the Colonial City into a tourist-cultural sector. Under the Balaguer government, hotel owners were pushed/advised to offer a night in the Colonial City to boost tourism development. Despite these specific past events, touristification of the historic centre is a recent phenomenon. The main push arose in 2000, supported by plans, programmes and strategies that sought to regenerate the historic centre through concentration on tourism and cultural activities. The Dominican tourism model is mainly built around sun and sand, and the Colonial City welcomes 13.01 per cent of overseas visitors to the Dominican Republic (2013).

The increasing touristification of the historic centre is measured here by two different indicators: tourism accommodation and number of visitors. The data are from internal statistics provided by the Ministry of Tourism. Firstly, the number and distribution of tourist establishments, where we see a trend for a higher number of hotels and beds. The Colonial City offered six hotels in 1989 (176 rooms); this rose to 24 in 2004 (586 beds, 440 rooms) and, by 2014, numbers had risen to 55 tourist establishments and 1,011 rooms (1,795 beds). This growth is combined with a gradual decline in resident numbers. We are therefore seeing a declining representativeness amongst the population in the historic centre in terms of the Distrito Nacional, whilst its economic specialisation in terms of tourism and leisure services is on the rise. A series of data illustrates this trend towards tourist exclusivity:

- The population of the Distrito Nacional increased by 53.11 per cent between 1981 and 2010, and by 30.62 per cent between 2002 and 2010. In the same period, the Colonial City lost demographic weight: –50 per cent (1981–2010) and –30.13 per cent (2002–2010) (ONE, 2010).

- Compared to the increase in housing numbers from 1993 to 2002 (16.83 per cent), recent years have seen a major decline (2002–2010: –12.89 per cent) (ONE, 2010). This is a result of the historic centre's model currently being put into action: more tourism, more leisure, more holidays and fewer permanent residents.

Although tourism establishments are spread across almost the entire Colonial City, there is a major concentration around three streets: the pedestrian shopping street of El Conde, the heritage and residential street of Padre Bellini and the administrative and heritage street of Isabel la Católica. The more run-down, working-class neighbourhoods (San Lázaro, San Antón and Santa Bárbara) and the streets around Avenida Mella fall outside this circuit. In turn, the presence of small hotels, condominiums and tourist apartments in the area around El Conde and the areas with greater heritage sites located in the south also stand out. The so-called tourism 'polygon', housing the main heritage buildings, serves as a recreational zone with an exclusive selection of accommodations, restaurants and shops (see Figure 2.2).

The historic centre also plays a role in the touristification and gentrification processes by consolidating another type of accommodation selection, converting residential apartments into commercial lodgings advertised on Airbnb and other sites. This new type of holiday residence covers not only empty homes taking over the traditional rental market, since this new type of accommodation market offers renters higher revenue. Airbnb had 123 homes for rent in October 2015. Airbnb has a presence across all city neighbourhoods, although the selection in those areas with greater heritage stands out. It has scant presence on Avenida Mella and in the lower income residential areas to the north, except San Antón, given its heritage and urban advantages (the ruins of San Francisco, Plaza de San Antón . . .) (see Figure 2.3).

A second indicator for measuring touristification is visitor numbers. Figures from the Ministry of Tourism show that 619,835 tourists visited the Colonial City in 2013, of whom 89.91 per cent (557,333) were from overseas. Nevertheless, only 12.99 per cent stayed in the Colonial City overnight. According to the Tourism Carrying Capacity Monitoring System for the Colonial City (2014), 53 per cent of those surveyed spent less than a day in the Colonial City. These figures confirm how the historic centre functions as a secondary or supplementary offer in resort packages and cruises.

This supplementary nature leads to intensive use of space, i.e. a concentration of pressure from tourism in just a few streets and for just a few hours a day. The tourists who come to the city on a cruise (the Don Diego, Sans Souci and Cristóbal Colón port terminals), on a planned day trip from a beach resort (Boca Chica, Bávaro, Punta Cana, La Romana . . .) or those who stay in the city itself have very localised routes that match the streets with the highest number of unique historical buildings, restaurants and shops aimed at tourists. This comprises three main axes: El Conde (Colón Park and Cathedral), Las Damas (Ozama Fortress and Casa de Ovando) and the Plaza de España (Puerta de San Diego and Alcázar de Colón). This type of day-trip tourism is likely to increase in coming years.

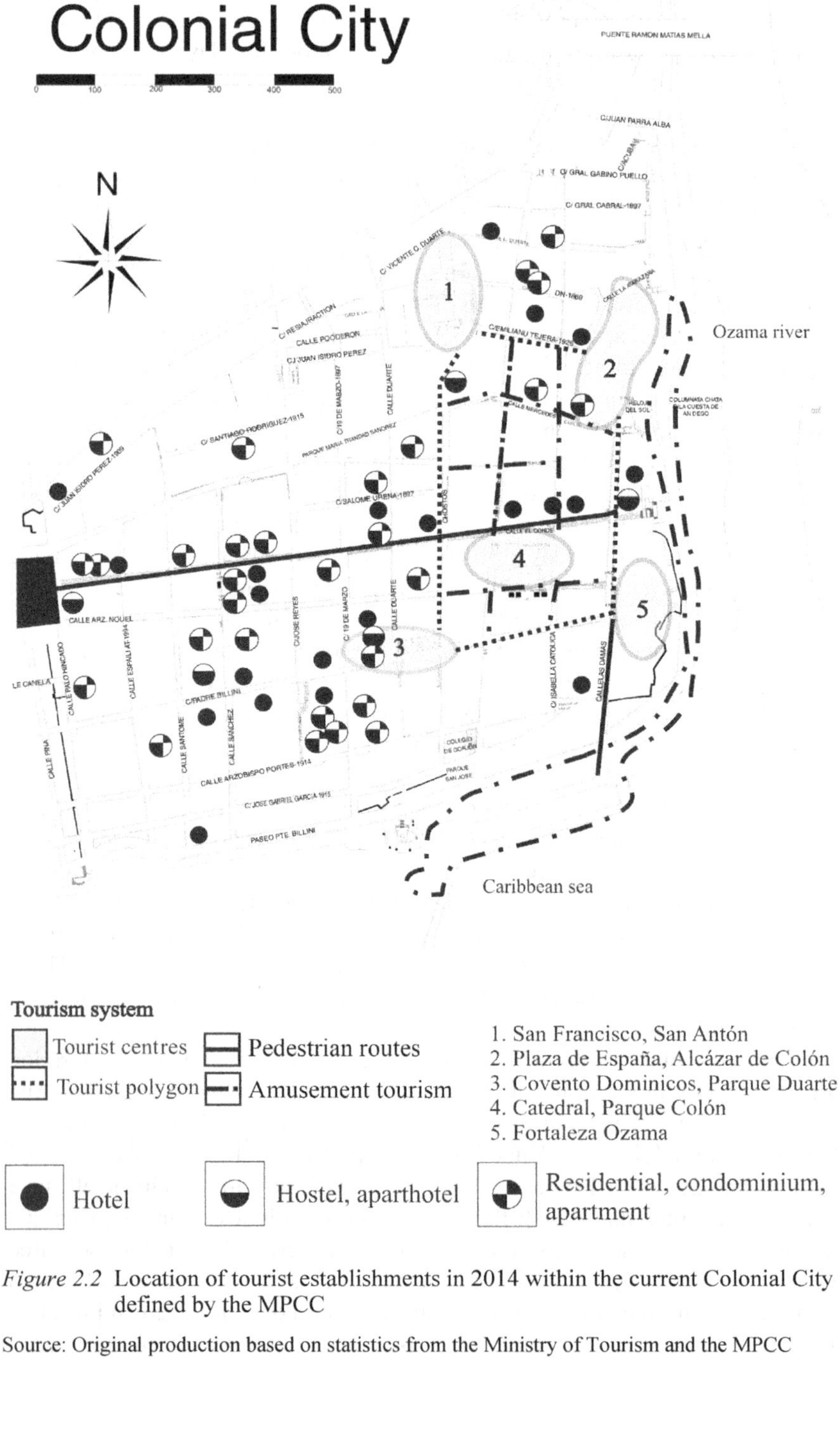

Colonial City

Figure 2.2 Location of tourist establishments in 2014 within the current Colonial City defined by the MPCC

Source: Original production based on statistics from the Ministry of Tourism and the MPCC

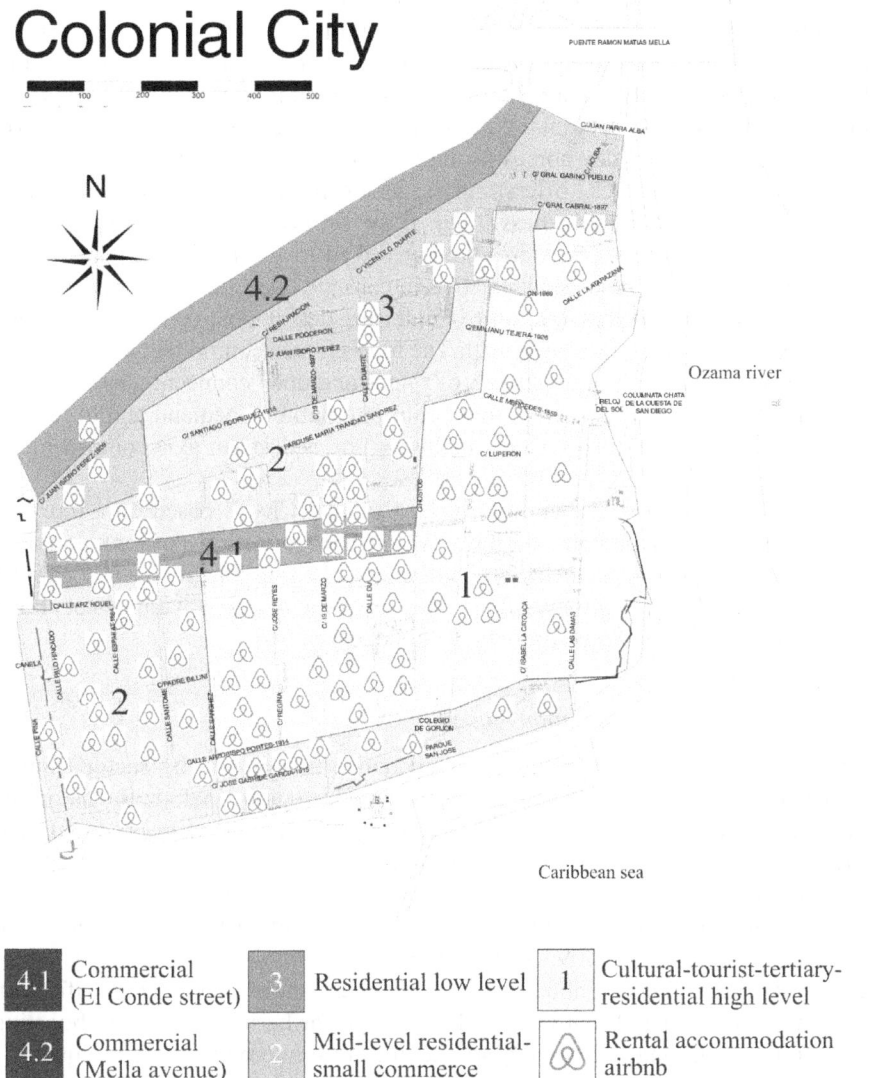

Figure 2.3 Location of Airbnb's holiday rentals (15 October 2015) within the social and urban structure of the Colonial City as defined by the MPCC

Source: Original production based on Airbnb and MPCC

For example, the Coral main road, which opened on 8 August 2012 connects La Romana and Bayahibe to Bávaro-Punta Cana, offering an area that houses around 60 per cent of the hotels in the Dominican Republic direct access to the capital. Around 20,000 vehicles are estimated to use this road, carrying around 600,000 tourists per year. Before it was built, it took three hours to travel the 215 kilometres between Santo Domingo and Punta Cana. The journey time was cut to just 60 minutes by this new road. One of the knock-on effects from this road will be people travelling in both directions. On the one hand, more visitors in Bávaro-Punta Cana will travel out for the day to Santo Domingo and the old town. On the other, it makes the Colonial City a suitable and accessible spot for property investment and for holiday homes for high-income social classes.

In this light, the MPCC offers tourism as a central force for urban regeneration and regional competitiveness. Four of the six priority actions set out in the strategic and regulatory plans are directly linked to tourism development. The strategic plan frames a tourism area covering around one fifth of the old town and promotes a waterfront area as a future fun leisure destination, currently in decline as an urban space and as a tourist spot. The tourism area comprises five main sites of attraction, supported by a network of routes and open connecting spaces. This area or 'polygon' will be used to meet both international demand and to join up with other complementary systems including the pedestrian axis comprising El Conde and the so-called 'port boulevard' (see Figure 2.4).

In short, the urban regeneration of the Colonial City is based on a tourism-heritage duality. Packaging up heritage, with its categorisation measures and risks of becoming trivial, is occurring whilst the city is losing inhabitants and gaining hotel beds, and higher visitor numbers: there is a ratio of 8.38 inhabitants per hotel bed (1989: 37.34) and 86.66 tourists per resident in 2014.

Gentrification as a regeneration model

The gentrification processes have spread both spatially and by sector (Smith, 2006). Tourism is one of the sectors affected and regions include towns on the periphery of tourism areas. Like other consumer spaces, tourism produces gentrification (Smith, 1996; Judd and Fainstein, 1999; Wilson and Tallon, 2011). Prepping historic centres in Latin America for symbolic gentrification destroys some of their features with high heritage value, leading to 'musealisation' of heritage (Monterrubio, 2009; Nelle, 2009). Santo Domingo is a spatially segregated city, split into neighbourhoods of different social classes. Gentrification is planned in the belief that it is the best option for heritage recovery, boosting the economy and making the historic centre more international. Consequently, property reclassification and tourism could lead to higher urban fragmentation and privatisation.

There is no scientific bibliography analysing in detail gentrification in the Colonial City. This is probably due to how new the process is as well as the general interest in boosting overseas investment as a way to kick-start the economy and renovate urban and social fabric. The effects these processes may have are not taken into account by current urban planning. The negative impact of property revaluation and its associated gentrifying processes make no appearance in the 132 pages of the Regulatory Plan or in the 562 pages of the Strategic Plan.

Colonial City

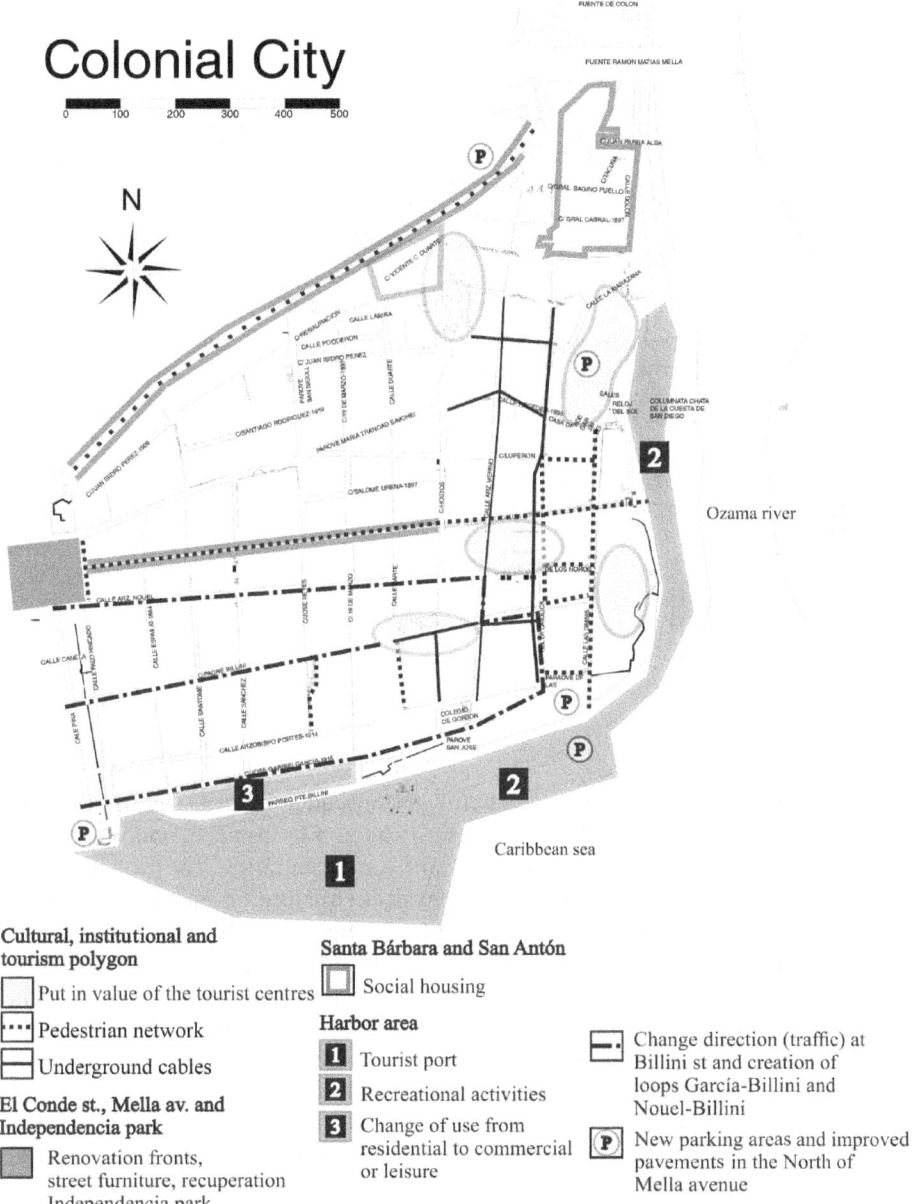

0 100 200 300 400 500

N

Ozama river

Caribbean sea

Cultural, institutional and tourism polygon

☐ Put in value of the tourist centres

┄ Pedestrian network

⊟ Underground cables

El Conde st., Mella av. and Independencia park

▨ Renovation fronts, street furniture, recuperation Independencia park

Santa Bárbara and San Antón

⊞ Social housing

Harbor area

1 Tourist port

2 Recreational activities

3 Change of use from residential to commercial or leisure

⊟ Change direction (traffic) at Billini st and creation of loops García-Billini and Nouel-Billini

Ⓟ New parking areas and improved pavements in the North of Mella avenue

Figure 2.4 Summary of the strategic plan proposal for the Colonial City

Source: Original production based on Lombardi and Associati data (2004)

Figure 2.5 The risks of tourism 'dumbing down' in the Colonial City

Source: Original production

These processes begin with apparently inconsequential urban policies; however, they have a huge effect for later gentrification. For instance, removing street sellers is a common practice in many Latin American cities. The Colonial City has seen 164 hawkers relocated into a physical structure or square, where they are regulated as per the items they sell. As with any suburban gated community, installing CCTV on streets and in squares is a common practice widely seen in historic centres under the premise of local and visitor safety. The Ministry of Tourism, through its Promoting Tourism in the Colonial City programme approved a monitoring centre and the installation of 123 CCTV cameras at 53 intersections in the Colonial City in 2015, with a budget of 8.6 million US dollars.

In these circumstances, the Colonial City is an ideal setting for gentrification: quality urban heritage; large rundown residential areas awaiting forced resettlement so as to be renovated; favourable urban planning, as well as domestic and, especially, overseas investment interested in the property sector. At present, it seems to have enjoyed great success as an activity linked to tourism and limited to highly specific areas – those with high heritage value (see Table 2.4).

The map showing the blocks with low-income households (income below 6,000 RDS, about 119 Euros) and a high percentage of rented homes (over 70 per cent of homes being rented in buildings on the street block) shows the location of poorer

Table 2.4 Indicators according to dominant use at street block level. Total (number of blocks in the area) and percentage (per cent of blocks compared to total street blocks in the area)

	Income (under 6,000 RDS)		Blocks with over 70% of homes being let		Blocks with over 50% empty homes		Tourist accommodation*		Airbnb accommodation	
	Total	%	Total	%	Total	%	Total	%	Total	%
ZONE 1	28	87.5	18	56.25	5	15.16	38	69.09	53	43.44
ZONE 2	34	79.06	16	37.20	7	16.27	14	25.45	58	47.54
ZONE 3	29	100	13	50.00	4	15.38	3	5.45	11	9.01

Source: Original production based on Lombardi and Associati data (2004, 2006) and Airbnb

Note: *The percentage of tourist lodgings and Airbnb by sector is in relation to the total amount in the Colonial City.

social classes: few owners and low income (see Figure 2.6). Although it is spread across all sectors, the localisation in sector 1 stands out, defined by the strategic plan as the cultural, tourist, tertiary and high-quality residential area. In 2004, the year the data referred to, the level of gentrification is low and many owners prefer renting properties to poorer families, being seen as a form of protection against expropriation and capitalisation for better sales' opportunities (to foreigners). This situation cuts the supply of homes, keeps the market at a standstill, pushes up property prices through speculation and perpetuates declining slum-like conditions. It is estimated that 25 per cent of buildings in the Colonial City at the start of the century were in these conditions (Toca, 2000).

Although renting is the most common trend seen in the Colonial City, there has been a marked decrease. In just eight years, it has fallen by ten points, especially affecting those areas most susceptible to gentrification. In 1998, 69 per cent of homes were rented; in 2002, the figure fell to 62 per cent and in 2010, to 52.99 per cent (total: 1,477). Homes that are owned and fully paid for are the second most common (total: 941; 33.76 per cent in 2010). Property is held by just a few owners and a single family may have more than eighty properties it neither sells nor repairs (Toca, 2000). According to the Colonial City Development Foundation Directive, the Vicini family had 78 properties and the Portela family held 42 in the mid-2000s. According to Forbes Mexico, the Vicini family is the richest in the Dominican Republic (sugar mills, manufacturing, finance, media . . .). In third place, we find transferred or borrowed homes (total: 226; 8.10 per cent). Leaving aside a small group comprising other factors (23 instances), in fourth place come those homes with an outstanding purchase loan or mortgage. They represent 0.43 per cent (120 homes), a very low percentage if we compare this situation to property markets in Europe or North America. In short, compared to an upper-class minority who own more than one debt-free property, most individuals (on low incomes) rent (Toca, 2000). Lombardi and Associati (2004) argue that the main reasons for the high number of empty homes and the bad condition of many are the low rents and the Dominican legislation that provides protection to tenants. However, this reasoning does not take into account

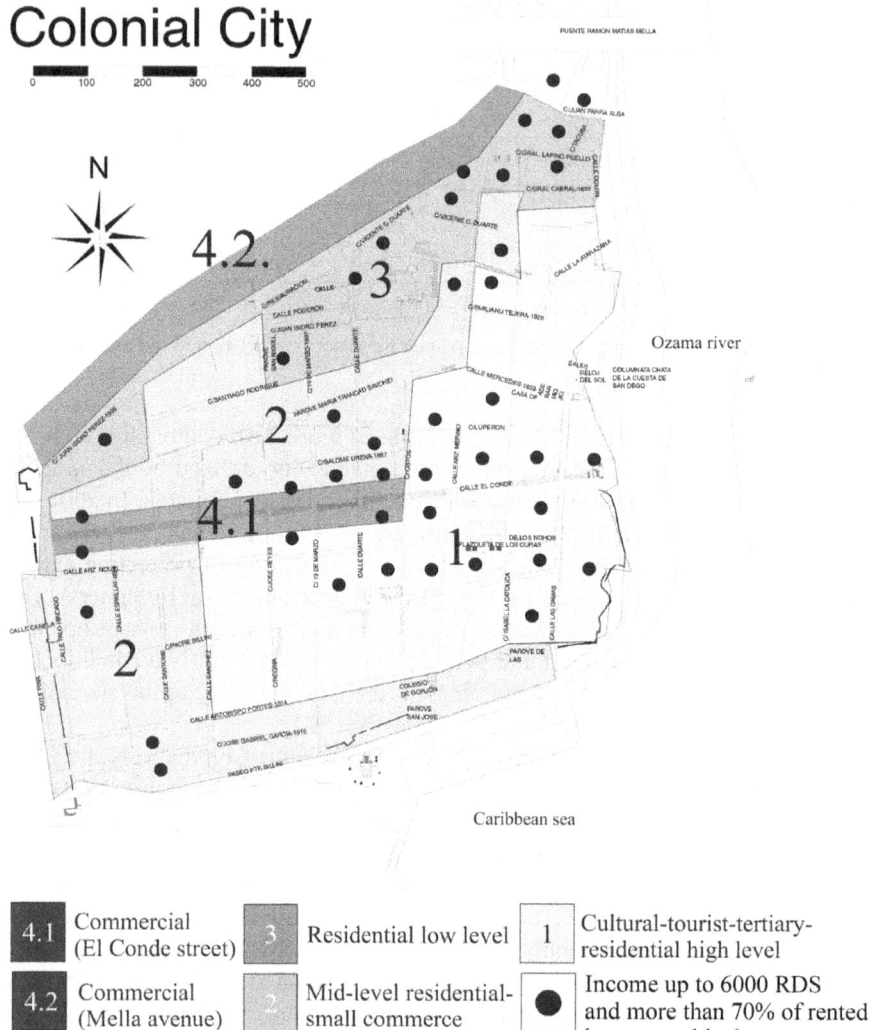

Figure 2.6 The location of low-income households and residents living in rented properties on the social and urban structural map of the Colonial City

Source: Original production based on the MPCC

the relationship between speculation and maintaining empty homes that are awaiting renovation, the arrival of investors and, in short, higher business expectations.

The fact that there are empty homes is an indicator not only of population drift from the historic centre, but also of interest in creating a large quality housing market likely to attract tourists or second homeowners aimed at new social classes. In 2004, 11 per cent of homes were empty and only ten blocks had all their homes occupied. In 2010, there were 692 empty homes, representing 20.01 per cent of the total. Although there are empty homes across the historic centre, the small concentration in sector 1 and 3 stands out. Sector 1 is home to low-income households and has a large proportion of rented properties. Although also home to low-income households, sector 3 is awaiting investment for regeneration (see Figure 2.7).

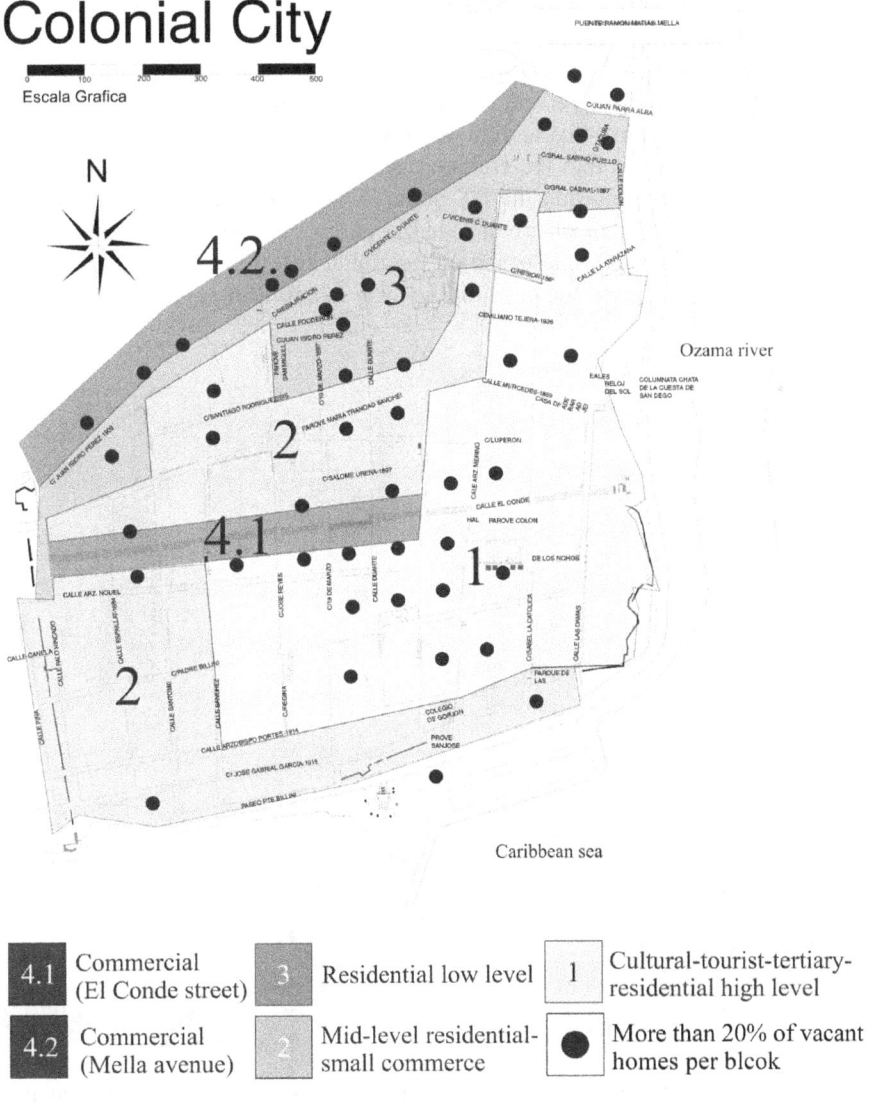

Colonial City

Figure 2.7 Location of empty homes (over 20 per cent of homes on a block) on the social and urban structural map of the Colonial City

Source: Original production based on the MPCC

These factors, alongside the function specialisation and large scheduled zoning programmes, favour higher real estate activity and the arrival of overseas investors. Some renovated homes are reaching prices above 1,500,000 dollars. As an example, the German real estate agent Engel and Völkers had five homes for sale in the Colonial City as of November 2015. The average price had reached 2,354.74 USD/m². The Castilla Luxury Residence is a completely renovated building containing thirty apartments in the style of a gated community with communal gardens and swimming pool. Defining itself as a 'project inspired by Mediterranean cities but with the tropical charm of the Caribbean', it offers homes with a price tag above 800,000 US dollars (see Figure 2.8).

El Conde is the backbone of the historic centre. It is the most important shopping and tourist thoroughfare, and major investments for regeneration are planned. By way of a conclusion to this section, we have mapped the main land uses. During our fieldwork we noted that the commercial establishments offered dual currencies: the Dominican peso and the US dollar. We believe the peso/dollar duality may represent an element of urban segregation by highlighting the existence of 'two parallel economies', one based on the dollar (centred around tourism from overseas) and the other on the peso for the local Dominican population. Except in

Figure 2.8 Construction of gentrified spaces: property revaluation and new residential condominiums

Source: Original production

one instance, all commercial establishments on El Conde and Las Damas streets allow customers to pay in both dollar and pesos. Euro payment is also accepted at jewellery shops.

Retailing mainly aimed at tourists represents the overarching business activity. This falls back in the eastern part and on Las Damas street where there are cultural establishments, restaurants and luxury hotels. El Conde is also the main area for buying souvenirs. We numbered 61 street sellers there. Other types of typical retail outlets in the area include tobacconists (3), art shops (2) and jewellers (5). There are two small shopping centres and some fashion and restaurant franchises, mainly with Dominican capital (Pollo Rey, KB Stores, Helados BON . . .), and to a lesser degree, from overseas (KFC). A total of 28 premises are empty (see Figure 2.9).

Conclusions

One idea summarises the urban processes being experienced in some historic centres in the Caribbean: heritage success is social failure. This idea should make us think about the impact of urban regeneration and renovation policies that are ever further removed from integration and social targets. Using tourism as a main focus for urban regeneration is consolidating historic cities that are spatially fragmented and socially fractured. Old areas in decline are moving backwards

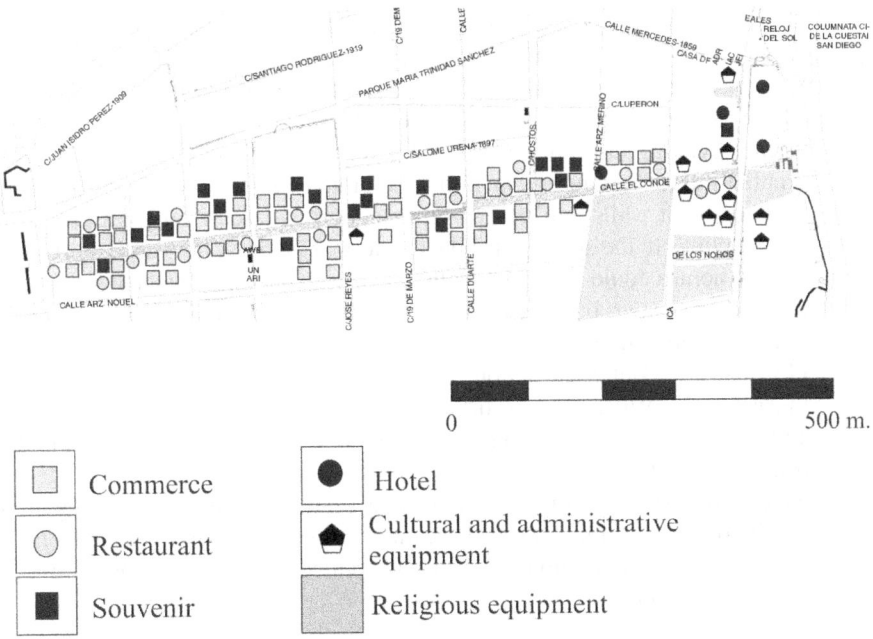

☐ Commerce	● Hotel	
○ Restaurant	⌂ Cultural and administrative equipment	
■ Souvenir	▨ Religious equipment	

Figure 2.9 Ground level land use on El Conde and Las Damas streets in 2014

Source: Original production

and new tourist/renovated real estate scenarios are progressing side-by-side. The result is a schismatic city.

Santo Domingo has a wide experience in planned evictions. This may not be happening at present, since democratic channels and procedures have improved. Eliminating poverty strongholds occurs through other mechanisms for driving out the population and later tourism and real estate specialisation. The MPCC outlines a residential area for working classes in the northeast sector of the old walled city, far from more promoted areas, whilst at the same time, zoning other more exclusive residential areas.

The MPCC documentation is high quality. The studies are exhaustive, highly detailed and well set out. Nonetheless, physical regeneration is prioritised, tourism and property revaluation are underscored as focal points for development and a commitment is made to planning policies for land use (zoning). Gentrification, homogenisation and contextualisation may lead to the city being understood as an enclave of consumerism, with new and pretty, gentrified reserves that are (re)built on property revaluation and social transformation. In this sense, protecting and recovering heritage is important; nevertheless, without the proper social and resettlement policies, the living memory of the Colonial City will become a tourist souvenir. In short, the following are the three main threats resulting from gentrifying the Colonial City:

- Economic: more speculation, higher cost of living in general and higher housing costs in particular. International investor groups see revaluing the Colonial City as a synonym of economic return. The gradual appearance of foreign estate agents, such as Remax from the US and Engel and Völkers from Germany, stand out in the property market.
- Social: evicting and failing to resettle former residents. The loss of demographic weight and the ageing population make revitalising the historic centre more difficult whilst elitism and repopulation with temporary residents, who have no links to the area, contribute to a loss of identity. Increased tourism and second homes could make the Colonial City a space that is solely inhabited a few months a year. In December 2014, Airbnb was advertising 61 rental properties. By November 2015, this figure had doubled to 122.
- Physical: the contextualisation of space and musealisation for high-income tourism and residential use. Imposing new architectural designs for homes and commercial premises, which match the tastes of new residents and have little to do with local culture and heritage. Luxury complexes change how the courtyards of colonial houses are used, removing plant life, changing the profile of the Colonial City and creating an imbalance in the socioeconomic make-up of the population. They also introduce different patterns of co-existence and customs.

Thus, the historic centre of Santo Domingo is an urban space undergoing transition. Many of the processes are part of the changes affecting cities in the era of globalisation and the dynamics that characterise neoliberal, postmodern cities.

Tourism and incipient gentrification are the most influential processes, given their ability to model landscapes, focalise the economy and introduce major transformations in the social make-up of the city.

Acknowledgements

This research has been funded by two research projects: 'Crisis and social vulnerability in Spanish island cities. Changes in the social reproduction spaces.' CSO2015-68738-P (AEI/FEDER, UE); 'Tourism and transformation of urban spaces. Comparative analysis and knowledge transfer to different cities of Cuba and the Dominican Republic,' Government of the Balearic Islands and OCDS (UIB).

Notes

1 Twelve kilometres northeast of Trinidad are three interconnected rural valleys that make up the 225-km² Valley de los Ingenios. The former plantations, mill buildings and archaeological sites represent the richest and best-preserved testimony of the Caribbean sugar agro-industrial process of the eighteenth and nineteenth centuries, and of the slavery phenomenon associated with it.
2 Many publications and institutions include Bermuda in the Caribbean due to its cultural and economic similarities to other islands in the region. The city of Santa Cruz de Mompox is inland but the region of Bolívar, where it is located, runs to the Caribbean Sea.
3 These shops have tables and chairs outside on the pavement, used as a terrace. The music is generally provided by cars that park along them. They operate as bars but at corner shop prices and offer a delivery service.
4 Information for this section is based on fieldwork and the MPCC. The 2010 census data are to Colonial City scale. The maps, which at block level represent average income, empty houses and the percentage of rented property, provide some keys underlying the analysed processes. This information is shown on the dominant use district map according to the MPCC which largely matches the sector breakdown provided in section 3.

References

Achtenberg, E. and Currents, R. (2013) *Gentrification in Cuba? The Contradictions of Old Havana*, Washington: NACLA Report on the Americas.
Arocho, A.I. (2007) *El fenómeno de la gentrificación en la comunidad Alto del Cabro, Santurce: análisis del problema y propuesta para el establecimiento de una estrategia de permanencia comunitaria*, San Juan: Master thesis, University of Puerto Rico.
Bailey, N. (2008) 'The Challenge and Response to Global Tourism in the Post-modern Era: The Commodification, Reconfiguration and Mutual Transformation of Habana Vieja, Cuba', *Urban Studies*, vol. 45, no. 5–6, pp. 1079–1096.
Cela, J. (1992) *Con los trastes en la espalda: remodelación estatal y movilidad intraurbana*, Santo Domingo: Ciudad Alternativa.
De Jesús, V.M. (2011) *Las actitudes de los residentes y las residentes hacia el realojo en una comunidad urbana pobre en San Juan: 2006–2007*, Salamanca: PhD thesis, University of Salamanca.
Dilla, A. (2010) 'Ciudades en el Caribe', *Umbral*, vol. 3, pp. 4–34.

Dürr, E. and Jaffe R. (2012) 'Theorizing Slum Tourism: Performing, Negotiating and Transforming Inequality', *European Review of Latin American and Caribbean Studies*, vol. 93, pp. 113–123.

Faxas, L. (2007) *El mito roto. Sistema político y movimiento popular en la República Dominicana, 1961–1990*, Mexico D.F.: Siglo XXI.

Franco, P. (2010) 'Como los pobladores enfrentan los desalojos en Barrio Valiente', in Cabannes, Y., Guimarães, S. and Johnson, C. (ed.), *Como los pobladores se enfrentan a los desalojos*, London: University College London, pp. 79–89.

González, J.M. (ed.) (2015) *Ciudades en transición. Procesos urbanos y políticas de rehabilitación en contextos diferenciados: centro histórico de La Habana y Ciudad Colonial de Santo Domingo*, Palma: University of the Balearic Islands.

González, J.M. and Lois, R.C. (2010) 'The Historic Centre in Spanish Industrial and Postindustrial Cities', *The Open Urban Studies Journal*, vol. 3, pp. 34–46.

Gutiérrez, R. (2008) 'Centros históricos en Iberoamérica. Experiencias y reflexiones', in Gutiérrez, R. (coord.) *El centro histórico de Santo Domingo*, Santo Domingo: AECID, pp. 7–18.

Hiernaux, D. and González C.A. (2014) 'Turismo y gentrificación: pistas teóricas sobre una articulación', *Revista de Geografía Norte Grande*, vol. 58, pp. 55–70.

Janoschka, M., Sequera, J. and Salinas, L. (2014) 'Gentrification in Spain and Latin America – a Critical Dialogue', *International Journal of Urban and Regional Research*, vol. 38, no. 4, pp. 1234–1265.

Jones, G. and Varley, A. (1999) 'The Reconquest of the Historic Centre: Urban Conservation and Gentrification in Puebla, Mexico', *Environment and Planning A*, vol. 31, no. 9, pp. 1547–1566.

Judd D.R. and Fainstein S.S. (1999) *The Tourist City*, New Haven: Yale University Press.

Lees, L. (2003) 'Super-gentrification: The Case of Brooklyn Heights, New York City', *Urban Studies*, vol. 40, no. 12, pp. 2487–2509.

Lees, L., Slater, T. and Wyly, E. (2008) *Gentrification*, New York: Routledge.

Lombardi and Associati (2004) *Plan Estratégico de revitalización integral de la Ciudad Colonial de Santo Domingo*, Santo Domingo: BID and Secretariado Técnico de la Presidencia de la República Dominicana.

Lombardi and Associati (2006) *Plan Regulador de la Ciudad Colonial de Santo Domingo*, Santo Domingo: BID and Secretariado Técnico de la Presidencia de la República Dominicana.

Lungo, M. and Baires, S. (2001) 'Socio-Spatial Segregation and Urban Land Regulation in Latin American Cities', Conference paper, Cambridge, Massachusetts: Lincoln Institute of Land Policy, pp. 2–21. Available at: http://citeseerx.ist.psu.edu/viewdoc/download?doi=10.1.1.200.8130&rep=rep1&type=pdf [accessed 31 Jan 2017]

Monterrubio, A. (2009) *Hábitat popular, renovación urbana y movimientos sociales en barrios céntricos de la Ciudad de México 1985–2006*, Mexico D.F.: PhD thesis, The Metropolitan Autonomous University of Mexico.

Morel, E. and Mcjía, M. (1998) 'The Dominican Republic: Urban Renewal and Evictions in Santo Domingo', in Azuela, A., Duhan, E. and Ortiz, E. (ed.), *Evictions and the Right to Housing: Experience from Canada, Chile, the Dominican Republic, South Africa and South Korea*, Ottawa: International Development Research Centre, pp. 83–143.

Nelle, A. (2009) 'Museality in the Urban Context: An Investigation of Museality and Musealization Processes in Three Spanish-colonial World Heritage Towns', *Urban Design International*, vol. 14, no. 3, pp. 152–171.

ONE (2010) *IX Censo Nacional de Población y Vivienda*, Santo Domingo: Oficina Nacional de Estadística.

Prieto, E. (2008) 'El aporte de la cooperación española en el rescate del centro histórico de Santo Domingo', in Gutiérrez, R. (ed.), *El centro histórico de Santo Domingo*, Santo Domingo: AECID, pp. 43–50.

Rodríguez, K.Y. (2008) 'Cultura y Acción Comunitaria: Resistencia comunitaria ante el Plan Maestro de Revitalización de Santurce Centro', in *Proyecto Mapa Cultural del Puerto Rico Contemporáneo*, San Juan: Universidad de Puerto Rico, pp. 1–13.

Scarpaci, J. (2000a) 'Winners and Losers in Restoring Old Havana', in *Cuba in Transition*, vol. 10, Miami: Association for the Study of the Cuban Economy, pp. 289–300.

Scarpaci, J. (2000b) 'Reshaping Habana Vieja: Revitalization, Historic Preservation, and Restructuring in the Socialist City', *Urban Geography*, vol. 21, no. 8, pp. 724–744.

Scarpaci, J. (2002) 'La transformación de los centros históricos latinoamericanos y el proceso de globalización', *Revista de Geografía*, vol. 1, pp. 15–33.

Scarpaci, J., Segre R. and Coyula, M. (2002) *Havana: Two Faces of the Antillean Metropolis*, Chapel Hill: University of North Carolina Press.

Semple, H.M. (1999) 'Downtown Revitalisation in Kingston, Jamaica: An Assessment', *Caribbean Geography*, vol. 10, no. 2, pp. 89–102.

Shaw, K. (2008) 'Gentrification: What It Is, Why It Is, and What Can Be Done About It', *Geography Compass*, vol. 2, no. 5, pp. 1697–1728.

Smith, N. (1996) *The New Urban Frontier. Gentrification and the Revanchist City*, London: Routledge.

Smith, N. (2002) 'New Globalism, New Urbanism: Gentrification as Global Urban Strategy', *Antipode*, vol. 34, no. 3, pp. 427–450.

Smith, N. (2006) 'Gentrification Generalized: From Local Anomaly to Urban "Regeneration" as Global Urban Strategy', in Downey, G. (ed.), *Frontiers of Capital: Ethnographic Reflections on the New Economy*, London: Duke University Press, pp.191–208.

Suárez, C.A. (2009) *Marketing the Gates: The Politics of Gated Communities in Guaynabo, Puerto Rico*, Dissertations, vol. 139, Amherst: University of Massachusetts-Amherst.

Súarez, C.A. (2010) *Gentrification as Municipal Policy? The Experience of Guaynabo, Puerto Rico*, Unpublished.

Thomas, G.A. (1991) 'The Gentrification of Paradise: St. John's, Antigua', *Urban Geography*, vol. 12, no. 5, pp. 469–487.

Toca, N. (2000) 'Dimensión socio-económica y política de la revitalización del centro histórica de Santo Domingo, República Dominicana', in *Requailification, revitalisation e auto-durabilité des Centres historiques: un projet urbain*, Salvador de Bahía: Sirchal.

Vergara, C. (2013) 'Gentrificación y renovación urbana. Abordajes conceptuales y expresiones en América Latina', *Anales de Geografía*, vol. 33, no. 1, pp. 219–234.

Verma, G. (2006) 'Haussmann in the Tropics', in Davis, M. (ed.), *Planet of Slums*, London-New York: Verso, pp. 95–120.

Wilson, J. and Tallon, A. (2011) 'Geographies of Gentrification and Tourism', in Wilson, J. (ed.), *The Routledge Handbook of Tourism Geographies*, London: Routledge, pp. 103–112.

3 Airbnb and tourism gentrification

Critical insights from the exploratory analysis of the 'Airbnb syndrome' in Reykjavík

Anne-Cécile Mermet

Introduction

After the dramatic collapse of the Icelandic economy in October 2008, Iceland is currently going through what the media describes as a "miraculous" recovery (O'Brien, 2015). In the years following the crisis, whereas the income from fisheries or from aluminium production has dropped considerably, the tourist influx has surged (+ 105 per cent between 2007 and 2014), and has played a crucial part in this ongoing upturn (Jóhannesson and Huijbens, 2010). It is particularly visible in the Reykjavík urban landscape, where building sites are increasing, often with a purpose related to tourism.

Beyond the impact on the urban landscape, the development of tourism in the city is tightly intertwined with the social production of urban space. Notably, many narratives consider tourism as a new dynamic fuelling the gentrification process in Reykjavík (Mathiesen *et al.*, 2014), especially through the development of a new parallel housing market dedicated to tourists, by way of short-term rentals (today it represents 2.5 per cent of residences in the urban area). Even if this is a less-known case study than San Francisco, Barcelona or Paris, Reykjavík is currently undergoing a significant 'Airbnb syndrome' in a very specific post-crisis context, which makes it a particularly instructive example for the analysis of the interaction between the emergence of short-term rentals and local housing markets. Indeed, the crisis has also had a considerable impact both on the local real estate market and on the social and economic situations of households which resulted in a new massive demand on the very small long-term rental market. Thus, the case of Reykjavík provides an interesting insight into the tension that is emerging between the growing presence of tourism in the production of urban space and the weakening of the economic situation of the inhabitants.

This contribution intends to complete the tourism gentrification literature, which has rather focused on the effect of tourism-oriented urban projects on residential gentrification. It proposes a more direct way of thinking the link between tourism and gentrification which turns tourists into gentrifiers due to the new short-term rental market. I argue first that short-term rentals must be considered as a new segment of the housing market, which tightly interacts with the more classic segments such as long-term rentals, and, second, that the emergence of this new real estate market fuels a *direct* form of tourism gentrification. This chapter presents the first results of the exploratory phase of research into tourism gentrification in Reykjavík.

(In)direct tourism gentrification

Contemporary gentrification corresponds to what scholars have called the "third wave of gentrification" (Hackworth and Smith, 2001), which is henceforth considered as "a global urban strategy" (Smith, 2002). It is argued that the process has been "generalized" (Smith, 2006) both spatially (with rural, suburban or variegated – in developing countries – gentrification) and sectorally. Indeed, gentrification is tightly interwoven with other urban dynamics such as industry (Curran, 2004; Yoon and Currid-Halkett, 2015), retail (Bridge and Dowling, 2001; Sullivan and Shaw, 2011; Zukin *et al.*, 2009), but also tourism, which has long been described as another major force that reshapes contemporary cities (Hoffman *et al.*, 2003; Judd and Fainstein, 1999; Maitland, 2010).

Literature has only recently begun to pay attention to the link between urban tourism and gentrification (Bures and Cain, 2008; Gladstone and Preau, 2008; Gotham, 2005; Herrera *et al.*, 2007; Liang and Bao, 2015; Vives Miró, 2011; Wilson and Tallon, 2011). The first study on the topic has been provided by Gotham, who has developed the concept of 'tourism gentrification' "as a heuristic device to explain the transformation of a middle class neighbourhood into a relatively affluent and exclusive enclave marked by a proliferation of corporate entertainment and tourism venues" (2005: 1102). In this paper, he demonstrates how the development of tourism in the Vieux Carré in New Orleans (through tourism-oriented projects, mega-events, but also the setting-up of dozens of hotels and the development of new retail structures catering to tourists) has not only resulted in a dramatic increase in tourist influx, but also in the transformation of the socio-demographic features of the area in favour of a wealthier and whiter population. Thus, the underlying theory of this paper is actually that the development of tourism in a city can be one of the driving forces which foster or intensify *residential* gentrification. Most of the more recent papers are based on the same idea, by analysing how tourism-oriented urban projects enhance urban environments, provide amenities and produce consumption spaces which are also highly sought-after by gentrifying populations (Füller and Michel, 2014: 1310; Herrera *et al.*, 2007: 277; Maitland, 2010: 178), and which can raise displacement issues (Gladstone and Preau, 2008; Herrera *et al.*, 2007: 140).

To sum up, literature has so far largely focused on *indirect* tourism gentrification. Surprisingly, much less attention has been paid to the more direct link between tourism and gentrification, through the existence of segments of the real estate market specifically and directly dedicated to tourists. Here, I argue that a more direct form of tourism gentrification, which makes tourists gentrifiers, can be found behind these segments of the real estate market.

The second-home is the best-known segment of this market targeting tourists. Indeed, this issue has been studied by a wide range of authors (e.g. Hall and Müller, 2004), and focuses mainly on the conflicts between locals and tourists (Farstad and Rye, 2013; Gallent, 2014) or on the way second-home owners negotiate their multi-spatialities (Chevalier *et al.*, 2013; McIntyre *et al.*, 2006; Norris and Winston, 2010). Nevertheless, housing studies have recently begun to link the growing development of second homes to the question of gentrification (Paris, 2009), and especially of rural gentrification (Paris, 2009; Solana-Solana, 2010). For instance, Paris proposed a pattern of the effects of the growth of second homes in rural areas which mirrors the classic features of residential gentrification (Paris, 2009: 300; 2010: 24): rising prices on the local real estate market, displacement of long-term inhabitants which are replaced by wealthier – albeit temporary – ones, and upgrading of the built environment. More recently, the rise of the sharing economy led to the creation of a new segment of real estate targeting tourists, namely, short-term rentals.

New urban tourism and Airbnb

Bursting the tourism bubble

This new kind of tourism accommodation needs to be recontextualised within the emergence of a "new urban tourism" (Füller and Michel, 2014), involving new tourism practices and new urban impacts. This tourism is based on tourists' quest for an 'authentic experience', by spreading throughout the entire city (admittedly with a predilection for gentrifying neighbourhoods), seeking 'off the beaten track' (but always secure) itineraries (Djament-Tran and Guinand, 2014; Maitland, 2013; Maitland and Newman, 2014), going for a walk with a local 'greeter' and, above all, staying in a local's home.

The development of websites such as Airbnb now allows tourists either to share locals' home for the time of their stay, by renting a room in an occupied apartment which permits "interaction between hosts and guests as part of the visitor experience" (Stors and Kagermeier, 2015: 11), or to rent an entire property and "live 'like a local'" (Kaplan and Nadler, 2015: 105) for the time of their stay. Eventually, it gives rise to the emergence of a new segment of the real estate market made by millions of listings targeting this new tourist demand, in addition to the classic real estate transactions for primary or second homes and to rentals based on long-term lease.

The symptoms of the 'Airbnb syndrome'

The emergence of this new kind of rentals is the object of a heated debate in tourism metropolises such as San Francisco, Paris, London or Barcelona (Kaplan and Nadler, 2015; Penn State School of Hospitality Management, 2016; Samaan, 2015; Schneidermann, 2014; Sperling, 2015; Zervas *et al.*, 2015) which articulates around three main cornerstones.

The first is related to the legal vacuum inherent in the sharing economy activities (Ranchordas, 2015) and sees this new kind of accommodation as unfair competition with the ordinary hotel trade (Fang *et al.*, 2015; Lehr, 2015; Neeser *et al.*, 2015; Packin, 2015; Zervas *et al.*, 2015), notably because it has long escaped legal taxes and licenses (even if numerous cities such as Paris or San Francisco have recently created new laws to tackle that problem).

The second recurrent issue is the question of the changing atmosphere in the streets and buildings where many apartments have turned into Airbnb listings: noises of suitcases in the entrance halls, the different lifestyles between tourists and locals (especially during the night), but also the difficulties which appear in the buildings where numerous owners rent their apartments to tourists and do not feel concerned by the management of collective ownerships (Chevalier *et al.*, 2013; O'Sullivan, 2014).

Last but not least, the central point of the debate is perhaps the issue of the effects of the multiplication of short-term rentals on local housing markets. On the one hand, Airbnb defends the idea according to which its activity is a way to provide supplemental income to households, and thus to allow them to stay in the place they live, notwithstanding the issue of mortgage repayment or of rising rents in a context of economic crisis and austerity policies (Sperling, 2015). To this end, Airbnb has launched several websites[1] which regularly publish data about the presence of the company in different cities. These reports invariably stress the fact that the majority of Airbnb rooms available on the platform are only occasionally rented to travellers, that hosts "use their Airbnb income to help pay their mortgage or rent",[2] or that a large majority of the listings are located throughout the cities, mainly "outside the main tourist area[s]".[3] But on the other hand, some argue that "Airbnb is embedded within neoliberal urbanism, fuelling the social and economic forces behind gentrification" (Thoem, 2015). Several critical websites[4] and recent reports indicate that those new accommodations increase the pressure on local real estate markets and fuel gentrification trends by removing hundreds of properties from the classic rental market (Stulberg, 2015), contributing to the rent increase (Samaan, 2015) or leading to speculation, especially in case of "multi-unit hosts" (Penn State School of Hospitality Management, 2016; Schneidermann, 2014). Furthermore, recent research in Barcelona has shed light on direct and indirect displacement of long-term inhabitants caused by the conversion of properties into tourist accommodation (Cócola Gant, 2016).

The cities involved have adopted various strategies to tackle those issues. Since October 2014, in San Francisco, Airbnb has been required to collect the tourist tax and then to transfer it to the City; only people living at least nine months a year in the city are allowed to rent out a room on a short-term basis; a room or an apartment can be rented up to 90 days a year maximum; only principal housing can be listed on Airbnb; finally, hosts have to register and obtain a business license for a $50 fee a year. Since May 2016, renting out whole properties without an official license is forbidden in Berlin. Some other cities are more open to short-term rentals, such as London, where residents have been allowed to rent their home without any permission from the local authorities since March 2015.[5]

The emergence of this short-time rental market can thus be seen as a new force which tightly interacts with neighbourhood changes and can foster direct tourism gentrification. It opens up a wide agenda for critical urban studies and raises several crucial questions concerning above all the impact of this new market on local housing markets. Are the assets available on Airbnb removed from the ordinary housing market? Does it give rise to an increase of real estate values? And what are its social impacts? To what extent does it allow locals to earn supplemental income which could allow them to stay in their home and support the increased cost of living in a context of austerity? Or, on the contrary, does it encourage speculation by making local real estate more strained for sales as well as rentals? Does it make it more difficult for households who have seen their incomes drop considerably to stay or find housing in the touristic parts of cities? Does it lead to displacement issues as argued by scholars of residential gentrification (Atkinson, 2000; Freeman and Braconi, 2004; Newman and Wyly, 2006)? To sum up, scholars need to establish whether this new market is a force fuelling a new and direct form of tourism gentrification, following the same process that has been emphasised for second homes. The aim is here to bring insight from the exploratory empirical analysis of the phenomenon in an emerging tourism metropolis, namely Reykjavík, the capital of Iceland.

Methods and sources of the study

To compare the development of the Airbnb market against the backdrop of the wider evolution of the local real estate market, I used three kinds of data sets. The first and most important one is an exhaustive database of the Airbnb listings available in the Reykjavík capital area in December 2015 (including the suburban municipalities of Kópavogur, Hafnarfjörður, Mosfellsbær Seltjarnarnes, Garðabær and Álftanes – see Table 3.1). I collected it by using a web-scraping technique which, thanks to the preliminary writing of a code, allows the harvesting of all the publicly available data from the source page of the Airbnb website. This technique is today mainly used in marketing and consumption studies (Polidoro *et al.*, 2015; Vargiu and Urru, 2012) but has also been used for a few journalistic analyses of Airbnb (Abbiateci, 2014) or even by the members or the Airbnb community to evaluate the price of an apartment on the platform.[6] I used the software OutWit Hub which enabled me to gather a database including 2,033 listings in this area including quantitative (prices, minimum time of stay, number of rooms available, number of comments, rating etc.), qualitative (ID number of the host, type of housing, type of property, municipality etc.) as well as geographic (latitude and longitude) variables.

I have also 'scraped' a geo-referenced inventory of the 122 hotels listed in the area by the Icelandic Yellow Pages.

The second set of data has been provided by Registers Iceland (an office of the Icelandic Ministry of Interior, whose main attribution is to register

Table 3.1 Statistical insight of the Airbnb database

Main quantitative variables	Maximum number of guests	Number of rooms	Minimum time of the stay	Number of comments	Price per night	Weekly prices	Monthly price
N Valid	1618	1924	1993	1441	1816	176	171
Missing	415	109	40	592	217	1857	1862
Mean	3.18	1.76	2.52	22.37	116.78	598.18	2268.79
Median	3.00	1.00	2.00	10.00	98.00	500.00	1900.00
Mode	2	1	1	1	100	500	1680
Standard deviation	1.178	1.003	1.888	30.822	108.742	365.248	1475.100
Minimum	0	1	1	1	16	100	475
Maximum	12	8	21	231	2687	2655	9001

Main qualitative variables	ID of the host	Municipality	Type of housing	Type of property
N Valid	1679	2032	2028	1993
Missing	354	1	5	40

Source: Airbnb website (December 2015)

"a range of information on Iceland's resident and real estate properties"[7]) who regularly publishes data on real estate. I mainly used three main databases from that source:

- The detailed database providing disaggregated data of the real estate transactions in the capital area from 2008 to 2014 (20,185 transactions registered), thus allowing to take into account the effects of the economic crisis and of the current economic recovery, including numerous variables (price, size, number of rooms etc.). The accurate geographical coordinates are not available but the district ('matssvæði'[8]) of each transaction is mentioned.
- The aggregated database providing data on the evolution of the real estate values at a district level from 1991 to 2014.
- The database on the evolution of the rental market from 2011 to 2014 in which data is aggregated at a mega-district level. It displays several variables for each category of housing (from studio to more than six rooms), such as the average size, the average price, the standard deviation of the price etc.

Finally, spatial data comes from two main sources: Reykjavíkurborg (Municipality of Reykjavík) which provides a wide set of open data related to the city (including the total number of properties per municipality) and Landmælingar Íslands (National Land Survey of Iceland). The maps and spatial analyses have been made thanks to QGIS.

A complementary and more qualitative approach has been provided by the constitution of a press corpus dealing with the issues both of development of

Airbnb and of the evolution of the real estate market from English-speaking newspapers (English version of *Morgunblaðið, Iceland Monitor,*[9] *IceNews, The Reykjavík Grapevine, Iceland Review Online*) with about 80 papers gathered.

Reykjavík: the emergence of a tourism metropolis in a post-crisis context

The tourism-oriented reshaping of Reykjavík

Since the 2000s and especially since the economic meltdown in 2008, Iceland has witnessed a growing influx of tourists (from 360,000 arrivals in 2004 to 998,000 in 2014 – see Figure 3.1). This economic activity is today the second-largest sector and is considered as key in the recovery of the country (Jóhannesson and Huijbens, 2010).

While this increase is mostly driven by the natural attractions of the island (Ólafsdóttir and Runnström, 2011), Reykjavík has emerged as a city destination on its own, promoted by the development of short stays from Europe or of stop-over on the way to North America and Western Europe. In the context of the economic meltdown that has affected the country since 2008, the Icelandic capital has thus taken advantage of this new increasing economic benefit. International tourism has become the driving force behind the current reshaping of the city, and especially the downtown area. After having been mainly shaped by the financial sector in the 2000s during the golden age of the "neo-vikings" (Benediktsson and Karlsdóttir, 2011) (see next section), the city centre is today being redesigned by the tourism industry through new tourism-oriented urban projects, aiming to draw the "visitor class" (Eisinger, 2000) to Reykjavík, such as the architectural icon Harpa, a concert hall and conference centre. The ongoing renewal of the 'Old Harbor' is also considered as one of the centrepieces of the new Reykjavík municipal plan (2010–2030), published in

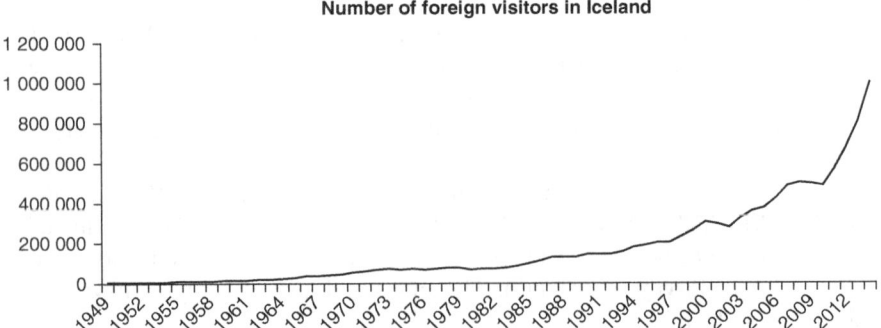

Number of foreign visitors in Iceland

Figure 3.1 Evolution of the number of foreign visitors in Iceland, 1949–2014

Source: Icelandic Tourist Board (2015)

June 2014. After the recent set up of the Saga Museum, the Maritime Museum and the Volcano House in this sector, this document plans the aestheticisation of public spaces, the replacement of the current mundane warehouses by new, carefully designed buildings, made of fine materials. Several cafés, restaurants, but also art galleries have recently set up there, in addition to the famous fish restaurants and small huts selling tickets for the whale watching tours which are already well established.

This increase of tourist influx also raises the issue of the amount of accommodation available for those visitors. The city is currently witnessing the construction of dozens of new hotels in response to this growing demand, which is a part of the recovery of the building sector after the economic meltdown, increasing the number of hotel rooms available in Reykjavík from 900 in 2000 to 5,700 expected in 2017 (Robert, 2014). The accommodation sector in Reykjavík is also being deeply reshaped by the emergence of a growing short-time rental market, notably due to the success of Airbnb which has quickly spread in the city since 2009.

Overview of the post-crisis real estate market in Reykjavík

Analysing the link between the expansion of Airbnb and the evolution of the real estate market is particularly interesting in the case of Reykjavík, because it happened against the backdrop of the economic crisis which has had considerable effects on real estate values (see Figure 3.2).

The Icelandic real estate market is the mirror image of the economic fluctuations of the country. Indeed, the economic boom of this "Nordic Tiger" (Bergmann, 2014) in the 2000s relied partially on the building sector, whose expansion led to a strong and rapid urban sprawl. Between 2000 and 2008,

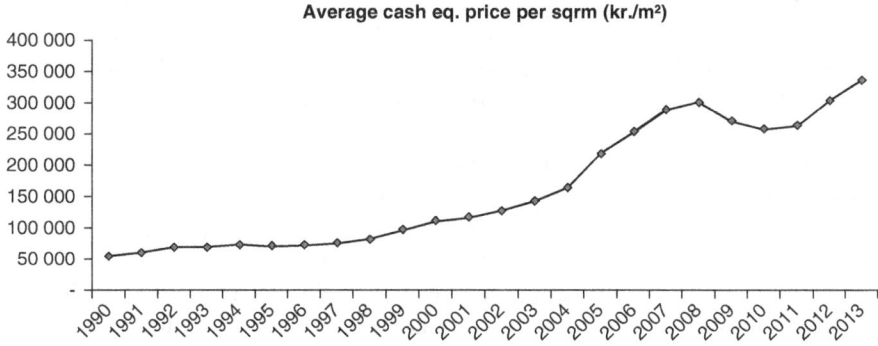

Figure 3.2 The evolution of real estate values in the 'Reykjavík West' districts, 1990–2014 (see Figure 3.3 for location)

Source: Registers Iceland (2015)

the built-up area of the city has risen by 25 per cent (Mathiesen *et al.*, 2014: 13). This urban development is actually the consequence of the "government policy to encourage homeownership, rather than renting" (Tulinius, 2013) in the 2000s (in a country which already culturally valued home ownership over rental), especially through the promotion of long-term mortgages provided by the Housing Financing Fund (HFF) founded in 1999. As a result, "by mid-2004, almost ninety percent of Icelandic households held an HFF loan" (Bagus *et al.*, 2011: 56). But the development of real estate at that time also gave rise to "radical urban transformations" (Mathiesen *et al.*, 2014: 24) in the city centre, aiming at demonstrating the new financial power of this country: building a business district located next to the city centre, transforming the seafront fishery warehouses into luxurious and modern residential complex, launching the building of an architectural icon Harpa (*supra*). Moreover, whereas the suburban real estate values used to be higher than downtown ones for decades, since the late 1990s, the trend has been reversed (Iceland Monitor, 2016).

The collapse of October 2008 suddenly stopped these urban developments and brought about significant social difficulties for Icelandic households. The cumulative effects of the fall of the Icelandic Krona, the rise of unemployment, the extent of households' debt accumulation, while the three main banks of the country went bankrupt (Ólafsson, 2011), caused increasing difficulties in mortgage repayments.[10] As a result, young people arrived on the housing market without having the means to obtain a mortgage anymore (Andersen, 2012a) and had to turn towards the rental market (in 2007, 9 per cent of them rented a property, compared to 28 per cent in 2013). Thus, as the rental market had been historically underdeveloped in the country, the situation soon became highly tense (Andersen, 2012b), especially downtown, as reported in the media:

> Since January 2011, the average rental price has risen by 22,8% in Iceland. The area worst afflicted by high rental prices is the older part of Reykjavík, the area roughly within postcodes 101, 105 and 107. (Tulinius, 2013)
>
> Rental prices in 101 Reykjavík have gone through the roof. The average monthly price for a 40-square-meter studio apartment has increased from ISK 88,000 (USD 734, EUR 548) to ISK 97,000 (USD 808, EUR 604) in one year (between 2012 and 2013). (Arnarsdóttir, 2013)

This situation, combined with the recovery of the sale prices of properties, especially in the central neighbourhoods (see Figure 3.3) resulted in a housing crisis as the demand on the rental market now exceeded the supply, with 50 possible tenants for one apartment, according to *Iceland Review* (Arnarsdóttir, 2013).

The recovery of the real estate market is actually not homogeneous in the whole city. Instead, it resulted in varied situations, between the central neighbourhoods which have globally reached superior values since 2012, and the outskirts where some districts (especially in the southern part of the urban area) still have inferior average values than before 2008.

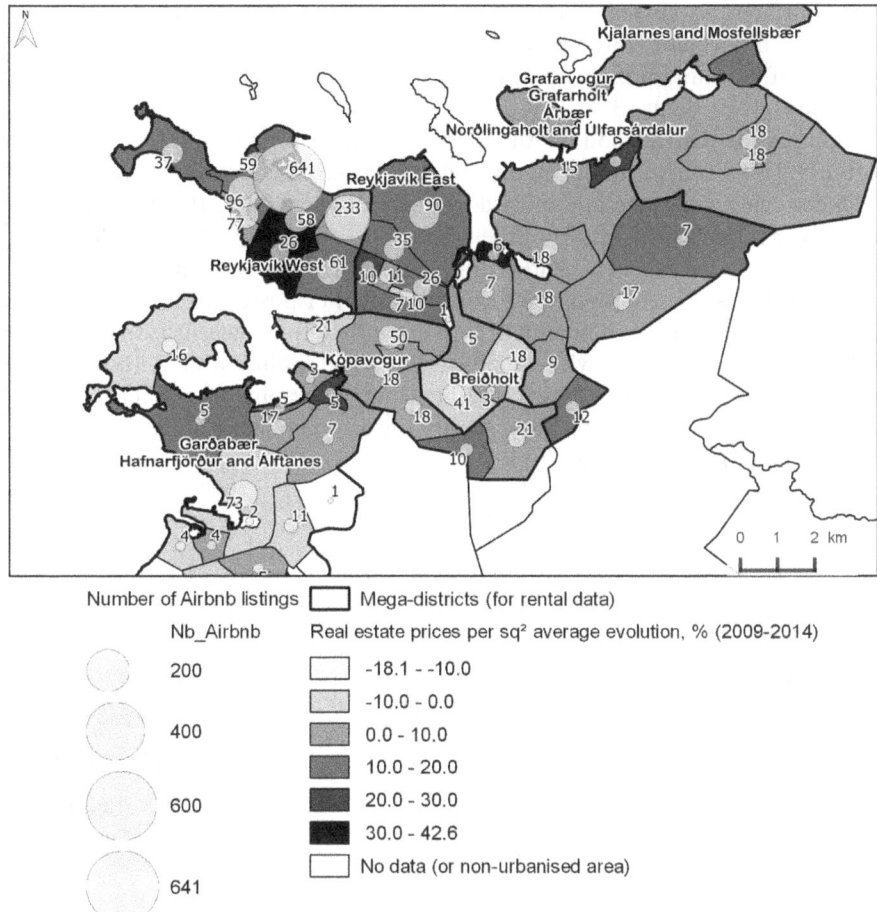

Figure 3.3 The evolution of real estate values (kr/m²) between 2009 and 2014 and the number of Airbnb listings

Sources: Airbnb website (December 2015), Registers Iceland (2015), Reykjavíkurborg (2015)

Small, central and entire homes: the three keywords of a highly concentrated market

According to the figures provided by the media, the number of Airbnb listings available in the Reykjavík capital area has constantly increased: from 600 in 2013 (Beckett, 2013) (i.e. 0.7 per cent of the total number of residencies in the city), to more than 1,000 in 2014 (Fulton, 2014) (1.2 per cent of the housings) and more than 2,000 in 2015 (2.5 per cent of residencies). In December 2015, one can find a wide variety of property types, ranging from standard apartments (1,551) or houses (334) to more original kinds of accommodation such as a yurt, a hut, or a ship. One- to three-room accommodation represents the overwhelming majority of the entire properties available (respectively 38, 35 and 18 per cent of the listings – see Figure 3.4).

The spatial distribution of the listings is also highly concentrated (see Figure 3.5), as a huge majority are located in the municipality of Reykjavík (80.8 per cent) and especially in its downtown districts (101, 105 and 107 post-codes, 65 per cent). Most parts of the urban area, especially on the outskirts, are almost ignored by this new market (only 1 per cent of the listings are located in the less well-off neighbourhood of Breiðholt (111 postcode) or 2 per cent in Mosfellsbær, a suburban municipality). It is worth noting that this map coincides with the one of the hotels, which tends to show that, unlike what Airbnb claims in its publications, it does not really reshape the geography of tourist accommo-dation in the city (see Figure 3.6) – similar results have been highlighted in the case of Barcelona (Guttierez, 2015).

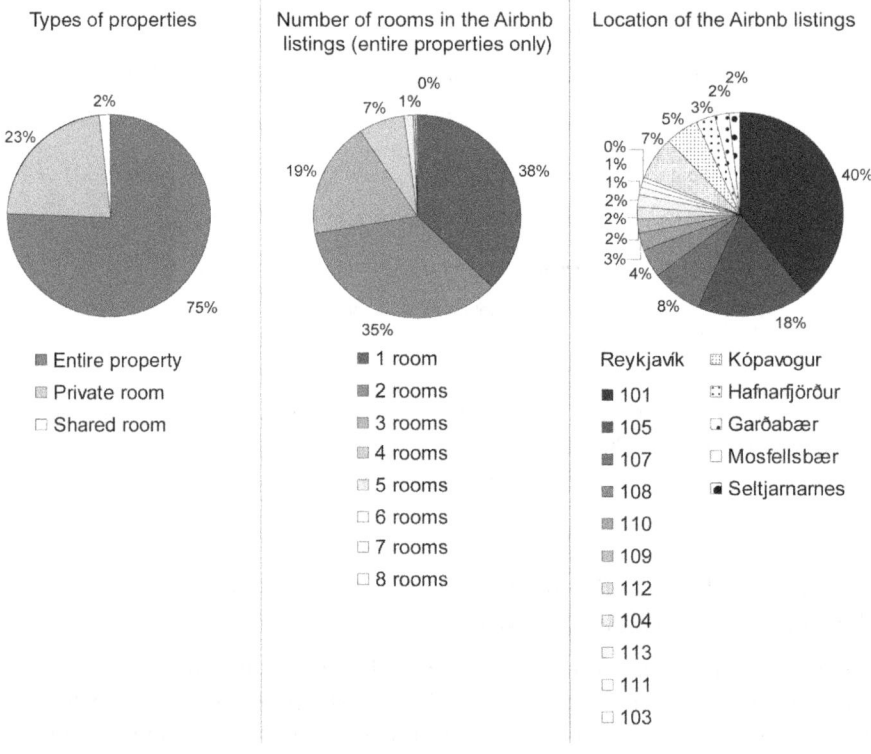

Figure 3.4 The three main features of the short-term rental market in Reykjavík

Source: Airbnb website (December 2015)

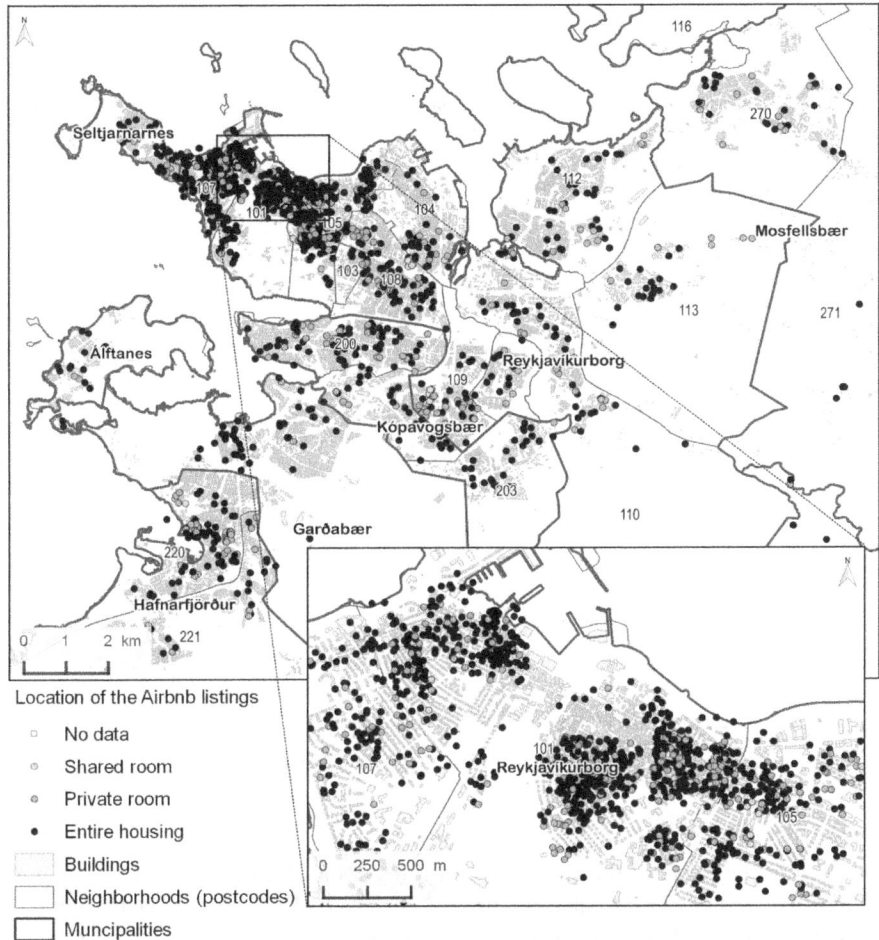

Figure 3.5 Location of Airbnb listings in the Reykjavík urban area

Sources: Airbnb website (December 2015), Reykjavíkurborg (2015), Landmælingar Íslands (2015). Cartography by the author

Another major feature of this market is that it concerns mostly entire homes (75.3 per cent of the listings) rather than rooms (whether private – 22.9 per cent – or shared – 1.6 per cent) in occupied properties. Admittedly, it does not amount to saying that all the 1,531 homes concerned (i.e. 1.9 per cent of the total number of houses in the urban area) have been removed from the ordinary market since they may be rented from time to time when their owners are away. Nevertheless, it is a higher rate than in most other tourism metropolises (62 per cent in San Francisco, 61 per cent in Berlin and Madrid, 59 per cent in New York, 52 per cent in London[11]) and some media sources claim that most of them are actually rented to tourists most

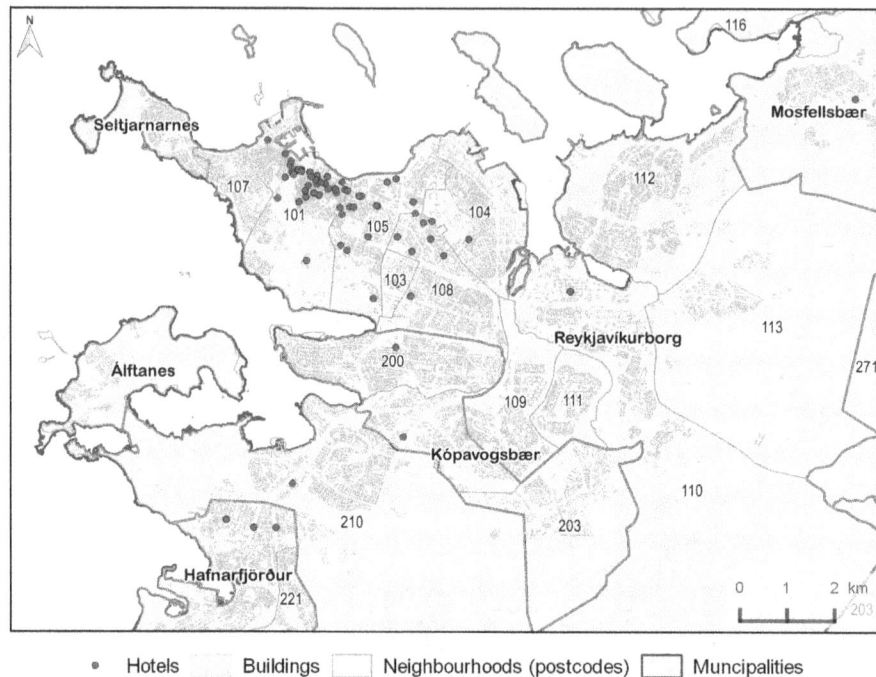

● Hotels   Buildings [ ] Neighbourhoods (postcodes) [ ] Muncipalities

Figure 3.6 Location of hotels in the Reykjavík urban area

Sources: Já.is (February 2016), Reykjavíkurborg (2015), Landmælingar Íslands (2015). Cartography by the author

of the time, especially during the tourism peaks (during the summer or the main international festivals organised in the city, such as Iceland Airwaves each autumn). For example, *The Reykjavík Grapevine* assesses that "in any given six-week period, less than 20 per cent of the Reykjavík listings in Airbnb are actually vacant, with the majority of them rented up by tourists" (Fontaine, 2015a). In fact, a last minute search made on 18 February 2016 for listings available from the 19th to the 21st showed that only 16 per cent of the listings were still free. Other media outlets point out that a significant number of the listings are only dedicated to tourists:

> Many listings are properties that have been bought and renovated explicitly for the purpose of renting to tourists. Many 'host' profiles even state something along the lines of 'I have numerous rentals available throughout 101' imploring would-be guests to be in touch for something that suits their specific needs. (Fulton, 2014)

For instance, behind the 13 listings proposed by 'Elísabet' in the database, there is a real company called 'Thomsen Apartments' who boasts on its website about being "located right in Reykjavík City Centre", close to the main tourist spots of the city. These apartments are not located in the same building but scattered in the very central neighbourhood of Þingholt (in the heart of the 101 district).

The database also gives an account of this sensitive question of 'multi-unit hosts' (see Table 3.2). Contrary to what has been observed in other cities such as Geneva (Conti, 2014), it is a relatively marginal phenomenon in the Reykjavík case (180 houses involved, i.e. 0.2 per cent of the total number of houses in the area). Nevertheless, the significant concentration of these listings in the central area (especially in 101 Reykjavík) makes this phenomenon rather less anecdotal there.

To sum up, the Airbnb 'syndrome' concerns entire small apartments (one to three rooms) located in the downtown area, which exactly match the market segment and the type of location that are most sought-after by locals who have to turn towards the rental market (students and young couples).

Towards a 'tourism gentrification rent gap'

Although there is no significant statistical correlation between the evolution of real estate prices between 2008 and 2014 and the number of Airbnb listings in the districts of the city,[12] most of the listings are located in the central neighbourhoods which have witnessed the most significant increase in real estate values in the urban area (see Figure 3.3). It is not unreasonable to assume that the development of these new short-term rentals has some disruptive effects on the classic rental housing market, at least in this very specific part of the city.

Figure 3.7 compares the theoretical average rental values in the short-term market and the actual average values on the long-term rental market in the central neighbourhoods (named as 'Reykjavík West' on Figure 3.7) and in a sample of

Table 3.2 An assessment of multi-unit hosts in Reykjavík

Number of rooms for the same ID	Number of IDs	Percentage
1	955	79.7
2	132	11
3	64	5.3
4	22	1.8
5	9	0.8
6	5	0.4
7	3	0.3
8	1	0.1
9	3	0.3
10	1	0.1
11	1	0.1
13	1	0.1
15	1	0.1
Total	1198	100

Source: Airbnb website (December 2015)

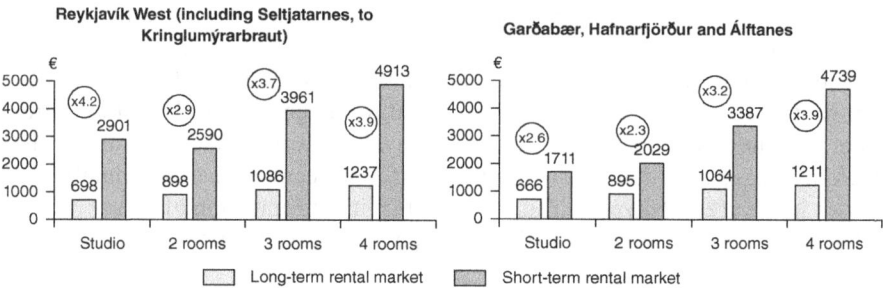

Figure 3.7 Comparison between the actual monthly average values for long-term rentals and the theoretical monthly average values for short-term rentals[13] (in €)

Sources: Airbnb website (December 2015), Registers Iceland (2015)

municipalities located on the outskirts (Garðabær, Hafnarfjörður and Álftanes – for location, see Figure 3.3). The common point between these two markets is that the rental values are always higher in the central part of the city, even if the difference is smaller for properties with four rooms or more. But the most striking point is the huge difference between the rental values for the two types of market (a ratio of 1:4 or 1:3 in the most represented segments of the market – i.e. one to three rooms). Even if the short-term values are purely theoretical, it shows that the value of an entire month's rent on the long-term market could be earned in only one week on average on the short-term market. Moreover, as the occupation rate of the Airbnb listings is very high (more than 80 per cent, see above), these theoretical values should be quite close to reality – at least during tourism peaks. This figure thus highlights the existence of a new kind of rent gap (Smith, 1979) based on this new segment of the tourism real estate market.

The fact that there are henceforth two kinds of rental markets running in parallel creates some disruptive effects on the classic housing market, and is expressed through the appearance of a completely new competition for properties between locals and tourists. Indeed, the demand for housing from both locals and tourists is concentrated on the same area, namely downtown. The downtown districts (and especially the famous 101 postcode) are not only the most appealing areas for tourists, but also the most sought-after neighbourhoods for locals who value its dense built-up environment (which is rare in this city), its unique atmosphere drawing on numerous cafés and designer shops, and where every amenity is within walking distance (Arnarsdóttir, 2013). It thus gives rise to unfair competition between locals who want to live or stay in the centre and tourists, as highlighted by Jóhann Már Sigurbjörnsson, the chairperson of the Renters' Association of Iceland in an interview with the *Reykjavík Grapevine*:

> Locals simply cannot compete with tourists on the rental market. In some cases, renters have been asked by landlords to vacate their properties during

high tourist season [. . .]. The government, local authorities, and even labour unions must respond to this. There are thousands of apartments going off the rental market and onto the tourist market. (Fontaine, 2014a)

Furthermore, the media echo some cases of displacement due to the turning of housing into Airbnb listings (with sometimes temporary displacement for the peak season), as was the case for Sarah, who had to find a new apartment following "her former landlord's decision to dive headfirst into the tourist rental market, leaving her and her young family with just 8-weeks to find a new place to live as he rushed to get their long-term rental tourist-ready in time for June 1" (Fulton, 2014), or Julia Mai, a local artist who had to face the same situation over the summer of 2014:

> For example, I am currently in the process of looking for a new apartment, which is a really rather hard and frustrating endeavour. I was given until September to evacuate my current apartment, which the owners plan to renovate this coming winter and renting and then letting it out on Airbnb for the 2015 tourist season. I have a couple of weeks to find something [note, this interview was conducted a couple of weeks ago. At the time of publication, it does not look like Julia has found a proper place to stay]. (Fontaine, 2014b)

Locals confronted with this situation do sometimes react, in the manner of Julia Mai, who decided to run a joke crowdfunding campaign in August 2014 aiming to raise $6,000 to renovate a cardboard box:[14]

> As the rent market in Reykjavík has become a nightmare in the past few years I decided to take (a slightly sarcastic) part in what made it so bad. Many apartments in the city center have turned in to Airbnb's, which means less available flats as well as sky high prices for the few flats left for permanent residents. As an art project I decided to ask for funding to renovate my future home, a cardboard box, that I will eventually rent out through Airbnb, starting on the Culture Night on the 23rd of August.[15]

The public authorities have also taken hold of both the question of the tense housing market and of the need for regulation of short-term rentals. Several measures have been taken to solve the housing crisis, and especially the issue of the saturated rental market, during the electoral mandate of the 'Best party', which held the City Hall of Reykjavík from 2010 to 2014. The building of "more than 7,500 residences of various configurations [. . .] with an average of 750 being added to the market yearly" (Iceland Review, 2013) has been planned up until 2022, with an objective of densifying the city (for instance, by planning the urbanisation of the strategic site Vatnsmýri, which is adjacent to the city centre and currently shelters the domestic airport whose closing is planned). A series of measures intended to "strengthen the Icelandic rental market" (Kyser, 2013) was also heralded in 2013, for example by exempting landlords from capital-gains tax for the income they get from rents, or by giving grants to rental companies who would provide long-term leases. The Socio-democrats, who took over City Hall in 2014, are pursuing the same aim, as one of their prominent campaign promises was "to add 2500–3000 new apartments to the city in the next 4 years" (Fulton, 2014).

Regulating the Airbnb market is another side to managing the problem. Actually, it is estimated that "close to 60% of these accommodations have no operating license" (Fontaine, 2015b) from the Icelandic Tourist Board, which can raise security issues, but also generate incomes which avoid the tourism and capital-gains taxes. In order to thwart this trend, since 2014, police have made several checks on apartments which are suspected of being illegal tourism accommodation. For instance, in September 2014, four businesses which used to rent apartments on Airbnb without a license had to close (Fontaine, 2014c). These controls should be reinforced in the future.

Conclusion

The exploratory analyses presented here have established that, as has been underlined for second-homes (Paris, 2009), in Reykjavík, the 'Airbnb syndrome' also leads to neighbourhood changes which are very close to the main features of gentrification. First, I have highlighted a significant rent gap between short-term and long-term rental values which encourages owners to turn their assets into Airbnb listings. Second, some cases of physical displacement of locals due to the transformation of housing into Airbnb listings have been identified, and some clues tend to indicate the existence of very specific forms of 'temporary' displacement during the peak season. Moreover, the housing demand from both tourists and inhabitants is concentrated on the same areas in the downtown districts, which produces new competition between these two populations on the rental market.

Beyond these exploratory results, the convergence of three factors makes Reykjavík both an original and a very instructive laboratory for exploring what is at stake locally and the effects of the arrival of this new segment of the rental market, that is, short-term rentals, on the social production of urban space. First, the arrival of Airbnb in Reykjavík coincided with the tourism boom witnessed by the city since 2009, a boom which has surely accelerated the spreading of Airbnb in a city which has a significant lack of tourism accommodation. Second, Airbnb also arrived against a post-crisis backdrop which has had some significant effects on the regular renting market, which has witnessed a growing demand due to increasing difficulties in mortgage repayments undergone by local households. Finally, and maybe most importantly, Reykjavík is a highly effective example because its small size has made the 'Airbnb syndrome' very visible very quickly. Indeed, while Reykjavík is not among the top cities in the world in terms of stocks of Airbnb listings available (it is far from the 53,000 listings available in Paris in 2016), in relative values, it represents a significant proportion of the total number of properties in the city (2.5 per cent of properties in the urban area and 3.2 per cent only for the Reykjavík municipality). Thus, the combination of these three factors (tourism boom, weakening of the socio-economic situation of locals, magnifying effect due to the small size of the city) has exacerbated the disruptive effects of the development of short-term rentals.

These exploratory results prompt several questions to further an understanding of the Airbnb syndrome. First, the extent and the forms (temporary or not) of the displacements need to be investigated, with a special focus on the legal framework which allows it. Second, the link between the post-crisis context and the extent of the 'Airbnb syndrome' should be clarified through a more qualitative approach of both the Airbnb hosts (has the crisis encouraged them to embark on short-term rentals?) and the local renters (who are the households experiencing difficulties finding a property? Did the crisis disrupt their residential trajectories?).

Notes

1 http://blog.airbnb.com/; www.airbnbaction.com/new-york/; http://publicpolicy.airbnb. com/ (last checked: January 2016).
2 Source: http://blog.airbnb.com/economic-impact-airbnb/ (last checked: January 2016).
3 Source: http://blog.airbnb.com/airbnbs-economic-impact-nyc-community/ (last checked: February 2016).
4 See: http://insideairbnb.com/ and www.airbnbvsberlin.com/, (last checked: February 2016).
5 Nevertheless, the debate is still fierce, as evidenced by the anti-Airbnb referendum organised in November 2015 which was termed 'City of San Francisco Initiative to Restrict Short-term Rentals, Proposition F', one of whose slogans was 'Keep San Francisco affordable for the locals' (the proposed law has been rejected).
6 See: http://hamelsmu.github.io/AirbnbScrape/ (last checked: February 2016).
7 Source: www.skra.is/english/english/ (last checked: February 2016).
8 A matssvæði is the spatial unit used specifically by Registers Iceland. The map of all the matssvæði of the area is provided online: www.skra.is/library/Samnyttar-skrar-/ Fasteignamat/2014%20matssv%20hofudb01%20A3.pdf (last checked: February 2016).
9 The English version of the daily newspaper *Visir*.
10 It is worth noting that the current Icelandic government launched a household debt-relief plan in 2014.
11 Source: Inside Airbnb website (last checked: February 2016).
12 Pearson's correlation coefficient = 0.134 (correlation between the total number of Airbnb listings and the evolution of real estate prices between 2009 and 2014 by matssvæði) and 0.160 just for 0–2 room properties.
13 The Airbnb average rental values have been computed only from the prices of entire properties. I have used the monthly prices (8.4% of listings) or multiplied by 4 the weekly prices (8.7%) when the data was not missing (cf. Table 3.1). Otherwise, I have multiplied the price per night by 30.
14 Incidentally, at the time of writing this chapter (January 2016), she published on her personal Facebook page that she was looking for a new property 'in the downtown area (101, 105, 107). [. . .] Long term rent would be ideal but at this point anything on dry (icy) land will be considered'.
15 Source: www.facebook.com/ReykjavíkGrapevine/posts/1461295024124351 (last checked: February 2016).

References

Abbiateci, J. (2014) 'Comment nous avons enquêté avec les données d'Airbnb', *Le Bac À Sable*, 9 November, available at: http://blogs.letemps.ch/labs/2014/11/09/comment-nous-avons-enquete-avec-les-donnees-dairbnb/ (accessed 29 January 2016).

Andersen, A. (2012a) 'A Black Box', *The Reykjavík Grapevine*, 13 April, available at: http://grapevine.is/mag/feature/2012/04/13/a-black-box/ (accessed 20 February 2016).

Andersen, A. (2012b) 'Renting In The Free World', *The Reykjavík Grapevine*, 22 March, available at: http://grapevine.is/mag/articles/2012/03/22/renting-in-the-free-world/ (accessed 20 February 2016).

Arnarsdóttir, E.S. (2013) 'Hipsters' Paradise (ESA)', *Iceland Review*, 23 August, available at: http://icelandreview.com/stuff/views/2013/08/23/hipsters-paradise-esa (accessed 13 February 2016).

Atkinson, R. (2000) 'Measuring Gentrification and Displacement in Greater London', *Urban Studies*, vol. 37. no. 1, pp. 149–165.

Bagus, P., Howden, D. and Baxendale, T. (2011) *Deep Freeze: Iceland's Economic Collapse*, Auburn: Ludwig von Mises Institute.

Beckett, R. (2013) 'Apartment Owners Warned Of Renting Without License', *The Reykjavík Grapevine*, 15 April, available at: http://grapevine.is/news/2013/04/15/apartment-owners-warned-of-renting-without-license/ (accessed 12 February 2016).

Benediktsson, K. and Karlsdóttir, A. (2011) 'Iceland Crisis and Regional Development – Thanks For All The Fish?', *European Urban and Regional Studies*, vol. 18, no. 2, pp. 228–235.

Bergmann, E. (2014) *Iceland and the International Financial Crisis: Boom, Bust and Recovery*, Houndmills, Basingstoke: Palgrave Macmillan.

Bridge, G. and Dowling, R. (2001) 'Microgeographies of Retailing and Gentrification', *Australian Geographer*, vol. 32, no. 1, pp. 93–107.

Bures, R. and Cain, C. (2008) 'Dimensions of Gentrification in a Tourist City', Paper presented at Meeting of the Population Association of America. Available at: http://paa2008.princeton.edu/papers/81623 (accessed 01 Feb 2017)

Chevalier, S., Lallement, E. and Corbillé, S. (2013) *Paris comme résidence secondaire – Enquête chez ces propriétaires d'un nouveau genre*, Paris: Belin.

Cócola Gant, A. (2016) 'Holiday Rentals: The New Gentrification Battlefront', *Sociological Research Online*, vol. 21, no. 3.

Conti, J. (2014) 'Avec Airbnb, ma petite entreprise de location ne connaît pas la crise', *Le Temps*, Genève, 9 November, available at: www.letemps.ch/suisse/2014/11/09/airbnb-petite-entreprise-location-ne-connait-crise (accessed 14 February 2016).

Curran, W. (2004) 'Gentrification and the Nature of Work: Exploring the Links in Williamsburg, Brooklyn', *Environment and Planning A*, vol. 36, no. 7, pp. 1243–1258.

Djament-Tran, G. and Guinand, S. (2014) 'La diffusion des grands équipements culturels, vecteur de métropolisation des quartiers populaires ?', *Belgeo. Revue belge de géographie*, no. 1, available at: https://doi.org/10.4000/belgeo.12737.

Eisinger, P. (2000) 'The Politics of Bread and Circuses Building the City for the Visitor Class', *Urban Affairs Review*, vol. 35, no. 3, pp. 316–333.

Fang, B., Ye, Q. and Law, R. (2015) 'Effect of Sharing Economy on Tourism Industry Employment', *Annals of Tourism Research*, Forthcoming paper, published online first, pp. 1–4.

Farstad, M. and Rye, J.F. (2013) 'Second Home Owners, Locals and their Perspectives on Rural Development', *Journal of Rural Studies*, vol. 30, pp. 41–51.

Fontaine, P. (2014a) 'Tourism Having Effects On Rental Market', *The Reykjavík Grapevine*, 10 August, available at: http://grapevine.is/news/2014/08/10/tourism-having-effects-on-rental-market/ (accessed 14 February 2016).

Fontaine, P. (2014b) 'Visitors And Locals', *The Reykjavík Grapevine*, 2 September, available at: http://grapevine.is/mag/interview/2014/09/02/visitors-and-locals-the-rental-market-in-a-nutshell/ (accessed 19 February 2016).

Fontaine, P. (2014c) 'Police Raid Illegal Tourist Accommodations', *The Reykjavík Grapevine*, 17 September, available at: http://grapevine.is/news/2014/09/17/police-raid-illegal-tourist-accommodations/ (accessed 20 February 2016).

Fontaine, P. (2015a) '1,800 Reykjavík Apartments In Airbnb', *The Reykjavík Grapevine*, 17 July, available at: http://grapevine.is/news/2015/07/17/1800-reykjavik-apartments-in-airbnb/ (accessed 13 February 2016).

Fontaine, P. (2015b) 'Most Tourist Rentals Illegal', *The Reykjavík Grapevine*, 6 March, available at: http://grapevine.is/news/2015/03/06/most-tourist-rentals-illegal/ (accessed 20 February 2016).

Freeman, L. and Braconi, F. (2004) 'Gentrification and Displacement New York City in the 1990s', *Journal of the American Planning Association*, vol. 70, no. 1, pp. 39–52.

Füller, H. and Michel, B. (2014) '"Stop Being a Tourist!" New Dynamics of Urban Tourism in Berlin-Kreuzberg', *International Journal of Urban and Regional Research*, vol. 38, no. 4, pp. 1304–1318.

Fulton, C. (2014) 'Help! I Need A Place To Live!', *The Reykjavík Grapevine*, 16 June, available at: http://grapevine.is/mag/articles/2014/06/16/help-i-need-a-place-to-live/ (accessed 12 February 2016).

Gallent, N. (2014) 'The Social Value of Second Homes in Rural Communities', *Housing Theory & Society*, vol. 31, no. 2, pp. 174–191.

Gladstone, D. and Preau, J. (2008) 'Gentrification in Tourist Cities: Evidence from New Orleans Before and After Hurricane Katrina', *Housing Policy Debate*, vol. 19, no. 1, pp. 137–175.

Gotham, K.F. (2005) 'Tourism Gentrification: The Case of New Orleans' Vieux Carre (French Quarter)', *Urban Studies*, vol. 42, no. 7, pp. 1099–1121.

Guttierez, D. (2015) 'Airbnb: Disruptive Innovation and the Rise of an Informal Tourism Accommodation Sector', *Current Issues in Tourism*, vol. 18, no 12, pp. 1192–1217.

Hackworth, J. and Smith, N. (2001) 'The Changing State of Gentrification', *Tijdschrift Voor Economische En Sociale Geografie*, vol. 92, no. 4, pp. 464–477.

Hall, C.M. and Müller, D.K. (2004) *Tourism, Mobility, and Second Homes: Between Elite Landscape and Common Ground*, Clevedon: Channel View Publications.

Herrera, L.M.G., Smith, N. and Vera, M.Á.M. (2007) 'Gentrification, Displacement, and Tourism in Santa Cruz De Tenerife', *Urban Geography*, vol. 28, no. 3, pp. 276–298.

Hoffman, L., Fainstein, S. and Judd, D. (2003) *Cities and Visitors: Regulating People, Markets, and City Space*, Malden, Mass: Blackwell Publications.

Iceland Monitor. (2016) 'Property 50% dearer in central Reykjavík', *Iceland Monitor*, 25 February, available at: http://icelandmonitor.mbl.is/news/news/2016/02/25/property_50_prosent_dearer_in_central_reykjavik/ (accessed 26 February 2016).

Iceland Review. (2013) 'Eight Hundred New Apartments Planned in Reykjavík', *Iceland Review*, 19 March, available at: http://icelandreview.com/news/2013/03/19/eight-hundred-new-apartments-planned-reykjavik (accessed 20 February 2016).

Jóhannesson, G.T. and Huijbens, E.H. (2010) 'Tourism in Times of Crisis: Exploring the Discourse of Tourism Development in Iceland', *Current Issues in Tourism*, vol. 13, no. 5, pp. 419–434.

Judd, D.R. and Fainstein, S.S. (1999) *The Tourist City*, New Haven and London: Yale University Press.

Kaplan, R.A. and Nadler, M.L. (2015) 'Airbnb: A Case Study in Occupancy Regulation and Taxation', *The University of Chicago Law Review*, vol. 82, pp. 103–115.

Kyser, L. (2013) 'Social Democrats Seek to Strengthen Rental Market', *The Reykjavík Grapevine*, 11 September, available at: http://grapevine.is/news/2013/09/11/social-democrats-seek-to-strengthen-rental-market/ (accessed 20 February 2016).

Lehr, D.D. (2015) 'An Analysis of the Changing Competitive Landscape in the Hotel Industry Regarding Airbnb', Master's thesis, Dominican University of California, available at: http://scholar.dominican.edu/masters-theses/188/ (accessed 8 December 2015).

Liang, Z.-X. and Bao, J.-G. (2015) 'Tourism Gentrification in Shenzhen, China: Causes and Socio-Spatial Consequences', *Tourism Geographies*, vol. 17, no. 3, pp. 461–481.

Maitland, R. (2010) 'Everyday Life as a Creative Experience in Cities', *International Journal of Culture, Tourism and Hospitality Research*, vol. 4, no. 3, pp. 176–185.

Maitland, R. (2013) 'Backstage Behaviour in the Global City: Tourists and the Search for the "Real London"', *Procedia - Social and Behavioral Sciences*, vol. 105, pp. 12–19.

Maitland, R. and Newman, P. (2014) *World Tourism Cities: Developing Tourism Off the Beaten Track*, London and New Nork: Routledge.

Mathiesen, A., Forget, T. and Zaccariotto, G. (2014) *Scarcity in Excess: The Built Environment and the Economic Crisis in Iceland*, New York: ActarD Inc.

McIntyre, N., Williams, D. and McHugh, K. (2006) *Multiple Dwelling and Tourism: Negotiating Place, Home and Identity*, Oxford and Cambridge: CAB International.

Neeser, D., Peitz, M. and Stuhler, J. (2015) 'Does Airbnb Hurt Hotel Business: Evidence from the Nordic Countries', Master thesis, Universidad Carlos III de Madrid, available at: http://www.researchgate.net/profile/David_Neeser/publication/282151529_Does_Airbnb_Hurt_Hotel_Business_Evidence_from_the_Nordic_Countries/links/5605310e08aea25fce322679.pdf (accessed 8 December 2015).

Newman, K. and Wyly, E.K. (2006) 'The Right to Stay Put, Revisited: Gentrification and Resistance to Displacement in New York City', *Urban Studies*, vol. 43, no. 1, pp. 23–57.

Norris, M. and Winston, N. (2010) 'Second-Home Owners: Escaping, Investing or Retiring?', *Tourism Geographies*, vol. 12, no. 4, pp. 546–567.

O'Brien, M. (2015) 'The Miraculous Story of Iceland', *The Washington Post*, 17 June, available at: https://www.washingtonpost.com/news/wonk/wp/2015/06/17/the-miraculous-story-of-iceland/ (accessed 21 February 2016).

Ólafsdóttir, R. and Runnström, M.C. (2011) 'How Wild is Iceland? Wilderness Quality with Respect to Nature-Based Tourism', *Tourism Geographies*, vol. 13, no. 2, pp. 280–298.

Ólafsson, S. (2011) *Iceland's Financial Crisis and Level of Living Consequences*, Social Research Centre, Reykjavík: University of Iceland.

O'Sullivan, F. (2014) 'Barcelona Organizes Against "Binge Tourism" – and Eyes a Street Protester for Mayor', *CityLab*, 27 August, available at: http://www.citylab.com/politics/2014/08/barcelona-organizes-against-binge-tourismand-eyes-a-street-protester-for-mayor/379239/ (accessed 17 February 2016).

Packin, E. (2015) 'The Manhattan Hotel Industry: A Changing Landscape The Effects of Airbnb and Land Values on Segmented Property levels', available at: http://dataspace.princeton.edu/jspui/handle/88435/dsp01kk91fn90r (accessed 8 December 2015).

Paris, C. (2009) 'Re-positioning Second Homes within Housing Studies: Household Investment, Gentrification, Multiple Residence, Mobility and Hyper-consumption', *Housing, Theory and Society*, vol. 26, no. 4, pp. 292–310.

Paris, C. (2010) *Affluence, Mobility and Second Home Ownership*, London and New Nork: Routledge.

Penn State School of Hospitality Management. (2016) *From Air Mattresses to Unregulated Business: An Analysis of the Other Side of Airbnb*, PennState, available at: http://www. ahla.com/uploadedFiles/_Common/pdf/PennState_AirBnbReport_.pdf (accessed 11 February 2016).

Polidoro, F., Giannini, R., Conte, R.L., Mosca, S. and Rossetti, F. (2015) 'Web Scraping Techniques to Collect Data on Consumer Electronics and Airfares for Italian HICP Compilation', *Statistical Journal of the IAOS*, vol. 31, no. 2, pp. 165–176.

Ranchordas, S. (2015) 'Does Sharing Mean Caring: Regulating Innovation in the Sharing Economy', *Minnesota Journal of Law, Science and Technology*, vol. 16, p. 413.

Robert, Z. (2014) 'Explosion in Apartments for Rent to Tourists', *Iceland Review*, 25 July, available at: http://icelandreview.com/news/2014/07/25/explosion-apartments-rent-tourists (accessed 12 February 2016).

Samaan, R. (2015) *Airbnb, Rising Rent, and the Housing Crisis in Los Angeles*, Laane, a new economy for all, available at: http://www.laane.org/wp-content/uploads/2015/03/AirBnB-Final.pdf (accessed 24 January 2015).

Schneidermann, E.T. (2014) *Airbnb in the City*, New York State Office of the Attorney General, New York, available at: http://skift.com/wp-content/uploads/2014/10/AIRBNB-REPORT.pdf (accessed: 11 February 2016).

Smith, N. (1979) 'Toward a Theory of Gentrification A Back to the City Movement by Capital, not People', *Journal of the American Planning Association*, vol. 45, no. 4, pp. 538–548.

Smith, N. (2002) 'New Globalism, New Urbanism: Gentrification as Global Urban Strategy', *Antipode*, vol. 34, no. 3, pp. 427–450.

Smith, N. (2006) 'Gentrification Generalized: From Local Anomaly to Urban "Regeneration" as Global Urban Strategy', in Downey, G. (ed.) *Frontiers of Capital: Ethnographic Reflections on the New Economy*, Duke University Press Books, Durham and London, pp. 191–208.

Solana-Solana, M. (2010) 'Rural Gentrification in Catalonia, Spain: A Case Study of Migration, Social Change and Conflicts in the Empordanet Area', *Geoforum*, vol. 41, no. 3, pp. 508–517.

Sperling, G. (2015) *How Airbnb Combats Middle Class Income Stagnation*, Airbnb, available at: http://www.cedarcityutah.com/wp-content/uploads/2015/07/MiddleClass Report-MT-061915_r1.pdf (accessed 7 December 2015).

Stors, N. and Kagermeier, A. (2015) 'Share Economy in Metropolitan Tourism. The Role of Authenticity-Seeking', *Metropolitan Tourism Experience Development: Diversion and Connectivity*, vol. 28, Budapest.

Stulberg, A. (2015) 'How Much Does Airbnb Impact Rents in NYC?', *The Real Deal New York*, 14 October, available at: http://therealdeal.com/2015/10/14/how-much-does-airbnb-impact-nyc-rents/ (accessed 11 February 2016).

Sullivan, D.M. and Shaw, S.C. (2011) 'Retail Gentrification and Race: The Case of Alberta Street in Portland, Oregon', *Urban Affairs Review*, vol. 47, no. 3, pp. 413–432.

Thoem, J. (2015) *Belong Anywhere, Commodify Everywhere.: A Critical Look into the State of Private Short-Term Rentals in Stockholm, Sweden.*, Degree projet in urban and regional planning, School of Architecture and of the Built Enviroment, Stokholm, available at: http://www.diva-portal.org/smash/record.jsf?pid=diva2:825261 (accessed 8 December 2015).

Tulinius, K. (2013) 'So What's This Overpriced Rental Market I Keep Hearing About?', *The Reykjavík Grapevine*, 16 September, available at: http://grapevine.is/mag/

column-opinion/2013/09/16/so-whats-this-overpriced-rental-market-i-keep-hearing-about/ (accessed 13 February 2016).

Vargiu, E. and Urru, M. (2012) 'Exploiting Web Scraping in a Collaborative Filtering-Based Approach to Web Advertising', *Artificial Intelligence Research*, vol. 2, no. 1, available at: https://doi.org/10.5430/air.v2n1p44.

Vives Miró, S. (2011) 'Producing a "Successful City": Neoliberal Urbanism and Gentrification in the Tourist City – The Case of Palma (Majorca)', *Urban Studies Research*, vol. 2011, pp. 1–14. Available at: http://www.hindawi.com/journals/usr/2011/989676/ (accessed 01 Feb 2017)

Wilson, J. and Tallon, A. (2011) 'Geographies of Gentrification and Tourism', in Wilson, J. (ed.) *The Routledge Handbook of Tourism Geographies*, Routledge, London, pp. 103–112.

Yoon, H. and Currid-Halkett, E. (2015) 'Industrial Gentrification in West Chelsea, New York: Who Survived and Who Did Not? Empirical Evidence from Discrete-Time Survival Analysis', *Urban Studies*, vol. 52, no. 1, pp. 20–49.

Zervas, G., Proserpio, D. and Byers, J. (2015) *The Rise of the Sharing Economy: Estimating the Impact of Airbnb on the Hotel Industry*, SSRN Scholarly Paper, no. 2366898, Social Science Research Network, Rochester, NY, available at: http://papers.ssrn.com/abstract=2366898 (accessed 7 December 2015).

Zukin, S., Trujillo, V., Frase, P., Jackson, D., Recuber, T. and Walker, A. (2009) 'New Retail Capital and Neighborhood Change: Boutiques and Gentrification in New York City', *City & Community*, vol. 8, no. 1, pp. 47–64.

4 Tourism gentrification in the cities of Latin America

The socio-economic trajectory of Cartagena de Indias, Colombia

Sairi T. Piñeros

Introduction

Gentrification in cities of developing countries is explained in part due to increased integration of these economies into the capitalist system economy (Betancur, 2014; Lees, 2014). The economic neo-liberalisation process (deregulation and privatisation) opens up investment markets in different sectors (exploitation of natural resources, industry and services). Thus, the process of gentrification occurs in areas with potential to capture rents (Smith, 2010) through redevelopment (Betancur, 2014). In this sense, the central areas constitute potentially profitable spaces due to their location, connection with other areas of the city, proximity to the administrative and business centres and, in the case of historical centres, their cultural heritage.

In some cities in Latin America, the state or municipal government, with private partners, created new projects to preserve or rehabilitate downtown areas. This is the case in Rio de Janeiro with *Corridor cultural* (Rubino, 2005), Mexico DF with *Programa de rescate* (Walker 2008)[1] and Chile (Inzulza-Contardo, 2012). These rehabilitation projects can produce demolition or rehabilitation for new uses. In the first case, historical buildings may be demolished to make way for the construction of high-rise buildings or lofts. In the second case, old buildings with cultural or historical value are rehabilitated for commercial (restaurants and shops), cultural (museums) or tourist (hotels) uses. As Rubino suggested, "gentrification processes in central and/or historical areas are not simply a matter of who lives in the housing, but also of tourism, leisure and cultural activities" (Rubino, 2005: 227).

In this sense, Gotham (2005) was one of the first authors to use the term "tourism gentrification" when he referred to the urban transformations that were occurring in the *Vieux Carré* (French Quarter) in New Orleans (USA). Gotham states that tourism gentrification "reflects new institutional connections between the local institutions, the real estate industry and the global economy" (Gotham, 2005: 1114).

Thus, the historical centres with tourist potential, especially those listed by UNESCO as a World Heritage Site, become attractive areas for capital investment and for consumption. In this chapter I use Cartagena de Indias, Colombia, to demonstrate how tourism in neighbourhoods located in the historical centre generates gentrification and how this manifests itself, affecting and transforming the urban landscape and urban functions.

Cartagena de Indias, the capital of the Bolivar department, is a city of more than 900,000 inhabitants located north-west of the Colombian Caribbean coast. In 1984, UNESCO recognised Cartagena as a World Heritage Site, thanks to its outstanding military architecture from the sixteenth, seventeenth and eighteenth centuries including the port, the fortresses and other monuments (UNESCO, n.d.) (Figure 4.1).

In recent years, the number of tourists in the city has increased[2] along with cruise passengers[3] and also tourism infrastructure (accommodation[4] and tourist services). The main attraction, in terms of tourist flows,[5] is the historical centre with its ramparts, historical monuments (churches, colonial homes, etc.), museums, accommodation (hotels, boutique hotels and hostels) and other services such as cafes, restaurants and boutiques. This increase in tourism supply and demand has had an impact on the city, not only in the beautification of buildings' facades but also through new commercial uses such as brand-name boutiques and the daily movement of the local population.

This chapter is divided into four parts. First, I present research on tourism gentrification in Latin American cities. Second, I describe the historical context of Cartagena as well as the conservation and restoration of its heritage. Third, I explain the process of gentrification from its initial stage of private investment to the tourism investment in recent years. Finally, I will focus on land use changes and the decentralisation of urban functions.

Methodology

One of the main problems facing researchers working on gentrification is the lack or inaccuracy of data. Colombia's population data is outdated. The most recent census data for the city dates from 2005. In Cartagena, official data (such as information concerning land use in the historical centre) is limited or out of date. The territorial ordering plan (*Plan de Ordenamiento Territorial*, POT), that provides planning directives for the city, dates from 2001. The POT should have been updated in recent years but this has not been done.[6]

My methodology is therefore based on a compilation of different empirical sources, the combination of which can provide interesting insight into on-goings processes. More precisely, I checked the Cartagena Chamber of Commerce web database,[7] mainly to identify the location, types of business (clothing, shoes, jewellery and crafts shops and hotels) and the date of commercial operating licences. In order to organise this information, I established GPS coordinates of hotels and commercial brands (shops) during my fieldwork in 2013–2015. I then used Google Historical Street View (historical images from past Street View collections) to complete my spatial database. One of this chapter's contributions is therefore to show that a combination of today's existing on-line resources can provide interesting insight, even in the context of a lack of accurate official data.

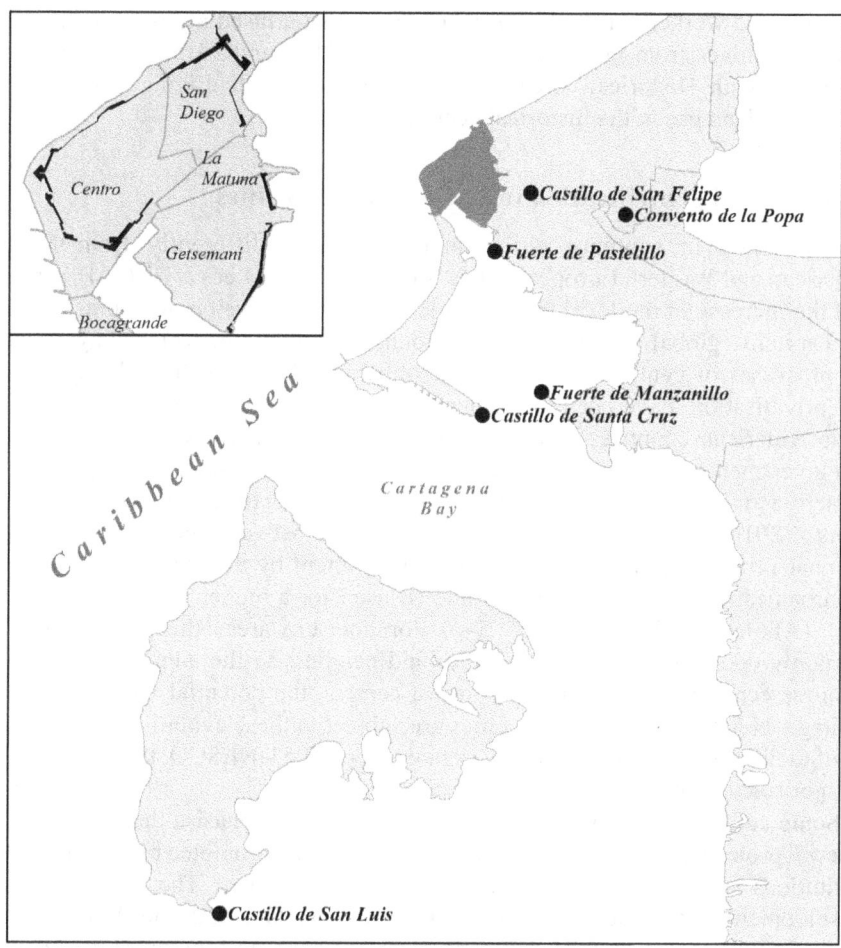

Cartagena UNESCO World Heritage Site

- • Monuments
- ▓▓ Historic Centre

░░ City of Cartagena de Indias

Figure 4.1 Cartagena UNESCO World Heritage Site
Source: Author, based on data from UNESCO

For educational and institutional functions, I consulted the spatial database of land use provided by Cartagena city's municipal government's Department of Planning[8] to acquire information on land use in the city in 2001. I superimposed this spatial database onto the GPS coordinates of commerce and tourism services to illustrate the transformation of the historical centre between 2001 and 2014.

In January 2015, I carried out observations in the field and interviews with residents and workers of the historical centre. I also utilised my personal photographic archives from visits to the city between 2007 and 2015. These pictures combined with Historical Street View are helpful in identifying evidence of changes in land use in the historical centre.

Tourism and gentrification in Latin American cities

Gentrification in southern cities[9] manifests itself differently from North American and Western European cities. Lees (2014) and Lees *et al.* (2016) argue that the increase in the level of economic insertion of developing countries into the capitalist global system brings profound restructuring, which may generate processes of gentrification. Thus, neoliberal policies, such as deregulation and privatisation, have been implemented gradually and to different degrees in Latin American countries and have progressively opened new capital markets and investment opportunities (Betancur, 2014). In this sense, central spaces are attractive for capital investment which makes it possible to capture the "rent gap" (Smith, 2010). This is defined as "the difference between the potential value of inner urban land (low value due to abandonment by deindustrialization and suburbanization) and its potential value (if used for a higher value and 'better' use)" (Atkinson and Bridge, 2005: 5–6). For inner-city areas, this potential value is mainly due to location, accessibility and nearness to the administrative and business centres. In the case of historical centres, the potential value is due to heritage. Not all Latin American cities are subject to these evolutions. The quality of their heritage, as identified by organisations like UNESCO, therefore plays a major role (Conti, 2016).

Some authors (Betancur, 2014; Lees *et al.*, 2016) emphasise that most urban renewal projects carried out in Latin American cities are promoted by international institutions such as The Organization of American States, The Inter-America Development Bank, The International Development Bank and The World Bank. These institutions encourage urban renewal for heritage tourism (Lees *et al.*, 2016: 101).[10]

In consequence, historical centres with high heritage value, such as those listed by UNESCO as World Heritage Sites, tend to attract global economic capital allowing for their restoration and maintenance, thus attracting tourists who, in their turn, attract capital to the places they visit. A circle of investment for conservation, of tourismification (due to the conservation) and profitability (due to the tourismification) is introduced. Virtuous or not, this circle generates changes in land use and urban functions in the historical centres.

National and local governments can play an important role in this process, since they implement urban renewal projects or create public policies intended to attract capital for renovation and rehabilitation of rundown areas. Thus, many case studies on gentrification in Latin America discuss the influence of different types of public policies implemented in these cities for renovation and/or generation of real estate investment and their impacts. This is the case for cities in Mexico (Hiernaux, 2003), Chile (Inzulza-Contardo, 2012) and Argentina (Herzer *et al.*, 2015).

Due to the complex urban structure of Latin American city centres, where different land uses – administrative, educational, residential (including lower-income residents), formal commerce and informal business (street vendors) – coexist in the same space, several authors are led (Inzulza-Contardo, 2012; Rubino, 2005) to suggest that Latin American cities have their own process of gentrification. It manifests itself differently in each city. Inzulza-Contardo (2012) called this process "Latino gentrification". The author defines it as:

> the replacement of the existing residential typology (one-, two- or three-storey terraced houses) by a new housing tendency (middle- and high-rise buildings), rather than the displacement of people of low income to outskirts by people of higher income, or *the return to the central areas of the middle class seeking to develop its artistic or cultural activities*. (Inzulza-Contardo, 2012: 2101) (emphasis added)

This "seeking" to develop artistic, cultural and leisure activities by the middle and upper-classes could lead to a change in land use. This change manifests itself in terms of the transformation of residential land use into commercial or office use (Carrion, 2005[11]) but also into tourism facilities. The manifestation of these changes, understood as part of "Latino gentrification" (Inzulza-Contardo, 2012), has been studied in different Latin American UNESCO cities (Delgadillo, 2015) such as Salvador Bahia (Nobre, 2002), Mexico City (Delgadillo, 2009; Hiernaux, 2003), Panama (Sigler and Wachsmuth, 2015) and Santo Domingo (Gonzalez, Chapter 2 in this book). Thus, the restoration in Latin American historical centres does "not produce residential class replacement or the return to the gentry but commercial heritage tourism" (Betancur, 2014: 5).

In the case of Cartagena, gentrification did not begin with urban renewal projects or commercial uses but with individual investors who restored colonial buildings as vacation homes or second residences in the neighbourhoods of Centro and San Diego (Rojas, 1999). Diaz de Paniagua and Paniagua (1994)[12] mentioned the changes that were taking place in the 1980s in the neighbourhood of San Diego, such as the purchase and restoration of houses and real-estate speculation through the construction of a five-star hotel. These changes resulted in the loss of spaces for inhabitants to socialise in.[13] Likewise, the authors claim that the conservation and protection of heritage in Cartagena is concentrated only in material aspects (technical restoration of buildings) and not in the preservation of social and intangible heritage.

Since 2011, the implementation of economic incentives (more specifically tax exemptions for tourist accommodation) has increased security measures and Cartagena's positioning as a tourist destination has encouraged national and international stakeholders, such as international hotel chains, to invest in the city. This fact, coupled with the increase in the flow of tourists, has linked Cartagena more directly with globalisation.

Real estate speculation was accentuated in the neighbourhoods of Centro and San Diego and expanded into Getsemaní. Posso (2015) argues that the conservation policies and tributary incentives created by the Colombian state encouraged companies to invest in this area and made it difficult for low-income local residents to stay there. However, the process of gentrification in the historical centre of Cartagena today is not only related to the changes in residential use to second homes but is also accompanied by the displacement of other vital urban functions (administrative, institutional and educational) and the transformation of buildings for tourist services. In order to understand this process, it is necessary to look at the historical context of the city and the process of its conservation and restoration.

Historical context of Cartagena

Cartagena was founded in 1503 and belonged to the 'Keys of the Indies', a set of twenty cities dispersed throughout the American territories conquered by Spain (Zúñiga, 1997). These cities were key points allowing not only a connection between Spain and its American territories but also its ability to exert control over the latter.

The Spanish built the Dique Canal between the main river of New Granada, *La Magdalena*, and Cartagena. The river channel made communication possible between Santa Fe de Bogotá, the capital Viceroyalty of New Granada, and the Caribbean coast (Lemaitre, 1983a). This allowed trade between the Viceroyalty and Spain. Because of its role in the release of goods from Peru and Bolivia and the slave-trade, the city was considered a central place of trade between America and Europe (Nichols, 1973). Consequently, the city was attacked many times by pirates and corsairs. In order to protect it, the Spanish kingdom built military structures between the sixteenth and eighteenth centuries, including many forts along the bay, the city walls and one fortress (Castillo de San Felipe de Barajas) (Lemaitre, 1983a).

According to Diaz de Paniagua and Paniagua (1994), the early urban layout of Cartagena had features resembling the urban structure of medieval Spain. By the sixteenth century, the current features of the city were formed: the city (what is known today as the Centro neighbourhood) was where the legal and administrative powers were concentrated; the *alfoz* or the immediate environment of the city (today's San Diego neighbourhood) was characterised by rural–urban interaction; and the *arrabal* or suburbs (Getsemaní) was where residents (the poor, blacks and mulattoes) not accepted by the Spanish lived. This initial urban configuration influenced the city's architecture, in terms of the types of homes (size and height), and land use (Melero, 2004).

Over time, both the *alfoz* and the *arrabal* were integrated into the city and joined the walled city. Today the walled city of Cartagena is composed of three neighbourhoods: Centro or Cathedral, San Diego and Getsemaní. It is important to keep this urban configuration in mind since the process of gentrification has occurred in different ways and at different times in the three neighbourhoods. I will explain this process in more detail in the third part of this chapter.

In the history of Cartagena, the nineteenth century is considered to be a "lost century" (Lemaitre, 1983b) for the city in both economical terms and population growth. This was due to the devastation caused by the wars of independence (1810–1825), the Spanish expeditionary fleet under Pablo Morillo (1815) and the closure of the Dique Canal. The city and its port lost its role of economic centre in the country's economy. Cartagena was also impacted by epidemics of tropical diseases such as cholera (1849), yellow fever (1900) and smallpox (1901–1902), which caused the city to lose much of its population (Samudio, 1999).

In the early twentieth century, US companies built pipelines between Cartagena and the interior of the country; with this, the city experienced new economic development and population grown. The walled city no longer had the capacity to accommodate all residential, commercial, industrial and institutional functions. Increasingly the centre became more and more dense, generating congestion and pollution problems. Many families decided to establish their residence in neighbourhoods outside the walls. According to Samudio (2001) "This process was not only of the bourgeoisie; the middle and working classes also settled elsewhere along the coast and in towards the mainland"[14] (Samudio, 2001: 135).

At the same time, following the example of other European and Latin American cities, some portions of the walls surrounding the city were destroyed. The walls were considered "as a material and psychological obstacle, like a useless inheritance that prevented the entry of modernity" (Cunin and Rinaudo, 2006, para. 8). This destruction facilitated not only movement between the centre and new neighbourhoods but also the process of developing hygiene and sanitation systems.[15]

Conservation and restoration process

Between 1880 and 1924, the Cartagena walls were dismantled. The local press called it "*murallicidio*" (Lemaitre, 1983b). The dismantling of the walls ended in 1924 due to the absence of economic resources for their demolition (Samudio, 2006: 4). The city's economic crisis allowed for the conservation of Spanish military architecture, as compared to other Latin American cities that experienced demolitions such as Havana (1863) or San Juan de Puerto Rico (1865–1897). At the same time, an appropriation, defence and conservation movement emerged, led by a group of citizens defending the historical value of the city. This led to the foundation in 1923 of the Society of Public Improvements of Cartagena (*Sociedad de Mejoras Públicas*). The Colombian state, through funds given directly to this organisation, financed the conservation and restoration of historical monuments: walls, forts, and the fortress of San Felipe de Barajas.

From the 1930s to the 1970s, the state and local public sector invested primarily in the restoration of the major monuments (walls and forts) and other historical landmark buildings. This was supported by national laws establishing measures of protection and the ratification of international agreements.[16] In 1959, the Colombian government declared the historical centre of Cartagena – and the other historical centres of Colombian cities – a national monument. The historical area is defined as the set of sixteenth-, seventeenth- and eighteenth-century streets, squares, walls, buildings, houses and historical buildings in the outskirts of the city (Congreso de Colombia, n.d.: National Law 163 of 1959). Despite this declaration, no protective measures were taken for the neighbourhoods in the historical centre. The historical centre gradually deteriorated due to the abandonment of most of the homes, congestion, pollution, street vendors and the subdivision of homes. The area was not perceived as very attractive to tourist development.

At that time (1970–80), the economic sector of the city and the country were more interested in the development of coastal tourism and less in the development of cultural tourism. In Cartagena, tourism development focused primarily on the peninsula of Bocagrande and Laguito. The historical centre was almost completely void of tourist activity. According to data from the Colombian National Registry of Tourism and the French guide MA (Bordier-Chêne and Bordier-Chêne, 1982), the historical centre had four hotels located in the proximity of the commercial area, *La Matuna*.

In the late 1960s, the process of the restoration of historical buildings began in the walled city, especially in Centro neighbourhood (Table 4.1). Many of these restorations were carried out by government agencies and private organisations such as banks.

This table shows that the major monuments such as the walls and forts were preserved by the state, through the Society of Public Improvements, while conservation of monuments and homes located in the neighbourhoods of the historical centre was to their owners. This meant that only owners with purchasing power could restore their homes. Thus, many homes in the historical centre deteriorated until the 1970s when the slow process of gentrification began.

The gentrification process

The process of gentrification in the historical centre of Cartagena emerged as a consequence of private sector investments to preserve urban heritage. I will

Table 4.1 Buildings restored and their use, 1960–1983

Building	Year of restoration	Use
Customs House	1969	Institutional – Town hall
Las Bovedas	1975	Handicraft Market
Casa del Marques de Valdehoyos	?	Institutional – National Corporation of Tourism
Wineries of Customs House	1979	Cultural – Modern Art Museum
San Diego Convent	1979	Cultural – School of Fine Arts
Chamber of Commerce	1981	Institutional
Old headquarters of the Bank of Bolivar	1981	Cultural – Bartolomé Calvo Library
Colonial House	1982	Cultural – Gold Museum
Inquisition Palace	1983	Cultural – Historical Museum of Cartagena

Sources: Data from Mendoza de Riaño (2000), Viloria de la Hoz (2005) and Téllez (2016)

outline the two phases of this process, individual private and tourist investments, and their consequences in terms of land use and urban functions in the historical centre neighbourhoods.

Private investment

The restoration of the walls, forts and government buildings, and the relative safety of the Centro and San Diego neighbourhoods, were key factors leading private players to invest in the restoration of colonial homes (Rojas, 1999). This began when Gloria Zea, the director of the Colombian Institute of Culture ('*Colcultura*'), and Fernando Botero's ex-wife, bought and restored a colonial home in the late 1970s. Then, the architect-restorer Alberto Barrera, who renovated Zea's home, along with a group of friends, bought a dozen colonial homes, which were re-sold to Colombian public figures (Wolinski, 2014). In the following years, several persons of note in the cultural, political and economic elite of Colombia – many of them from the interior, Bogota and Medellin – started buying and restoring colonial homes as vacation residences (Scarpaci, 2005). This process gave social prestige to individuals who restored homes in the historical centre. The local elite who lived in Bocagrande and Castillogrande also started buying homes in the historical centre to establish their residence (Diaz de Paniagua and Paniagua, 1994). For these elites, the walled city was perceived as a refuge, a good place to relax, far from the mass tourism that had developed in the Bocagrande area (Rojas, 1999; Scarpaci, 2005).

At that time, in the Bocagrande and Laguito area, mass tourism increased due to the construction of several high-rise buildings,[17] most of them rented out as vacation apartments to the middle-classes. The main avenues were transformed into commercial corridors with boutiques, bazaars ('*sanandresitos*'), restaurants and bars. This intensity of tourist activities came with various problems, such as street vendors, drug trafficking, prostitution and super saturation of public services like electricity, water, sewerage and parking (Samudio, 1999).

The greatest amount of individual private investment was focused on restoring homes in the Centro neighbourhood. Some colonial homes in the San Diego and Getsemaní areas continued to deteriorate during the 1980s. According to Rojas (1999: 54), private investment intensified in the 1980s when an upper-middle-class clientele from the interior of the country, mainly from Bogota, Medellin and Cali, was attracted by the exclusive atmosphere of the historical city centre. This upper-middle-class invested in smaller homes in Centro and San Diego (1980s and 1990s). During these same years, real estate investors (national and international) started projects with multi-family homes; they bought big houses, restored facades and built private apartments inside, which they sold to members of the upper-middle-class (Rojas, 1999).

Cartagena's inscription process as an UNESCO World Heritage Site began in 1983 when the Colombian government proposed the inscription of the historical centre. The document only included domestic colonial architecture located in three neighbourhoods: Centro or San Pedro, San Diego and Getsemaní.[18] Following the recommendation of ICOMOS,[19] the government included military architecture from the sixteenth, seventeenth and eighteenth centuries located in Cartagena's bay and changed the site's name to "port, fortresses and other monuments".[20] In 1984, UNESCO declared Cartagena a World Heritage Site. In August 1985, this declaration was publicly celebrated with the installation of a commemorative plaque. According to a tourism official, this inscription as an UNESCO World Heritage Site had no immediate impact on tourism (Piñeros, 2013) for two reasons. First, the media coverage was quickly overshadowed by two events that same year: the siege of the Palace of Justice in Bogota by the guerrilla group M-19 (5 and 6 November 1985) and second, the Armero disaster caused by the eruption of the Nevado del Ruiz volcano (13 and 14 November 1985). However, local players (hotels, travel agencies, restaurants and bars, among others) continued to develop tourism on the Bocagrande peninsula.

Nevertheless, according to Diaz de Paniagua and Paniagua (1994) and Posso (2015) one effect of the UNESCO inscription was an increase in housing pressure in the historical centre, mostly in the Centro and San Diego neighbourhoods. This effect was not immediate but was felt gradually in the following years. In Getsemaní purchasing pressure was low because the neighbourhood was physically and socially run down.[21]

Starting in 1988, local authorities began to think about the different issues the historical centre had been experiencing: the saturation of commercial functions, degradation of heritage and the absence of checks and balances to limit the architectural transformation and rehabilitation of historical properties. At the request of the Mayor's Office of Cartagena, The Fund for the Protection of the World Cultural and Natural Heritage financed four programmes between 1988 and 1991. These programmes were intended to provide technical assistance for the preservation and restoration of the city's architectural heritage as well as assistance in the development of protection and conservation norms for the historical centre. At the same time, some developments were prohibited in the historical centre, such as mass storage and truck traffic.

In 1992, the municipal government established Ordinance Number 6 which defined the administrative structure and regulatory procedures for preservation and rehabilitation based on a system of architectural typologies. At the same time, it established incentives in the form of exemptions from real estate and related taxes. Thus, the work of restoration and rehabilitation in Centro, San Diego and Getsemaní obtained exemptions from land taxes (Ordinance Number 6, 1992):

- Heritage site and typological restoration (100 per cent exemption).
- Facade restoration and interior adaptation (75 per cent exemption).
- General adaptation (50 per cent exemption).
- New construction (25 per cent exemption).

The procedures outlined in Ordinance Number 6 to obtain licenses, permits and certificates of accepted use were very costly. This meant that only owners with substantial resources could restore their properties (Diaz de Paniagua and Paniagua, 1994).

Simultaneously, real estate speculation was taking place in San Diego in the form of the construction of a luxury hotel in the abandoned municipal hospital building. This real estate pressure, along with the purchases and restorations carried out by wealthy families, generated, in the long run, a recalculation of public utilities and taxes. This meant that low-income residents could not continue living there; they either had to sell their property or move to leased property that better suited their disposable income. This change in ownership and residents was visible to the inhabitants themselves and has been referred to by multiple authors (Diaz de Paniagua and Paniagua, 1994; Scarpaci, 2005) but there are no official statistics showing population change.

Individual investments did not initially occur for economic reasons but rather to acquire prestige inside the same group of elites (Rojas, 1999). This created a 'trend' that attracted the local elite and upper-middle classes to buy and restore colonial homes. The arrival of new temporary residents brought new businesses such as restaurants and bars. This increased real estate and commercial demand in nearby areas. Thus, many families, unable to keep their homes due to the increase in real estate speculation, decided to sell. The gentrification process intensified in Centro and San Diego and extended to Getsemaní when investments in tourism accommodation began there as well.

Tourism investment

Tourism in Cartagena was initially oriented towards the beach areas, especially on the Bocagrande peninsula. Tourism infrastructure (hotels, casinos, restaurants, etc.) was concentrated in that area between the 1970s and 1980s.

Because of the increased density in Bocagrande, as mentioned before, the municipal authorities suspended construction permits in this area (Samudio, 1999). This pushed for the decentralisation of the hotel supply in the city. Hotel investors sought out new areas for their projects. In the early 1990s, tourism investment moved to the north with the construction of the Hotel Las Americas in 1993. In the historical centre, tourism investment began when two ancient monasteries were renovated to become luxury hotels – Santa Clara Sofitel Luxury Hotel (opened in 1994) and Santa Teresa Charleston (opened in 1996). They were the pioneers of a new urban function in the historical centre offering tourist accommodation which was previously exclusively located on the Bocagrande peninsula. This marked the beginning of the process of gentrification by tourism in Cartagena.

Tourist accommodation

Beginning in the 1970s, the Colombian government allowed the possibility of turning old convents, which were the property of the state, into luxury hotels (CIE and CNT, 1978). Investment in restoration and management would be provided by private investors because the state did not have sufficient funds. However, this project was never carried out and the monuments remained abandoned until the late 1980s.

In the early 1990s, the state auctioned off some historical properties, such as old convents in Cartagena (Correa, 1998), so that they could be restored and given new uses. The Old Convent of Santa Clara, built in 1621, located in San Diego was the first to be restored for use in the tourism industry. The work was carried out between 1992 and 1994 when a series of historical and archaeological studies were conducted. The hotel opened its doors in October, 1994. It is now a luxury hotel (122 guestrooms) belonging to the French chain, Sofitel Luxury Hotels. The other major monument restored was the Convent of Santa Teresa, built in 1610, located in Centro. Between 1995 and 1996, the Colombian hotel chain Hotels Charleston carried out restorations.

Similarly, other tourist activities and services appeared in these areas: antique shops, boutiques, jewellery, crafts, restaurants and bars. However, the investment in restoration to develop tourism was interrupted by Colombia's economic crisis in 1998 (Scarpaci, 2005).

Beginning in 2002, the Colombian government drew up a set of incentives to attract foreign investment to the country. Specifically, Article 207 paragraphs 3 and 4 of the Colombia Tax Statute provides tax exemptions for a period of thirty years for the renovation, expansion and construction of new hotels made between 1 January 2003 and 31 December 2017. Figure 4.2 illustrates the evolution of accommodation in the historical centre between 1990 and 2014.

Since 2004, these tax incentives have promoted the transformation of colonial homes into boutique hotels and hostels. The boutique hotels are characterised by a limited number of rooms and personalised services. Their decor is also an attraction. In the case of Cartagena, most of these hotels have colonial style decor combined with other elements of Indian, Mediterranean or other modern styles. They are located mostly in Centro and San Diego, but since 2006 some boutique hotels have popped up in Getsemaní as well.

Another style of tourist accommodation that has increased in the historical centre is the hostel. These are situated mainly in the Getsemaní neighbourhood. In the 80s and 90s low-priced accommodation such as motels were located in Getsemaní. Many of these forms of accommodation were re-modelled and converted into hostels. These housed mostly young tourists and backpackers. The hostels are mainly concentrated on the streets of Tripita Media and Media Luna. In 2013, when I talked to some of the tourists who were staying in this area, they told me that they preferred to stay in this neighbourhood because of the low-cost but also because it was more 'authentic'. They saw more of the daily life of the locals in Cartagena, where there are still small shops, artisans and children playing football in the streets.

Through fieldwork and spatial data, it emerged that there is a spatial differentiation in the accommodation supply (Figure 4.2). On the one hand, the high-quality hotels (boutique hotels such as the Santa Clara and Santa Teresa Hotels) are located in Centro and San Diego while the hostels and smalls hotels are situated in Getsemaní. However, this might change in Getsemaní. In 2014, a Viceroy Hotel Group project to build a six-star hotel was announced by the press.[22]

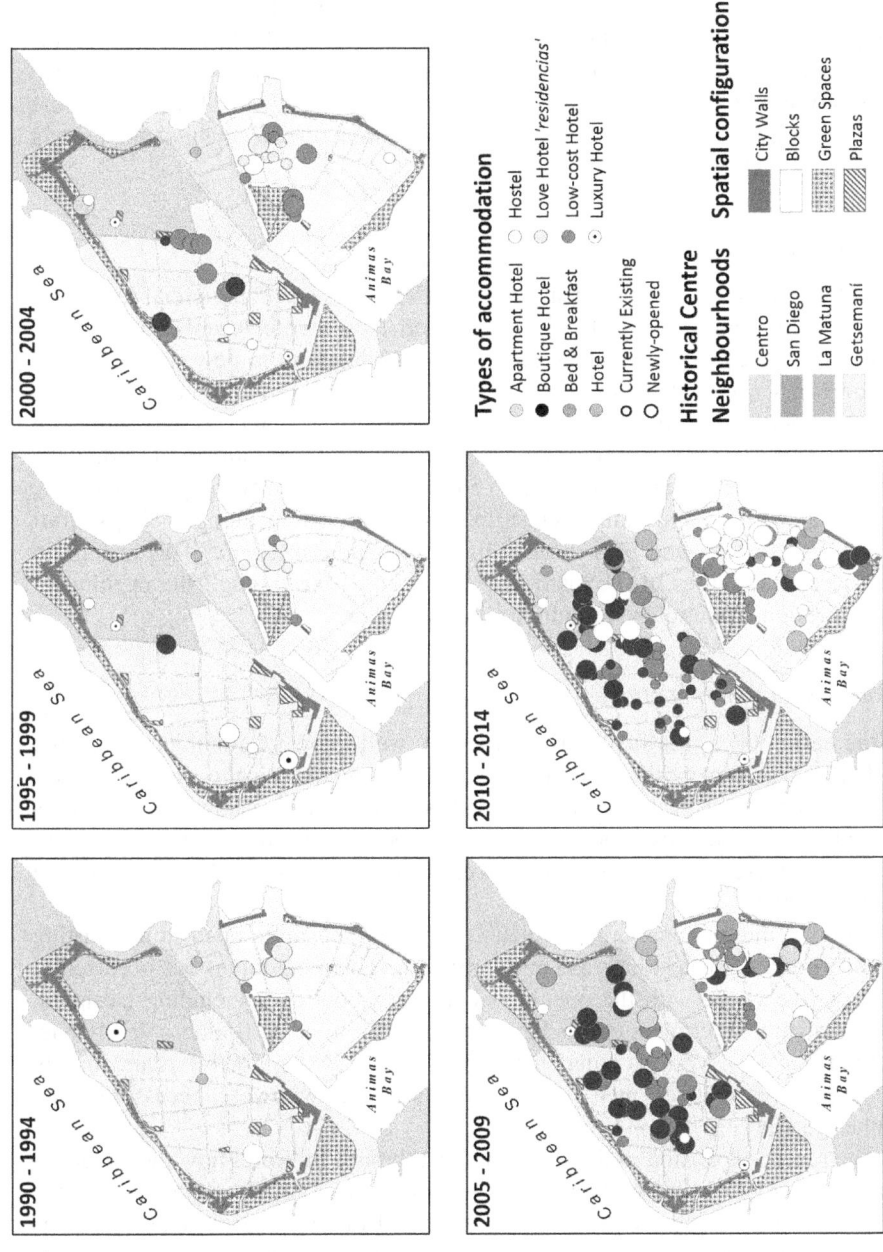

Figure 4.2 Accommodation growth statistics, 1990–2014

Sources: Author's spatial database, data from Cartagena's Chamber of Commerce and the Tourism National Registry

The site is located in the Old Convent, Obra Pia. Four Seasons will open a hotel in the old Club Cartagena in 2017.[23] The gentrification process may accelerate, turning towards luxury markets. In fact, an owner of one of the most popular hostels in Getsemaní stated online (Parker, 2014) that the implantation of these hotels will completely change the neighbourhood and they (i.e. hostels) have two possibilities: make investments in the hostels to improve their quality (meaning a price increase) and convert them into boutique hotels or sell them. Accordingly, the increase in hotel quality in the sector will generate a displacement of the hostels and small hotels in the sector.

Commercial activities

From the beginning, the historical centre of Cartagena had commercial functions, most of them geared towards local consumption. However, in recent decades many commercial premises have changed their use giving way to new shops focused on tourism services. This process started when the restoration of colonial homes by the national elite intensified. Several restaurants, art galleries and bars were opened. Rojas (1999) mentioned that the seasonal nature of tourism in Cartagena, at the time, caused a number of them to close. However, the opening of the luxury hotels Santa Clara and Santa Teresa in the mid-1990s encouraged shops, craft outlets, restaurants, and bars to open in Centro and San Diego. This change in commercial activities in the historical centre is visible through the opening of stores, designer boutiques, restaurants, bars and cafes.

Shops and designer boutiques

In the early 2000s, the first stores owned by well-known Colombian designers, such as Silvia Tcherassi or Jon Sonen, opened in the historical centre. In the following years other designers also opened boutiques in the same area. Jewellers specialising in emeralds relocated to this neighbourhood and the number of jewellery stores increased; four opened between 2000 and 2004, and twelve between 2005 and 2009.

The first international brands, such as Benetton, began to open stores starting in 2005 but the greatest increase in international chain boutiques has occurred in the last five years, with the opening of fourteen boutiques including Desigual, Bosi, Salvatore Ferragamo, Hackett London, Harmont & Blaine and Michael Kors (Table 4.2). Various other stores have opened in the last five years, mainly clothing stores that sell men's and women's attire, swimwear, accessories and leather goods.

The recent expansion of these international brands indicates how the historical centre of Cartagena participates in globalisation. This is largely due to the increase in tourist activity and the internationalisation of the destination. However, it is important to mention that the majority of international stores, national brands, craft outlets and jewellers are concentrated in the Centro neighbourhood.

Table 4.2 Store types by neighbourhood, 1990–2014

| | | Time Period | | | | | | |
Neighborhood	Store type	Before 1990	1990–1994	1995–1999	2000–2004	2005–2009	2010–2014	Total
Centro	Antique shop					1	1	2
	Handcrafts			1	2	9	6	18
	Other stores			1			4	5
	Other clothes stores	1		1	2	4	14	22
	Jewellery store	2	1		4	12	4	23
	Colombian designers			1	5	5	5	16
	International brands		1			2	14	17
	National brands				4	3	7	14
	Total	**3**	**2**	**4**	**17**	**36**	**55**	**117**
San Diego	Antique shop		1					1
	Handcrafts					1	1	2
	Jewellery store						1	1
	Total		**1**			**1**	**2**	**4**
Getsemani	Other clothes stores						1	1
	Colombian designers					1		1
	Total					**1**	**1**	**2**
	Total	**3**	**3**		**17**	**38**	**58**	**123**

Sources: Author's spatial database, data from Cartagena's Chamber of Commerce

Restaurants and bars

After the construction of the convention centre and the restoration of colonial homes during the 1980s, some restaurants and bars opened in the historical centre. Some of them are located on Arsenal Street which leads to the convention centre. However, the process began to gain real momentum when the Santa Clara and Santa Teresa hotels opened. Some restaurants around the San Diego Square were located in front of the hotel. This caused a series of conflicts between local residents and restaurant owners, who took up much of the public space in the square. This was mentioned by Scarpaci (2005) and the local press (Agámez, 2015; Álvarez, 2013).

In recent years, many bars and restaurants have opened in the streets of San Diego and Getsemaní, as well as San Toribio, San Diego and La Trinidad squares (Figure 4.3). Pictures taken in 2011 in Trinity Square, Getsemaní show a quiet place with small local shops around the square. At night, the square was used by residents to talk with neighbours or just take a breath of fresh air. Today, many cafes and restaurants are located around this square. At night it is increasingly filled with tourists. The first comment on TripAdvisor was made in June 2012; in 2014 this square was chosen by Travellers' Choice™ as the 2014 winner of the Top 10 Attractions Landmarks – Colombia. Two years later, it became a very popular place for young tourists (backpackers) and for tourists searching for a 'more authentic' experience, wanting to see a 'real neighbourhood'.

Sightseeing tours

Sightseeing tours are also changing. The horse-drawn carriage tour was introduced in the 1950s when tourism began on the Bocagrande peninsula. When the Santa Clara Hotel opened in 1995, some of these horse-drawn carriages began to make tours of the historical centre.[24] Today, most of these tours take place in this area.

The Tourism Corporation of Cartagena's map, distributed at tourist information points, proposes a walking tour which is focused mainly in Centro and includes some streets in San Diego.

In recent years, with the increase in visitors, the ways to visit the historical centre have also diversified. Now there are various options for touring the old town. Since 2012, a service for renting bicycles, Segways and scooters is available. Recently, a free walking tour has been made available. These tours coexist with more traditional tours: guided tours, horse-drawn carriages and *chiva*.[25] Compared to twenty years ago, visitors can now move more and more freely in these neighbourhoods. They like to get 'lost', discovering for themselves, by intuition or with the aid of a map, the streets of the historical centre, admiring its heritage and other aspects that may go unnoticed on a guided tour.

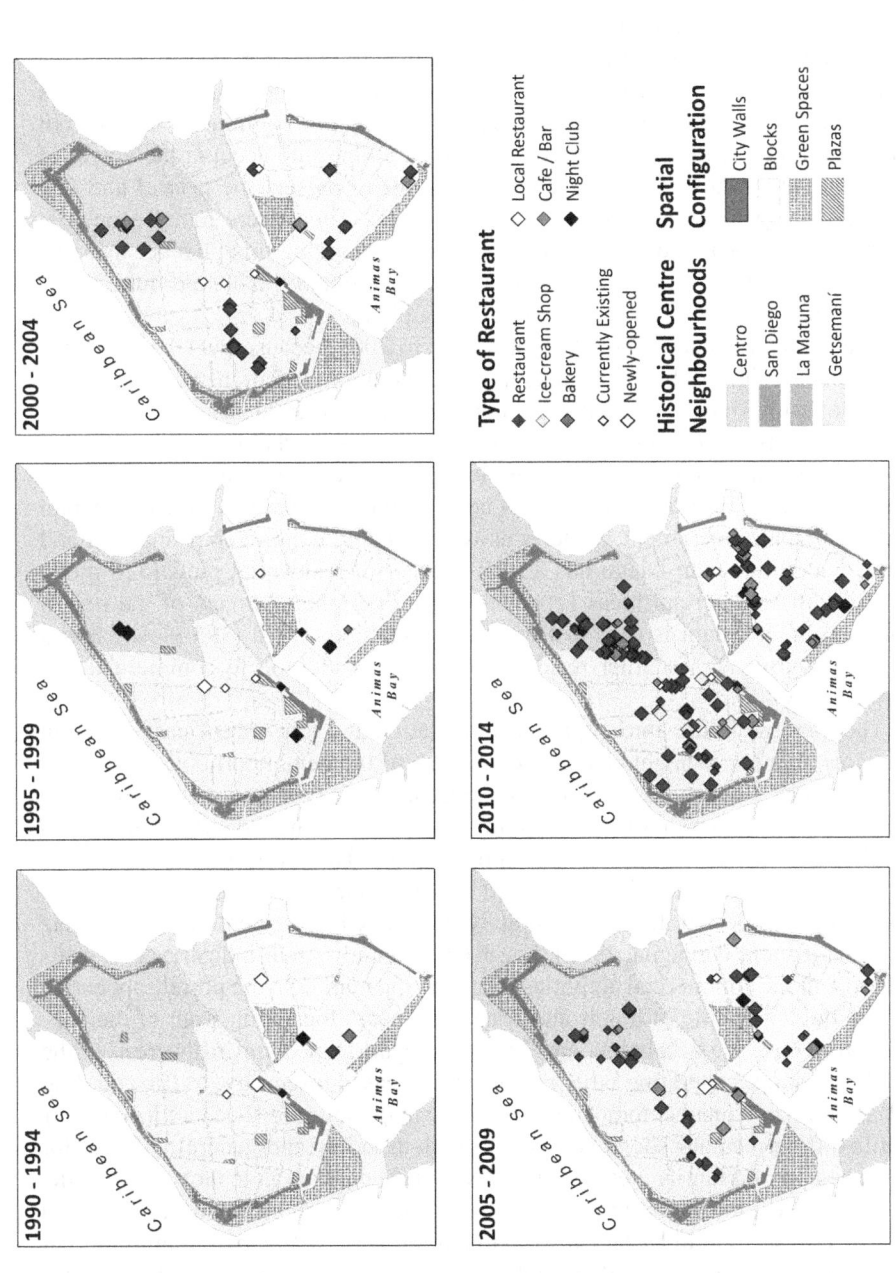

Figure 4.3 Restaurants growth statistics, 1990–2014

Sources: Author's spatial database, data from Cartagena's Chamber of Commerce

Changes in land use

Land use in the neighbourhoods of the historical centre is regulated by the territorial ordering plan – POT 2001. Since 2008, National Law 1185 and Ordinance 763 of 2009 have required a special management and protection plan for areas of cultural interest called PEMP (*Plan Especial de Manejo y Protección*). This is the case for the historical centre of Cartagena. In 2015 however, this plan was not approved by the national government. According to a community leader from Getsemaní, this plan was conceived of without taking into account the voices in the local community. There were only meetings to present the project and few people were given access to the document itself. There were many economic interests, both local and international players, surrounding this plan because this document regulates land use in the historical centre and this would outrank the regulations set forth by the territorial ordering plan.

The available data on land use dates back to 2001 because POT has not been reviewed or adjusted. I take this data into consideration to demonstrate changes in land use that have been generated by tourist activities over the last fourteen years.

Land use is not homogeneous (Table 4.3). Centro is characterised as having a very important institutional use (29 plots), compared with San Diego (6 plots) and Getsemaní (11 plots). This area is home to the headquarters of departmental and municipal government agencies as well as many commercial, financial and service activities. San Diego has a high number of residential plots (328 plots) mixed with commercial areas (26 plots) in the peripheral streets of the neighbourhood. Getsemaní is the most residential neighbourhood (395 plots). It is a popular quarter with smaller residential homes, in addition to commercial and service activities.

The increase in the amount of accommodation and tourist services located in the same area has brought changes to land use of the neighbourhoods of the historical centre. The process of change in land use due to tourism activities began in 1978 when the central market situated in Getsemaní was moved outside the neighbourhood in order to build the convention centre. This impacted on Getsemaní in numerous ways, including the loss of a major source of employment for local people who worked or had business activities related to the central market (Diaz de Paniagua and Paniagua, 1993). For example, many small industries and firms moved and the commercial activities related to the market in the port disappeared.

Although a couple of hotels and restaurants were located in front of the convention centre, bars, cabarets and brothels began to pop up in the rest of the neighbourhood, generating other problems: increased insecurity, traffic and drug sales. As a consequence, some families left the neighbourhood because it no longer fulfilled the necessary social conditions (such as safety and general welfare) for the cohesion of a family (resident 1 interview). The situation in Getsemaní at the time kept it off tourist routes until the early 2000s, when backpackers, attracted by low-cost accommodation (in Colombian terms '*residencias*'), stared to visit this neighbourhood. Gradually hostels and small modest hotels opened in the area (as was shown in Figure 4.2).

Table 4.3 Land use by neighbourhood

| | Historic centre neighbourhoods | | | |
	Centre	San Diego	Getsemaní	Total
No. blocks	30	17	22	69
No. plots	516	377	596	1,489
Land use				
Commercial	31	6	15	52
Commercial/institutional	4	0	1	5
Commercial/tourism	0	0	4	4
Institutional	29	6	11	46
Institutional/tourism	0	3	2	5
Institutional/mixed	15	0	0	15
Mixed	222	26	160	408
Mixed/tourism	0	0	2	2
Housing	194	328	395	917
Housing/institutional	19	2	2	23
Without use	2	6	4	12

Source: Adapted from *Plan de Ordenamiento Territorial del Distrito de Turístico y Cultural de Cartagena de Indias.* Decreto N° 0977 de 2001: 179–220

However, since 2004, with the increase of tourism in the city, there have been other changes in urban functions in the historical centre neighbourhoods. One of them has been the change of use of some buildings which used to be utilised for institutional, administrative and educational functions but are now used as tourist facilities.

In the case of educational use, in the 1970s, the National Tourism Corporation of Colombia promoted the transfer of primary and secondary schools out of the historical centre colonial buildings, arguing that they sped up the deterioration of these historical buildings (CIE and CNT, 1978: 32). In the 1980s, some of these institutions were transferred outside the city walls in order to spread the education system around the city.

From the 1990s to around 2010, other primary and secondary schools in the historical centre were relocated to other neighbourhoods and their buildings were transformed to house tourist services. In 2006, a real estate investor bought La Esperance School's six plots to build a multifamily complex with fifty-eight apartments. Another example is the Presentación School, which relocated in 2008. This school had two locations, one was transformed into another multifamily complex, '*La Casa del Boquetillo*' and the other is a cultural centre, '*Casa Museo Arte y Cultura la Presentación*'. Inside this centre, there is a boutique hotel. The last use of the Obra Pia Convent was as the Central School of Cartagena. It is now the six-star Viceroy Hotel. Today, two schools remain, '*La Milagrosa*', in Getsemaní and '*La Salle*', in San Diego. A local resident told me that the latter will be purchased in the coming years for the construction of a hotel.

Today, some universities, such as Rafael Núñez University and the University of Cartagena, remain in the area. The latter has already begun to decentralise some academic programmes. This process began in 1989 when the faculty of health science was transferred to the new headquarters in the Zaragocilla neighbourhood. The decentralisation process continued in the early 2000s with the relocation of the faculties of science and engineering, and economics. Today, at the historical centre headquarters, called San Agustin, only three faculties remain: humanities, law and political science, social sciences and education. These processes of decentralisation result in a decrease in the number of students who come to the historical centre on a daily basis.

The superposition data on land use as regulated in the territorial ordering plan of 2001 and the locations of accommodation and tourist services shows that land use is changing in the historical centre (Figure 4.4). The figure shows how many hotels, restaurants and shops are located in residential blocks. This has also happened in the areas formerly for institutional use.

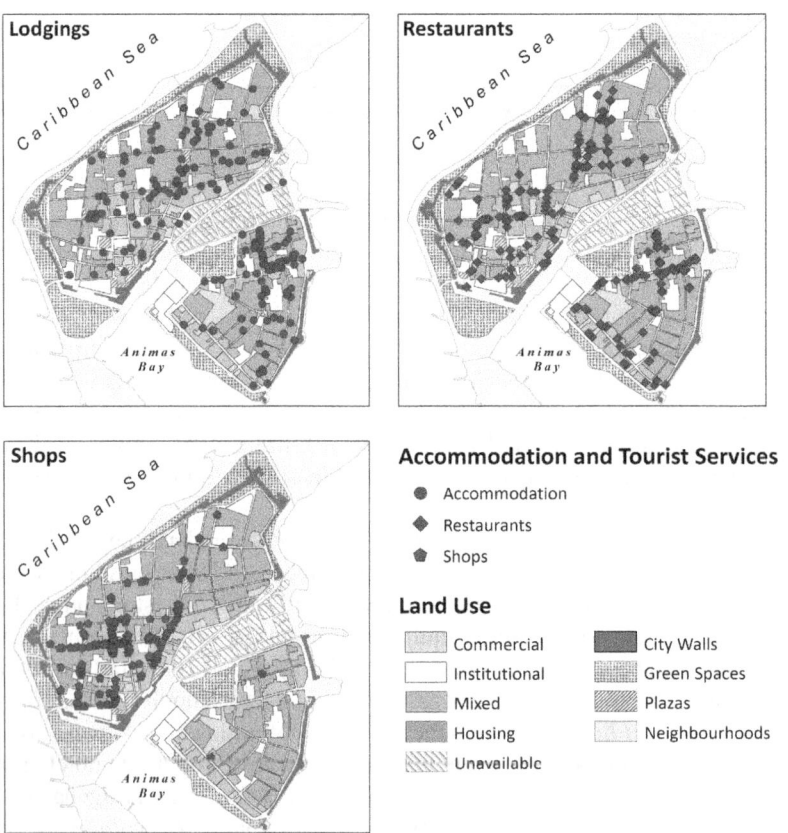

Figure 4.4 Superposition of land use data and the location of accommodation and tourist services

Sources: *Plan de Ordenamiento Territorial del Distrito de Turístico y Cultural de Cartagena de Indias* (2001) and author's spatial database

In recent years, some municipal departmental administrative functions were decentralised. In my 2007 field study, I observed that Centro was characterised by very local dynamics because of the urban administrative, commercial and institutional functions. Residents use to come to the centre for their administrative dealings with departmental and municipal authorities. Around 2010, several municipal offices were relocated outside the city walls. In 2013, the departmental headquarters no longer operated at Proclamation Place. The building has been converted into the Cultural Centre for Arts and Crafts of the Colombian Caribbean.

Today, one can still find some institutions such as departmental courts, notaries and the Mayor's office but these institutions are also being pressured to relocate so that their buildings can be converted for tourist services. The presence of the headquarters of the departmental government encouraged the revitalisation of the historical centre because many people came from different areas of the department. With the relocation of this institution, this population does not frequent the area as it did before. A resident told me:

> I only come to the centre when I have to do some kind of administrative procedure or buy something that I can only get here but I would not come to the centre, it is very far away from my house (more than one hour by bus) and rather expensive. (Interview with a local inhabitant in the historical centre, 2015)[26]

Commercial activities with no connection to tourist activity have also been affected. For example, during my 2013 fieldwork, I met an entrepreneur who had an art school. At that time, he was organising everything to move to another neighbourhood. He said that the incremental introduction of tourism affected him not only through the increasing cost of commercial space but also the restricted mobility within the centre.

During the last few years, the number of events and conferences in the city has increased and many plazas are being used for special dinners, weddings and private parties. The city is home to Colombian diplomats. It is in the convention centre that the most important political and economic meetings in the country (for example, the 6th Summit of the Americas in 2012) take place. The historical centre is used for private tours or special dinners in Plazas for international celebrity visitors. When events such as these are being held, access to the historical centre is restricted or closed to vehicles and pedestrians. When the visitor is a very special international celebrity, like President Obama or Prince Charles, safety standards are more stringent and often employees are unable to reach their workplaces. This causes constant discomfort for residents, workers, students and temporary visitors to the area.

Another problem has been the decline in the number of shops for the local population and the increase in tourist shops:

> Today, there are few places for us (the local population) that we can visit or attend. In fact, the services offered and the prices are more for tourists than for us. Before, you could find services for us. *Now everything is for tourists.*

There are only some small merchants where you can buy something to drink or eat but they could soon disappear like the others. (Interview with a worker in the historical centre, 2013)[27]

This process is continually happening today. For example, Figures 4.5 and 4.6 show an old local restaurant called '*La Loncheria*' that still existed in 2012. By 2014, this was transformed into an international cafe. This is demonstrated through the menu, the decor, WiFi service and the prices.

The changes in land use and increased tourism activities and services have also affected the social spaces of the inhabitants of San Diego and Getsemaní. In San Diego, for example, squares, streets and shops used to be social spaces where children played, families sat together in rocking chairs and celebrations were held on important dates like 7 December ('day of the candles') or 31 December (Diaz de Paniagua and Paniagua, 1994). Some of these activities and gatherings in the Fernandez de Madrid square or neighbourhood stores even existed in the early 90s. These social activities in San Diego changed with the opening of the Hotel San Clara in front of San Diego Square. Today, these social activities of local residents are almost non-existent because they have been replaced by areas for tourists and commercial purposes. Now, Fernández Madrid square is surrounded by restaurants and bars. Neighbourhood stores have practically disappeared, with only two remaining. San Diego Square is also surrounded by restaurants and shops where established artisans sell their products to tourists.

In the case of Getsemaní, changes in Trinity Square not only affected its commercial use, as mentioned before, but also its social use. In my interviews

Figure 4.5 Local restaurant, 2012

Source: Google Street View

Figure 4.6 International cafe, 2014

Source: Author's photograph archive

in 2013, residents told me that this plaza was a place for sport were children played football and that it was also a place to socialise for parents and neighbours. In recent years, these social practices have been lost. Fewer children play in the square and are now supplanted by street performances aimed at tourists where the young dance traditional Colombian dances such as the '*Cumbia*' or do acrobatic shows to the sounds of rap and hip-hop. At night, walking through the square, different languages are heard and many tourist groups eat street food and drink beer. This place is changing every day with the increase in tourism. It is losing its authenticity because the locals do not like it anymore and prefer to move away.

In recent years, these changes brought on by gentrification in Getsemaní have been exhibited in the local press (*El Universal* 29 October 2014), in blogs (Tyor, 2013) and in the international media (Caparros, 2014). This has also generated local social movements, such as the Tu Cultural foundation,[28] community meetings about the issue and publications by social researchers (Posso, 2015).

In short, the increase in accommodation and tourism services is transforming urban functions and social areas for residents, workers, students and temporary visitors to the area.

Conclusion

This chapter aimed to demonstrate that the process of gentrification in Cartagena is marked by two phases. The first began in the 80s with individual private investment in vacation homes and the second with private investment in tourism (accommodation and tourist services). Both phases of gentrification occurred in spatially differentiated patterns in the three historical centre neighbourhoods.

The process of gentrification by tourism in Cartagena has increased in recent years. This is due to the number of visitors, implementation of economic incentives (tax exemptions for construction of new hotels and the renovation/expansion of old hotels) and the international visibility of Cartagena through events and international meetings that make the city attractive for national and international investment.

The presence of five-star hotels, luxury stores and six-star hotel projects, such as Viceroy or Four Seasons, indicates how the historical centre of Cartagena has been integrated into global capital flows. This area has a higher potential value; in Smith's terms, it can capture a better 'rents cap' (Smith, 2010), and this potential value can generate a new stage in the process of gentrification by tourism in the city; hostels and small hotels could be replaced by boutique hotels, as well as four- and five-star hotels. Could this be tourism gentrification by tourism itself?

On the other hand, tourism has become increasingly present in the daily dynamics of the historical centre. It has gradually lost the urban functions that gave it vitality. If municipal, provincial and national authorities do not create the public policies necessary to prevent displacement of urban functions (administrative, institutional, educational and residential functions), it is possible that in the coming years the historical centre of Cartagena will become nothing more than a theme park. For some of Cartagena's residents, "Cartagena is a theme park" already:

> It is the open space where I can marry, where I can have my private party and other events. The functions of the historical centre are changing totally. I am in Disneyworld . . . if I don't have one or several businesses related to this theme park [hotel, restaurant, boutique . . .] I cannot stay in the historical centre. (Interview with an entrepreneur in the historical centre, 2014)

The process of gentrification by tourism manifests itself in various ways, from physical changes to the transformation of spatial relationships. This case study in Cartagena has shown that this process is not uncommon in the cities of Latin American countries. It is a process that is transforming all urban functions (residential, institutional and commercial) in the historical centre neighbourhoods and the spatial interaction between inhabitants of the city.

Notes

1 Cited by Lees (2014: 516).
2 For example, the number of foreign travellers who reported Cartagena as their main destination in Colombia in 2007 was 133,172 and 256,805 in 2014. Source: Cartagena Tourism Corporation.

3 31,063 cruise passengers were reported in 2003 and 310,957 in 2014. Data: Regional Port Society of Cartagena (*Sociedad Portuaria Regional de Cartagena*). In addition, since 2008, the cruise terminal has been a boarding point of passengers with Royal Caribbean and Pullmatur. Thus, Cartagena has become the principal port of arrival and departure for cruise passengers in Colombia.

4 The amount of tourist accommodation in Cartagena in 2008 was 8,111 rooms and 10,366 rooms in 2014. Source: Tourism Corporation Cartagena.

5 According to Zuleta and Jaramillo the historical centre is the most frequented place by both domestic (32 per cent) and international (21 per cent) tourists (Zuleta and Jaramillo, 2006: 123).

6 This has not been possible due to the crisis of governance in the city (six mayors in less than a year), not being allowed to draw roadmaps for the revision of POT and that creating large management plans for Adaptation to Climate Change and Risk Management is required by the new national regulations (Otero, 2012; El Universal Cartagena, 2012; El Espectador, 2013; extracartagena.com, 2013)

7 *Registro Único Empresarial y Social* – RUES (Confecamaras, n.d.).

8 *Mapa Interactivo de Asuntos del Suelo* – MIDAS (Alcaldía Mayor de Cartagena de Indias, 2010).

9 For Lees, South or global South is 'the places outside of the "usual suspects" in gentrification literature—that is cities in North America and Western Europe' (Lees, 2014: 506).

10 In this case, the Quito Lettre (*Normas de Quito*) of 1967 has underlined the importance of tourism for both preservation and maintenance of historic places.

11 Cited by Inzulza-Contardo (2012).

12 Diaz de Paniagua and Paniagua (1993) also conducted a study in Getsemaní about its history, heritage and social welfare which speaks of social deterioration (loss of cultural values) and exclusion, issues that dominated this neighbourhood during the 1990s.

13 The authors refer to the loss of cultural activities, such as the celebration of civil and religious holidays and socialising in spaces like traditional shops that are non-existent today.

14 Author's translation to English.

15 The city had poor public health, due to the absence of drinkable water and sewage systems and the threat of epidemics.

16 In 1936, Colombia ratified the Treaty of Montevideo on the protection of buildings of historical value from pre-Columbian, colonial, independence wars and the republican era, as well as the Roerich Pact concerning the protection of artistic and scientific institutions and historical monuments.

17 At that time, many investments in tourism or the construction of apartments were related to laundering drug money.

18 Délégation Permanente de Colombie auprès de l'UNESCO (1983).

19 ICOMOS letter, 4 April 1984.

20 Délégation Permanente de Colombie auprès de l'UNESCO (1984).

21 Illegal trade (selling stolen goods and drugs) and prostitution.

22 "Hospitality Net - Viceroy Hotel Group to Open Convento Obra Pia, Viceroy Cartagena", (Hospitality Net, 2014).

23 Corporación Turismo Cartagena de Indias (2015).

24 Interview in 2013 with the President of the horse carriage drivers' union, ASOCARCOCH.

25 This is a typical, colourful, Colombia bus.

26 Author's translation to English.

27 Author's translation to English.

28 A project on community empowerment called *Tu Acción para tu Barrio* (Your Action for your Neighbourhood). In this project they propose 'GETSEMA-NI-FI-CATION' instead of gentrification, to be understood as a socially just and integrative alternative form of socio-urban transformation.

References

Agámez, C. (2015) 'Polémica en San Diego por instalación de bolardos frente a la plaza', *El Universal Cartagena*. [Online] 28 April. [Accessed 27 June 2016]. Available from: www.eluniversal.com.co/cartagena/polemica-en-san-diego-por-instalacion-de-bolardos-frente-la-plaza-192010.

Alcaldía Mayor de Cartagena de Indias (2010) *Mapa Interactivo de Asuntos del Suelo - MIDAS*, [Online]. [Accessed 15 March 2015]. Available from: http://midas.cartagena.gov.co/.

Álvarez, R. (2013) 'Casas ruidosas en San Diego', *El Universal Cartagena*. [Online] 11 September. [Accessed 27 June 2016]. Available from: http://www.eluniversal.com.co/cartagena/casas-ruidosas-en-san-diego-134209.

Atkinson, R. and Bridge, G. (eds) (2005) *Gentrification in a Global Context: The New Urban Colonialism*, London: Routledge.

Betancur, J. (2014) 'Gentrification in Latin America: Overview and Critical Analysis', *Urban Studies Research*, vol. 2014, Article ID 986961, 14 pages. doi:10.1155/2014/986961.

Bordier-Chêne, J.-J. and Bordier-Chêne, M. (1982) *Pays andins: Colombie, Équateur, Pérou, Bolivie, Chili*, (Éd. 1982–83), Paris: M.A. Editions.

Caparros, M. (2014) 'A gentrificar, a gentrificar', *EL PAÍS*. [Online] 29 April. [Accessed 16 February 2016]. Available from: http://elpais.com/elpais/2014/04/23/eps/1398283121_375738.html.

CIE and CNT (eds) (1978) *Cartagena, zona histórica*. Bogotá: Centro de Investigaciones Estéticas e Historicas de la Universidad de Los Andes, Corporación Nacional de Turismo.

Confecamaras (n.d.). 'RUES - Registro Único Empresarial y Social, Cámaras de Comercio'. [Online]. [Accessed 27 February 2015]. Available from: www.rues.org.co/RUES_Web/.

Congreso de Colombia (n.d.). 'Ley 163 de 1959 Diciembre 30 Por el cual se dictan medidas sobre defensa y concervacion del patrimonio historico, artistico y monumentos publicos de la Nacion.'

Corporación Turismo Cartagena de Indias (2015) *Retos y realidades. El sector turistico de Cartagena de Indias*, [Online] Cartagena de Indias: Corporación Turismo Cartagena de Indias, [Accessed 1 November 2015] Available from: www.cartagenadeindias.travel/descargas.

Conti, A. (2016) 'The Impact of Tourism on Latin America World Heritage Towns', in Bourdeau, L., Gravari-Barbas, M. and Robinson, M. (eds), *World Heritage Sites and Tourism: Global and Local Relations*, London: Routledge, pp. 175–188.

Correa, F. (1998) 'Hotel Santa Clara Cartagena de Indias', *eltiempo.com*. [Online] 12 September. [Accessed 10 April 2013]. Available from: www.eltiempo.com/archivo/documento/MAM-832673.

Cunin, E. and Rinaudo, C. (2006) 'Entre patrimoine mondial et ségrégation locale : Cartagena et ses murailles', *Cahiers de la Méditerranée*, No. 73. [Online] [Accessed 24 November 2012] Available from: http://cdlm.revues.org/index1623.html.

Délégation Permanente de Colombie auprès de l'UNESCO (1983) 'Inclusión del Centro Historico de Cartagena a convención Patrimonio Cultural y Natural Mundial', UNESCO.

Délégation Permanente de Colombie auprès de l'UNESCO (1984) 'Proposition d'inscription sur la liste du PM soumise par la Colombie', UNESCO.

Delgadillo, V. (2009) 'Patrimonio urbano y turismo cultural en la ciudad de México: Las Chinampas de Xochimilco y el Centro Historico', *Andamios. Revista de Investigación Social*, vol. 6 no. 12, pp. 69–94.

Delgadillo, V. (2015) 'Patrimonio urbano, turismo y gentrificación', in Delgadillo, V., Díaz, I., Salinas, L. (eds), *Perspectivas del Estudio de la Gentrificacíon en México y América Latina*, Coyoacán: Instituto de Geografía, UNAM, pp. 113–132.

Diaz de Paniagua, R. and Paniagua, R. (1993) *Getsemaní. Historia, patrimonio y bienestar social en Cartagena*, Cartagena de Indias: Coreducar.

Diaz de Paniagua, R. and Paniagua, R. (1994) *San Diego. Historia, patrimonio y gentrificación en Cartagena*, Cartagena de Indias: Coreducar.

El Espectador (2013) 'El desafío de los POT'. [Online] 13 May [Accessed 20 July 2016]. Available from: http://www.elespectador.com/noticias/nacional/el-desafio-de-los-pot-articulo-421768.

El Universal Cartagena (2012) 'Lo primero es salir de la crisis'. [Online] 1 October. [Accessed 26 July 2016]. Available from: www.eluniversal.com.co/cartagena/editorial/lo-primero-es-salir-de-la-crisis.

extracartagena.com. (2013) 'Así avanza el Plan de Ordenamiento Territorial (POT)', [Online], [Accessed 26 July 2016]. Available from: www.extracartagena.com/index.php/noticias/sucesos/370-asi-avanza-el-plan-de-ordenamiento-territorial-pot.

Gotham, K.F. (2005) 'Tourism Gentrification: The Case of New Orleans "Vieux Carre (French Quarter)"', *Urban Studies*, vol. 42 no. 7, pp. 1099–1121.

Herzer, H., Di Virgilio, M. and Aguilera, M. (2015) 'Gentrification in Buenos Aires: Global Trends and Local Features', in Lees, L., Shin, H.B., López-Morales, E. (eds), *Global Gentrifications: Uneven Development and Displacement*, Bristol: Policy Press, pp. 199–222.

Hiernaux, D. (2003) 'La réappropriation de quartiers de Mexico par les classes moyennes: vers une gentrification', in Bidou-Zachariasen, C., Hiernaux, D. and Rivière d'Arc, H., *Retours en ville: des processus de "gentrification" urbaine aux politiques de "revitalisation" des centres*, Paris: Descartes & Cie., pp. 205–240.

Hospitality Net (2014) "Hospitality Net - Viceroy Hotel Group To Open Convento Obra Pia, Viceroy Cartagena". [Online]. [Accessed 25 July 2015]. Available from: www.hospitalitynet.org/news/4065593.html.

Inzulza-Contardo, J. (2012) '"Latino Gentrification"?: Focusing on Physical and Socioeconomic Patterns of Change in Latin American Inner Cities', *Urban Studies*, vol. 49, no. 10, pp. 2085–2107.

Lees, L. (2014) 'Gentrification in the Global South?', in Parnell, S. and Oldfield, S. (eds), *The Routledge Handbook on Cities of the Global South*, London & New York: Routledge, pp. 506–521.

Lees, L., Shin, H.B. and López-Morales, E. (2016) *Planetary Gentrification*, Cambridge: Polity Press.

Lemaitre, E. (1983a) *Historia general de Cartagena*, Vol. II La Colonia, Bogotá: Banco de la República.

Lemaitre, E. (1983b) *Historia general de Cartagena*, Vol. IV La Republica, Bogotá: Banco de la República.

Melero, N. (2004) 'Los centros históricos de Cartagena de Indias y La Habana Dos hitos del patrimonio colonial español en el Caribe', *Apuntes. Revista de estudios sobre patrimonio cultural*, vol. 17 no. 1–2, pp. 30–41.

Mendoza de Riaño, C. (2000) *Así es Cartagena de Indias*, Bogotá: Ediciones Gamma S.A.

Nichols, T. (1973) *Tres puertos de Colombia: estudio sobre el desarrollo de Cartagena, Santa Maria y Barranquilla*, Bogotá: Banco popular.

Nobre, E. (2002) 'Urban Regeneration Experiences in Brazil: Historical Preservation, Tourism Development and Gentrification in Salvador da Bahia', *URBAN DESIGN International*, vol. 7, no. 2, pp. 109–124.

Otero, C. (2012) 'Crisis de gobernabilidad en Cartagena', *Ola Política*. [Online]. [Accessed 25 July 2015]. Available from: www.olapolitica.com/content/crisis-de-gobernabilidad-en-cartagena-0.

Parker, J. (2014) 'Cartagena's Locals Brace for High-End Hotel Invasion', *Bloomberg.com*. [Online] 9 September. [Accessed 27 February 2015]. Available from: www.bloomberg.com/news/articles/2014-09-09/cartagena-s-locals-brace-for-high-end-hotel-invasion.

Piñeros, S. (2013) [Unpublished] *L'influence d'une inscription au Patrimoine Mondial (Unesco) dans le développement du tourisme: le cas de Carthagène des Indes, Colombie*, Mémoire M2 Spécialité Développement et Aménagement touristique des Territoires, Université Paris I Panthéon Sorbonne, Paris.

Posso, L. (2015) 'Patrimonialización, especulación inmobiliaria y turismo: gentrificación en el barrio Getsemaní', in Delgadillo, V., Díaz, I., Salinas, L. (eds), *Perspectivas del Estudio de la Gentrificacion en México y América Latina*, Coyoacán: Instituto de Geografía, UNAM, pp. 175–190.

Rojas, E. (1999) *Old Cities, New Assets: Preserving Latin America's Urban Heritage*, Washington, D.C: Inter-American development bank.

Rubino, S. (2005) 'A Curious Blend? City Revitalisation, Gentrification and Commodification in Brazil', in Atkinson, R. and Bridge, G. (eds), *Gentrification in a Global Context: The New Urban Colonialism*, London: Routledge, pp. 225–239.

Samudio, A. (1999) *El crecimiento urbano de Cartagena en el siglo XX: Manga y Bocagrande*, Cartagena de Indias: Universidad Jorge Tadeo Lozano, Seccional del Caribe, Departamento de Investigaciones.

Samudio, A. (2001) 'Los primeros barrios extramuros de Cartagena', *Patrimonio y urbanismo*, Cartagena de Indias: Universidad Jorge Tadeo Lozano, Seccional del Caribe, Facultad de Arquitectura, pp. 131–156.

Samudio, A. (2006) 'Cartagena veintiún años después de ser declarada patrimonio mundial', *Memorias. Revista Digital de Historia y Arqueología desde el Caribe.*, vol. 3 no. 6. [Online]. [Accessed 3 January 2013]. Available from: http://rcientificas.uninorte.edu.co/index.php/memorias.

Scarpaci, J. (2005) *Plazas and Barrios : Heritage Tourism and Globalization in the Latin American Centro Histórico*, Tucson: The University of Arizona Press.

Sigler, T. and Wachsmuth, D. (2015) 'Transnational Gentrification: Globalisation and Neighbourhood Change in Panama's Casco Antiguo', *Urban Studies*, vol. 53 no. 4, pp. 705–722.

Smith, N. (2010) 'Toward a Theory of Gentrification: A Back to the City Movement by Capital Not People', in Brown-Saracino, J. (eds), *The Gentrification Debates*, New York: Routledge, pp. 71–85.

Téllez, G. (2016) 'De las renovaciones autoritarias y sus efectos', [Online] [Accessed 26 July 2016]. Available from: http://imaginabogota.com/columna/de-las-renovaciones-autoritarias-y-sus-efectos/.

Tyor, A. (2013) 'Cartagena's Gentrification, But For Whom?', *Colombia Politics*. [Online] 20 December. [Accessed 20 June 2015]. Available from: www.colombia-politics.com/cartagena-gentrification/.

UNESCO (n.d.) 'UNESCO Centre du patrimoine mondial - Port, Forteresses et ensemble monumental de Carthagène'. [Online]. [Accessed 10 October 2012]. Available from: http://whc.unesco.org/fr/list/285/documents/.

Viloria de la Hoz, J. (2005) 'Historia del Banco de la Republica en Cartagena, 1923–2005: Fomento productivo, proyectos culturales y estudios économicos', *Cuadernos de Historia Economica y Empresarial Banco de la República, Centro de Investigaciones Económicas del Caribe Colombiano,* no. 4 [Online] March. [Accessed 29 April 2013]. Available from: www.banrep.gov.co/documentos/publicaciones/regional/cuadernos/14.pdf.

Wolinski, N. (2014) 'Carthagène des Indes vitrine de la Colombie d'hier et du futur', *The Good Life France,* pp. 176–179.

Zuleta Jaramillo, L.A. and Jaramillo Giraldo, L. (2006) *Cartagena de Indias: impacto económico de la zona histórica,* Bogotá: Convenio Andrés Bello.

Zúñiga Ángel, G. (1997) *San Luis de Bocachica, Un Gigante Olvidado En La Historia de Cartagena de Indias,* 2nd edn, Cartagena de Indias: Punto Centro Forum.

Part II

Tourism gentrification and the urban cultural and creative turn

5 The Barrio Chino as last frontier

The penetration of everyday tourism in the dodgy heart of the Raval

Alan Quaglieri Domínguez and
Alessandro Scarnato

Introduction

The Raval has always been a neighbourhood of change and transformation. If a single word could define it, it might well be 'complexity'. Arguably, this central Barcelona *barrio* has maintained a contradictory relationship with the rest of the city throughout its history, while offering the main stage for representation of the urban conflicts and transformations in the modern era, and for the embodiment of Barcelona's relations with the rest of the world.

The transition to the post-Fordist economy, the rise of the consumer society and the boom of international mobility made the Raval both a paradigm laboratory for intervention in the framework of the celebrated Barcelona model of urban transformation and a prime observatory of processes affecting the Catalan capital in the context of a 'certain global convergence to a more neo-liberal urban policy' (González, 2010: 17). The development of the neighbourhood's capacity for tourism and the changes that this requires and entails are key elements for enhancing Barcelona's competitiveness within what Harvey defines as the 'spatial division of consumption' (Harvey, 1989).

In post-Olympic Barcelona, the Raval, the most populated neighbourhood of Ciutat Vella, the old town, has been affected by converging processes: the intense transformation of its urban fabric and the 'steady amassing of symbolic capital' (Harvey, 2001: 405) around the Raval's cultural cluster (Subirats and Rius, 2006; Rius Ulldemolins and Zarlenga, 2014), the development of its capacity for tourism, and far-reaching sociocultural change with the arrival of intense flows of international migration.

Starting with a consideration of these dynamics, this chapter aims to cast light on the circular relationship between the tourism phenomenon and the process of urban transformation in a residential area of a metropolis. Analysis focuses particularly on South Raval, the area bordering the city's old port that gave rise to the mythical Barrio Chino, the red light district, whose stigmatized image of perdition and 'container of social problems' (Fernández, 2012: 51) is still rooted in Barcelona's collective consciousness.

Despite significant interventions in its physical fabric since the 1990s, in this part of the neighbourhood – dubbed *Ravalistan* due to the settlement here of intense flows of Asian migrants, particularly from Pakistan – the penetration of tourism and its connection to the rest of the city seemed slower than in the northern sector, and its supposed 'revitalization' was frustrated by the persistence of marginalization and the resurgence of old problems (Abella, 2004) related to street prostitution and drug-dealing. Now, however, in its current phase of real-estate recovery from the sharp economic downturn and the steady expansion of touristic demand in Barcelona, the area is experiencing an acceleration of its socio-economic transformation. What appeared to be the last frontier in the inner city has now been reached by this wave of 'touristification of everyday life' (Franklin, 2003: 206; Larsen, 2008: 26) that overloaded mass tourist precincts (Hayllar and Griffin, 2005), now increasingly penetrating and transforming the mundane fabric of "cosmopolitan" residential areas.

This trend promotes an epistemological shift towards a consideration of tourism as part of a wider process of urban change in which tourists and other populations culturally and spatially converge to integrate the local demand of goods and services, thereby enhancing the appeal (and the value) as urban destinations of the areas concerned. In this sense, the chapter focuses on the housing issue and the real-estate market, whose recent dynamics seem to confirm the capacity of tourism to fuel the 'foundational contradiction' between use and exchange value (Harvey, 2014), and the polarization of socio-economic development in the neoliberal city.

The chapter is organized as follows: the first section proceeds to frame the case of the South Raval into a theoretical discussion about current trends and new perspectives on tourism in urban destinations. We then outline the historic development of the South Raval and its evolution from a peripheral area of ancient Barcelona into a marginal slum in constant decay. In the second section we examine the massive urban intervention that took place after the 1992 Olympic Games and the socio-morphological changes in the urban landscape of the area. After this historical framework, the text focuses on the socio-cultural dynamic of the South Raval in order to proceed in outlining the main housing market features of the neighbour-hood. To this end, considerations about population movement, population origins, building conditions and the property market are proposed as a result of the analysis of historical data on different statistical scales (Municipal, district, neighbourhood and census sections) from the Barcelona Municipality statistical database (www.bcn.cat/estadistica). In the final section we propose an interpretation of the described dynamics of gentrification pointing out the role of the new global consumer as the key figure of the Raval's ongoing deep socio-demographic transformation.

Our investigation is based on an extensive corpus of studies regularly led and published by local public institutions (Regional Government, Regional Departments of Territory and Public Works, Municipality, Chamber of Commerce, municipal companies) and on a deep survey of local newspapers (particularly *Barcelona Metropolis Mediterrània, El País, La Vanguardia* and *El Periódico*). We also rely on local administrative archives of the Barcelona Central District and semi-directive interviews with official and local actors involved in the redevelopment projects.

Tourism mobility and everyday life

The rising concern at the development of tourism in a considerable sector of Barcelona's neighbourhood associations and citizen activism (Novy and Colomb, 2017; Russo and Scarnato, 2016) is explained by its mounting pervasion of the city's urban, socio-economic and cultural fabric. Emblematic examples of grass-roots mobilizations are the well-attended protests that have been taking place in the beachfront neighbourhood of La Barceloneta, a touristic epicentre during the summer, and the controversy surrounding restricted access to the public Park Güell (Arias Sans and Russo, 2017), one of the main hotspots on the Gaudí itinerary.

A growing number of issues such as liveability, the right to housing and saturation of public space are linked with the growth of tourism in a city, Barcelona (Ajuntament de Barcelona, 2016), and a country, Spain, where tourist numbers reach record levels year after year (Delgado, 2016). Traditionally, the flows of tourism had headed mainly towards the sun, sea and sand resorts along the Catalan coast, with Barcelona as a point of entry. But the city's incredible urban development in the last three decades has made it one of the top global destinations, reversing the regional tourism structure (Jiménez and Prats, 2006; Garay Tamajón and Cànoves Valiente, 2010).

In addition to effectual strategies to promote Barcelona as a monumental city – the branding of Gaudí's legacy is emblematic – the widespread exposure on the 'global catwalk' for tourism and investment (Degen, 2003) cannot be taken apart from the successful projection of an appealing image of a 'cosmopolitan Mediterranean metropolis' (Richards and Wilson, 2007).

In the great wave of the neoliberal Urban Renaissance, Barcelona, like other Western metropolises, underwent a long and intense process of transformation of its physical and symbolic landscape to ferry its economy out of the industrial crisis. In the transition 'from the work ethic to the aesthetic of consumption' (Bauman, 2005: 22), 'new urban economies' (McNeill and While, 2001) demand and promote a new relation with their city-users whose self-identities, in turn, are increasingly shaped by consumption (Giddens, 1991; Paddison, 2001; Bauman, 2005).

Two municipal campaigns are emblematic of this process: *Barcelona, posa't guapa* (Barcelona, get pretty), to enhance the physical landscape by rehabilitating building façades, and *Barcelona, la millor botiga del món* (Barcelona, the best shop in the world), aimed at promoting local commerce.

The adoption of a global urban and architectural discourse to spectacularly improve the skyline and tame the old urban fabric responds to the need to improve the 'city's legibility' (Lynch, 1960) and develop attractive 'new experiential milieus' (Degen, 2003) for a global audience of investors and consumers. In this sense, Francesc Múñoz (2008) used the term 'urbanalization'.

Tourism emerges, though, as a fundamental asset for providing the local supply of goods and services with further contingents of generally wealthier users bent on higher spending. At the same time, the growth of 'tourism reflexivity' in the city's policies (Sheller and Urry, 2004) and the proliferation of urban destinations suggest the increasing capacity of contemporary urban milieus to connect with the desires and expectations of a new generation of travellers (Russo and Quaglieri, 2013) that challenge epistemological assumptions in mainstream theories on the motivations and geographies of tourism.

For today's increasingly transnational, 'mobile lives' (Hannam, 2008; Elliot and Urry, 2010), travel is less and less an extraordinary break from daily routines (Edensor, 2001; Cohen, 2010), and interest in the exotic 'other' seems to be losing its traditional primacy as a pulling factor, especially to global cities (Sassen, 1991). Here, the idea of tourism in opposition to everyday life (Urry, 1990) is challenged by practices that are losing interest in gazing at extraordinary, 'rooted' attractions and discovering the quality of the 'commonplace' (Maitland, 2008).

The 'modern' consideration of tourism *per se* as a discrete activity or a specific consumer product (Franklin and Crang, 2001) separated from other urban spheres and other forms of travel has lost much of its analytical usefulness (Cohen and Cohen, 2015). The very category of tourism has been questioned by researchers (Franklin, 2004) who advocate the adoption of post-structural perspectives in the sense of the 'de-exoticization' of tourism (Larsen, 2008) and its de-differentiation (Rojek and Urry, 1997) from the quotidian. Moreover, tourism is an increasing part of everyday life and important to the understanding of wider socio-economic and spatial processes (Edensor, 2007; Hannam, 2008).

Awareness of the paradox of the staged authenticity of conventional tourist spaces and the quest for individualised experiences brings more expert urbanites (or post-tourists) to extend their spatial range beyond predictable "tourist bubbles" (Judd, 1999) and immerse themselves autonomously in supposedly more genuine atmospheres. The hierarchical and "enclavic" (Edensor, 2001) logic of the tourism industry's territorial organization, then, gives way to dynamic "tourismscapes" (Van der Duim, 2007; Edensor, 2006), overflowing "McDisneyized" (Ritzer and Liska, 1997) precincts and pervading the city's everyday fabric. In this sense, many post-industrial cities have witnessed the rise of "heterogeneous tourist spaces" as environments with blurred boundaries where different activities and populations coexist, and where tourists mingle with locals (Edensor, 2001).

As Jane Jacobs and Ruth Fincher (1998: 1) pointed out, "difference is undoubtedly a sustained feature of urban space" and the co-presence in the tourist space of different uses and users acts as a "marker of the real" (Maitland, 2008: 23), becoming essential in the building of urban imaginaries and the narration of "contrasting aesthetic contexts" (Edensor, 2001), where tourists emerge as "transforming agents" (Russo and Quaglieri, 2013). Instead of an extraordinary intrusion that contaminates the rooted essence of place, tourists therefore become contributors, alongside other actors, in the steady definition of the "global sense of place" (Massey, 1991). The mobility paradigm stresses the fluid and dynamic character of places as crossroads of complex networks by which different peoples and objects are 'contingently brought together to produce certain performances in certain places at certain times' (Hannam *et al.*, 2006: 13).

The development of global and cosmopolitan skills, tastes and morphologic elements on the urban scene and the pervading support of the Internet allow mobile "global cultural consumers" (Russo and Quaglieri, 2013) to autonomously deal with destinations and mundane "consumption landscapes" (Maitland, 2008). Once sanitized, previously dreary districts become the ideal stage where visitors can engage in local-like lifestyles among and with locals. This interest in informal interaction with the increasingly familiar "other" (Hottola, 2004) is confirmed by the boom of "network hospitality" platforms such as Couchsurfing, Airbnb and the various home-swapping communities (Germann Molz, 2011; Bialski, 2012; Russo and Quaglieri, 2016).

But if we consider recent socio-economic and even cultural divergences, a question arises: what do we mean by local? Again, research into networking-hospitality (Farbrother, 2010; Steylaerts and O'Dubhghaill, 2011; Bialski, 2012) stresses the interest of travellers in like-minded people as ideal counterparts.

Due to these globalization processes now penetrating everyday life and the "omnipresent cultural and linguistic diversity" (Vertovec and Cohen, 2002) in Western societies, the mainstream anthropological contraposition of tourism and resident populations seems unable to grasp the complexity of the contemporary host-guest relation. A binary discourse has pervaded public debate about the impact of tourism on the urban ecology and fostered increasingly "tourismophobic" (Palou i Rubio, 2011) attitudes, while the consideration of internal conflicts seem to have been relegated to second place.

According to Hall (2005: 129), tourism conceptualization should be considered by means of a more "comprehensive approach that involves the relationships between tourism, leisure and other social practices and behaviours related to human movement". Tourism may come to be considered just one form of (mostly) leisure-motivated mobility, intertwined with others, to and within the destination. In the context of the "mobilization of the world" (Elliott and Urry, 2010) and the related transnationalization of people's social networks, this complexity is highlighted by the growing importance of movements such as diasporas and return tourism (Coles and Timothy, 2004) or Visiting Friends and Relatives (VFR) as a direct consequence of other mobility. Several researchers (Williams and Hall, 2002; Vertovec and Cohen, 2002; Hall, 2005; Hannam, 2008) point out the blurring between migration and travel in a continuum of mobility (Hall, 2005). This includes, for instance, the hybrid approach to the destination of "migrant tourist-workers" conceptualized by Bianchi (2000) and the large body of work developed in the conceptual framework of retirement or lifestyle migration (O'Reilly, 2003; Haug *et al.*, 2007; O'Reilly and Benson, 2009; Thorpe, 2012; Huete *et al.*, 2013).

In fact, even as lifestyle has become pivotal in the constitution of "self-identity" (Giddens, 1991; Cohen *et al.*, 2013), it is increasingly emerging as a critical issue in the promotion of different mobility at different scales. From an epistemological perspective then, this calls for a shift in the contextualization of tourists in the complex and fluid social morphology of the post-industrial city, as well as their relations with other populations in terms of consumption and spatial patterns, and, therefore, their role in the processes of urban transformation.

In this sense, the rise of heterogeneous tourist areas in cities such as Barcelona is linked with spatial confluences of different urban users (including tourists) involved in shared practices of consumption in the context of converging urban imaginaries and lifestyles. It might be helpful to think in terms of a global and mobile "cosmopolitan consuming class" (Fainstein *et al.*, 2003) or an international postmodern middle-class (Sheller and Urry, 2004) "comprising residents, workers and visitors alike who want to consume amenity and culture, and enjoy familiar landscapes of consumption" (Maitland, 2007: 31) and which "globally moves very quickly from one place to another but anywhere requires [and does] more or less the same things" (Martinotti, 1993). Therefore, as Maitland suggests (2008: 15), it is in these areas that cosmopolitan lifestyles emerge as part of a physical and symbolic refashioning, and the development of an appealing commercial offer where tourism grows "as part of a wider process of change, regeneration and gentrification".

In other words, it is in these cosmopolitan milieus that tourism provides more contingents of consumers with the localized demand of lifestyle consumption and further peoples the cosmopolitan scene so as to improve local "people's climate" (Florida, 2002) and its attractiveness for more mobile populations. According to this circular logic, then, rather than distorting rooted, endogenous dynamics, the penetration of tourism into cities' mundane landscapes would directly and indirectly feed existing socio-spatial conflicts with other urban users who have different socio-cultural status, economic means and/or diverging interests and priorities in terms of liveability. Housing and public space emerge as emblematic fields that clearly represent competition in access to urban resources and the contradictions and growing inequalities promoted by the neoliberal city.

History matters

Barcelona's remarkable success as a global tourist destination in the last two decades has had a lot to do with the municipal strategy of overriding the typical institutional management of the Olympic Games with an eschatological tale of urban renewal, a strategy converting what until then had been the world's most important sports event into an extraordinary opportunity for a thorough renovation of the whole city, both physically and socially. However, the reconstruction of Barcelona (Bohigas, 1985) started well before the event of 1992 and could ultimately be considered as the completion of the famous 1865 project of urban engineer Ildefons Cerdà, when the city was allowed to expand beyond its medieval walls, which were eventually demolished. Thanks to the Cerdà project, the new city spread all around the old town, which, abandoned by the middle and upper classes, quickly embarked on a severe process of decay.

The Raval has been a marginalized part of the city for centuries, due to its peripheral position, and has suffered this abandonment more than anywhere else in the old town of Barcelona.

Between the eighteenth and the early twentieth centuries, as part of the social conflict between the new bourgeois city designed by Cerdà and the wretched old one, there were several uprisings in the area in consequence of its critical living conditions. In the 1920s, the southern sector was popularly known as the Barrio Chino (Chinese Quarter) thanks to its many bars, theatres and cabarets known throughout Europe for their bohemian atmosphere, nurtured by the literary descriptions of writers like Ernest Hemingway and Jean Genet. The bombings suffered during the Spanish Civil War (1936–1939) put a brutal end to that season, and, under the dictatorship of Generalissimo Franco (1939–1975), very few urban interventions were made, turning the former Barrio Chino into a social waste dump for all the realities that it was convenient for the Regime to hide there. The official image of a prosperous, peaceful, sober Spain would have been tarnished if prostitution, petty crime and dropouts had been visible anywhere but in specific streets obviously to be avoided by "decent citizens".

When, in 1981, the first democratically elected Barcelona City Council embarked on the "reconstruction of Barcelona" (Bohigas 1985), the situation in the former Barrio Chino was the most urgent problem to be faced. A detailed plan (Plan Especial de Reforma Interior, PERI) for the whole Raval, passed in 1985, designed an articulated system of public spaces to be created in the dense urban fabric of the area, from the South to the North, where a cultural hub of museums and university faculties was to be built, reusing and transforming the many abandoned religious buildings.

The urban fabric as a canvas to play with

The 1985 PERI envisaged linking the interventions in the south and north sectors of the Raval by opening up a huge central space called the Pla Central del Raval (PCR). The declared intention of respecting the existing urban fabric, local residents and popular atmosphere was stated as a priority in the operation, but it was no obstacle to the destruction of over 400 dwellings in a significant change of scale from the surrounding neighbourhood. Residents' associations pointed out in several street protests that the overall design of the PCR operation coincided with the location of blocks already affected by a process of expropriation started at the time of Cerdà and repeatedly proposed in the following decades, as in the frustrated plans from the 1950s and 1960s.

In 1990, the University of Barcelona began negotiations with the City Council to move the humanities faculties to new facilities to be built in North Raval. The municipal offices welcomed the proposal and used it as an opportunity to rethink the whole 1985 PERI. The new plan redistributed open spaces originally planned in North Raval, taking into account another operation promoted between 1991 and 1993 by the Catalan regional government in South Raval, substantially at odds with the 1985 PERI.

As regards the future space's morphology, the new square would have opened onto the commercial streets of Sant Pau and Hospital, and, to the east, a new intervention was planned between the project site and the historic Carrer d'en Robador, one of the streets traditionally identified with prostitution and petty crime. As a result of these changes, the area covered by the square (58 × 317 m) would be almost two acres (Figure 5.1).

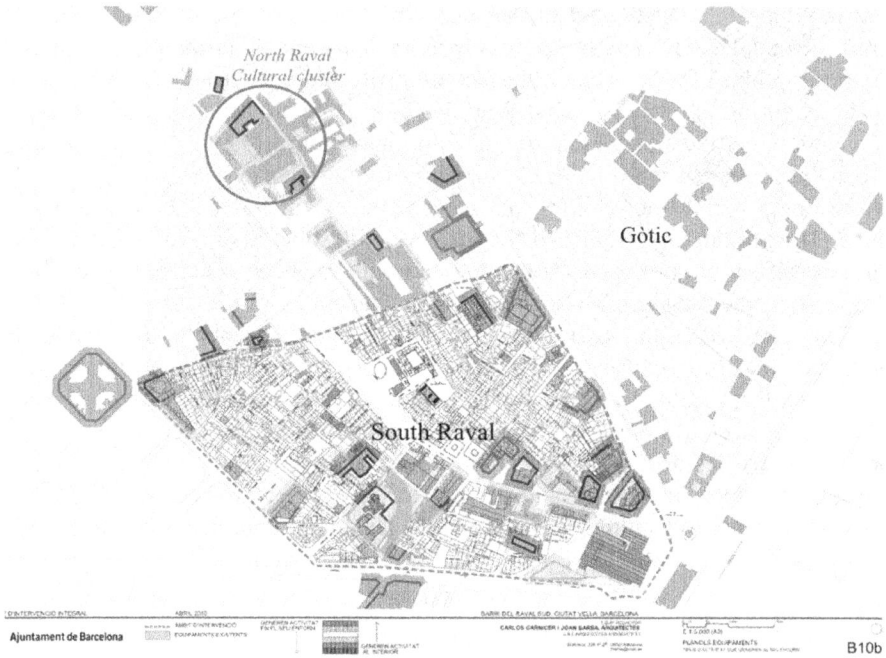

Figure 5.1 The system of cultural facilities (in darker colour) of Raval. In evidence the
         South Raval and the cultural cluster in the northern sector

Source: Authors' own adaption of Ajuntament de Barcelona data (2010)

The amended plan received its final approval in 1994, coinciding with European
Union grants that were to fund up to 85 per cent of the operation. The new square
emerged as a convincing architectural composition but caused considerable
controversy as an urban operation, as stated by local architects and urbanites in
several conferences and public declarations to local media (Soja, 2001). Criticism
was initially levelled at real-estate issues; to give the project its final dimensions,
the number of demolitions within the PCR area was extended, affecting some
listed Modernista buildings.

Difficult as the proposed morphologic change of the area was, the process
of expropriation was even more complex. Council administrators and local residents' associations had agreed on a fair model of compensation, with residents
being required to demonstrate a minimum of five years' continued occupation.
Those entitled were offered the choice of reallocation in the Raval or elsewhere
in the city, or financial compensation. Although everyone who met the require-
ments and expressed the desire to stay in the area received satisfaction (2016
interviews), conflict was inevitable.

First, there was the attitude of the municipal employees, generally described
as uncooperative in defining who was actually eligible for the compensation and

who was not. In addition, there was the unchecked economic trend. By late 1995 (coinciding with the start of demolition), there was a dramatic increase in real-estate prices, generating major friction between the Administration and residents about expropriations. Council planners did not want to postpone the construction once announced, since property values quickly outstripped the appraisals carried out in 1993, making them impracticable.

In operational terms, the granting of EU funds came as a boost and by early 1999 the first project phase was complete. On 16 September 2000, the last building was demolished and the now empty space was inaugurated with a festive event and newspaper chronicles signalling the end of "centuries of marginalization" (Aisa and Vidal, 2006) of the Raval. General enthusiasm momentarily took precedence over complaints about expropriation, demolition and social conflict.

Leaving aside the celebratory tone of press and officials, there were inevitably pros and cons to the Rambla del Raval operation. The pros included innovative design and private investment in remodelling buildings, which occurred surprisingly fast in comparison with other city-centre sectors. As regards the cons, real-estate prices had risen far more than expected. In the case of new warehouses or apartments rehabilitated by individuals, the prices tripled.[1] This contradiction in the economic sphere (market stagnation coinciding with price rises) was only apparent; the administration had not considered, in the original plan, that some-one outside of the Raval (owners or inhabitants) might be interested in investing in the area. The stagnation was, then, the consequence of the enlargement of the potential range of buyers, since many owners were just waiting to see where the new limit of the real estate market would lead.

Despite the appearance of the new Rambla, there were no perceptible signs of improvement of the social problems. Drug-dealers and addicts, petty crime and prostitution, poverty and insalubrity picked up where they had left off. Thus, initial enthusiasm gave way to general disenchantment.

Doubts and criticism gained intensity when work began on the project to remodel the area adjacent to Carrer d'en Robador in spring 2000. Unlike the Rambla del Raval, the Robador plan was purely – and quite explicitly – an initiative on the part of the Administration to raise funds to avoid bank debt. In fact, in the 1985 PERI, no significant intervention had been planned for the two blocks between Carrer d'en Robador and the new Rambla. Five years later, however, following the approval of the project's new layout, political voices were raised about the risks of such a large operation in terms of time and cost (Cia, 2001). Reports in municipal archives (ADCV, *passim*) testify that several members of the City Council called for a parallel operation to produce capital gains that would compensate for the dreaded rise in costs and secure a reserve of land that could, partially at least, be used for public infrastructures.

The solution seemed to lie in and around Carrer d'en Robador, one of Barcelona's most conflictive streets, famous for its round-the-clock prostitution and attendant problems of drugs, petty crime, STDs and lack of investment in pri-vate properties. A shock intervention here was therefore seen as an effective way to consolidate the success of the neighbouring Rambla del Raval. In 1997, a new detailed plan for the area was approved under the name "Pla Jardins Robador", including 10,000 m² of social housing and another 10,000 m² to be sold on the free market, plus 9,500 m² for trade or services, possibly with a catchy architectural appearance (Figure 5.2).[2]

The dynamic multicultural landscape of Raval

The complex, changing nature of the Raval's socio-cultural landscape is significantly related to its dynamic resident population and the far-reaching citywide urban interventions mentioned above.

The neighbourhood continues to be the main welcome area for newcomers to the city (6.6 per cent of all registered internal and international inbound migration in Barcelona 2014),[3] and, generally, the area most affected both by flows related to residential mobility and the migratory movement to/from Barcelona.[4]

Despite being a mainly working-class area, the Raval has continuously been marked by a degree of social mix until the second half of the nineteenth century, when most of the bourgeoisie moved to the new urbanized areas built after the demolition of the town walls, following the 1865 Cerdà Plan. The only remarkable exception to this dynamic was the present-day tourist hotspot centring on Gaudí's Palau Güell (1886–1990), a huge Modernist townhouse built for a local magnate in the South Raval, declared a UNESCO Heritage Site in 1984.

Figure 5.2 The impressive Filmoteca de Catalunya building (on the right) and a popular residential block in the Illa Robador Area

Source: A. Quaglieri Domínguez, 2015

Thereafter, intense migratory waves from the rest of Spain, associated with the need for labour for the Universal (1888) and International (1929) Expositions, emphasized the working-class profile of the Raval. This trend was reinforced after the Spanish Civil War (1936–1939), when the population peaked at 107,473 inhabitants (1950) concentrated in 1.1 km^2, one of the highest densities in Europe (Sargatal, 2001) After the industrial crisis of 1973 and for the next two decades, the entire Ciutat Vella district experienced an intense depopulation process that lasted until the mid-1990s.

By that time, the historic centre of Barcelona was undergoing severe economic decadence, steady deterioration of its housing stock, and deep-seated social decay related to the boom in heroin consumption. These factors, along with the social alarm raised by growing petty crime, pushed many families out of the Raval towards the residential suburbs. The population registered its low point of 34,871 inhabitants in 1996 at the same time as the neighbourhood re-emerged as the main focus of attraction for major waves of migration which, in the context of the acceleration of international flows to Europe (Cabré and Domingo, 2002), significantly affected the city of Barcelona in the mid-1990s.

During this period Ciutat Vella, and the Raval in particular, emerged as "moorings" (Hannam *et al.*, 2006) for transnational mobility in the framework of the establishment of the Schengen Area and the acceleration of south–north migratory trajectories. While the settlement strategies of these two macro-groups of migrants promoted differing geographies within Barcelona and strengthened the city's residential segregation (Bayona and López-Gay, 2011), the Raval emerged as the primary node for populations characterized by diverging mobile trajectories, push-and-pull factors and lifestyle choices.

Bearing in mind the growing socio-economic heterogeneity of national communities in contemporary cities, the consideration of nationality is still useful as a proxy indicator of migrant status and migratory pattern. Particularly in a context like the Raval, characterized by the weight of foreign communities, a comparison of size and growth rates of macro-groups such as those mentioned above could provide useful information for inferring considerations about the social fabric of the area and its evolution.

The affordability of its run-down, partly abandoned (22.6 per cent in 1991 housing census) housing stock, along with its centrality, made the Raval an enabling milieu for the settlement of intense flows of migrants from developing countries. In fact, the 2001 housing census indicated the condition of about a quarter (23.4 per cent) of the buildings as "ruinous" or "bad", and barely half (52.6 per cent) were considered to be "in good condition", as opposed to the rates for Barcelona as a whole, which were, respectively, 3.9 per cent and 84.8 per cent. In addition to age (66.6 per cent of the residential buildings were built before 1901), the decay of the housing stock was also related to low maintenance in a context historically characterized by a high rate of rented properties (68.1 per cent as opposed to 35.6 per cent for Barcelona as a whole in 1991), often with rent-controlled contracts. This aspect goes some way to explaining the low level of investment in maintenance and repair by landlords, leading to a residential down-filtering that affected the Raval, particularly the south-western sector, more than any other neighbourhood in the city.

Meanwhile, the Raval's growing cultural cluster gave rise to a vibrant scene with its related circuit of cultural and leisure consumption, and an increasing constellation of independent shops, libraries, art galleries, bars and restaurants attracted creative people, students and visitors (Figure 5.3). These atmospheric and commercial features further emerged as primary elements of liveability, thereby becoming relevant factors in the settlement strategies of the local or transient cosmopolitan consuming class.

Rather than boosting dynamics of economic substitution, the confluence of new populations in the context of mobility trajectories from the metropolitan to the intercontinental scale had been filling residential gaps, thereby reversing the depopulation trend. Whereas the Spanish citizen community was still immersed in the suburbanization process (−18.4 per cent between 1996 and 2005), the population of the Raval registered a 41.5 per cent growth in the same period, thanks particularly to an increase in migratory flows from non-European regions, which heightened both the working-class profile and the multicultural tag associated with the neighbourhood. Nevertheless, during this period the foreign Western, mostly European, community grew considerably, reinforcing the

Figure 5.3 Flea market in South Raval

Source: Authors, 8 June 2014

construction of the cosmopolitan imaginary associated with the area that played such an important part in boosting the touristic appeal of the Raval.

Whereas previously most of this group had settled in upper- and upper-middle-class districts, the rise of 'regenerated' areas in post-Olympic Barcelona such as the Raval and the whole historic centre seemed to match the lifestyle expectations of new waves of young middle-class Europeans, colloquially referred to as *guiris*.[5] Historically used in Spain as a somewhat pejorative term for foreigners, this word now largely refers to both tourists and migrants from Western countries (Subirats and Rius, 2006). Arguably, despite its various interpretations and uses in Barcelona (Monnet, 2001), the word *guiri* might represent an individual defined by specific socio-economic (middle-class), cultural and even generational conditions, regardless of their status as visitor, temporary or permanent resident. From the viewpoint of the resident population, then, this urban user profile suggests a continuum between tourism, professional, education-related mobility (internships abroad, Erasmus experiences) and certain migratory patterns, while at the same time stressing a separation from the rest of the migrant community, similarly to the internationally widespread term "expat".

The number of non-Spanish European residents increased almost nine-fold between 1996 and 2005, more than any other regional group, reaching 2,517 units, and representing 5.4 per cent of the area's population and 12.5 per cent of its foreign community (Figure 5.4). Moreover, two-thirds of this community was made up of citizens of the EU-15.[6]

Although the Western European community in the residential fabric was spread throughout the neighbourhood, during this period its weight and force in relation to other communities varied greatly between different zones of the Raval.

Unsurprisingly, where the Spanish population seemed less concerned by suburbanization dynamics and its average housing conditions were better, the weight of EU-15 citizens in the local community was significantly higher. This was the case of the areas more directly affected by the Raval cultural cluster, towards the north-east edge of the neighbourhood, bordering on the central Plaça Catalunya and the Ramblas – the city's touristic heart – where some one-third of the foreign community was represented by the European Community, with 86 per cent of EU-15 citizens.

Generally, although the macro-groups of migrants had grown between 1996 and 2005 in all areas of the Raval, the different settlement strategies gave rise to a complex social geography in keeping with the existing duality between northern and southern sectors. In order to proceed with the analysis of the resident population in the two mentioned sectors, data referring to North and South Raval were obtained by grouping Zones of Study (until 2009) or Census Sections (from 2010), the smallest territorial divisions with available statistical data respectively before and after the last new territorial division of Barcelona came into effect in 2009.[7]

The weight of non-Europeans within the overall resident population highlighted the exotic profile of the lower Raval. While the area had been linked with Barcelona's Moroccan community since the early 1990s, it was the arrival of

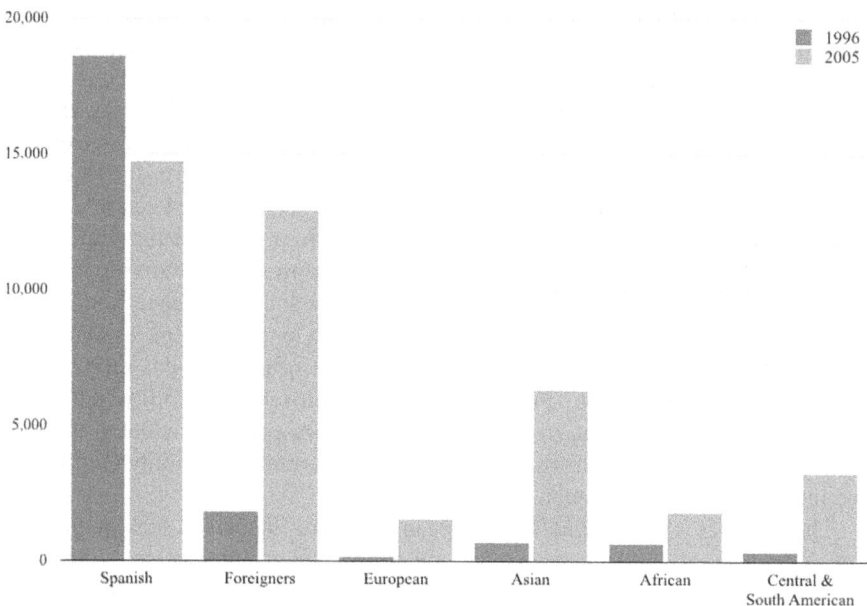

Figure 5.4 South Raval resident population by nationality (aggregated by world region),
comparison between 1996 and 2005

Source: Authors' own elaboration of data from www.bcn.cat/estadistica

intense migratory flows from the Indian Subcontinent that marked a demographic
upturn in South Raval. The rapid growth of the Pakistani community in particular
was significant, mainly from the Punjab Province of Gujrat, in the first half of the
2000s becoming the main foreign national group in the Raval and one of the two
largest in Barcelona (Scarnato, 2006). In addition to the Pakistani community,
Bengalis and Indians showed similar settlement patterns in terms of temporality
and spatial concentration, in Barcelona and within the Raval. The settlement of
these communities, in addition to their significant role in the area's demographic
recovery, saw an economic blossoming of the area thanks to the proliferation of
convenience stores, butcher's shops and greengrocers, as well as mobile device
shops and repair services, and travel agencies, which "transformed and redefined
the face" of the main commercial axes (Moreras, 2004). This, along with the presence in public space of their members, often wearing colourful traditional clothes,
gives these groups high visibility in the area's everyday landscape, thereby
reinforcing the multicultural imaginary associated with the Raval.

The Spanish financial crisis that began in 2008 marked a turning point in the
demographic revitalization of the city and coincided with a significant new phase
of de-population of the historic centre, with the loss of some 10 per cent of its
population between 2008 and 2015 (Table 5.1). Although this dynamic was more
evident in the neighbouring touristic Gòtic area, the Raval was also affected,

its growing trend first slowing and then overturning. In this sense, two different periods can be identified in the course of the last five years, as shown in Table 5.1.

While the Spanish population was still undergoing a deceleration (probably related to its natural movement), the foreign community registered two clearly opposing trends. Moreover, these demographic fluctuations reflected contrasting dynamics between migrant communities. While most of these groups registered negative rates between 2010 and 2012, larger than the city average, the overall population of South Raval grew thanks to the increase in the Asian community, particularly Pakistanis (+32.8 per cent) and Bengalis (+34.3 per cent). These groups, along with Filipino and Indian communities, presented significant concentration rates and location quotients in the Raval, above all in its southern sector (with the exception of the Filipinos), so the area continued to be the main reference for settlement of newcomers from these countries. Conversely, the deepening crisis seemed to affect other national groups such as Moroccans (−4.8 per cent), Dominicans (−21.6 per cent) and French (−7.8 per cent).

The following period (2012–2015), conversely, shows a degree of 'Westernization' of South Raval, bringing it into line with the rest of the neighbourhood and the surrounding city centre areas. Among the Europeans, the growth of communities such as Britons (+32.3 per cent), Russians (+36.3), Swedes (+25.7 per cent) and Italians (+20.4) stands out. Moreover, the growth of the EU-15 groups registered an upsurge that doubled that of the supposedly more 'European' North Raval, bringing the area into line with increasingly touristic and middle-class areas such as the Born, Eixample Dreta and Vila de Gràcia. Meanwhile, there was an acceleration of outbound flows of the Latin American and African communities from the city, especially from South Raval.

Finally, in its mature stage, the crisis represents a hiatus in the relation between Barcelona and the international migration fluxes that demographic trends in the South Raval seem to magnify. Increasing socio-economic inequalities and return migration boosted by the economic and political crisis, above all

Table 5.1 Evolution of resident population by nationality (aggregated by world region)

| Nationality | South Raval | | | | | Barcelona |
	2005	2009	2012	2015	Δ 2012–2015	Δ 2012–20152
Population	30,056	27,764	27,571	26,360	−4.4%	−1.0%
Spaniards	14,732	14,040	13,201	13,071	−1.0%	0.3%
Foreigners	15,324	13,724	14,370	13,289	−7.5%	−7.4%
European	1,816	2,576	2,238	2,667	19.2%	11.9%
African	2,007	1,417	1,281	1,177	−8.1%	−5.7%
Asian	7,828	6,917	8,490	8,010	−5.7%	−2.5%
C & S. American	3,255	2,728	2,110	1,356	−35.7%	−26.6%

Source: Authors' own elaboration of data from www.bcn.cat/estadistica

from South America and North Africa (Domingo *et al.*, 2014; García Ballesteros *et al.*, 2014), along with Barcelona's renewed appeal for fresh contingents of European expats and the steady growth in tourist flows to the city are outlining a new social geography.

The population of the analysed area is still characterized by high complexity in socio-economic and cultural terms, but the balance of forces within the foreign community seems to be significantly changing in favour of progressive Westernization and, therefore, gentrification, like other sectors of the historic centre and the middle/upper-middle class areas that make up Barcelona's 'cosmopolitan circuit'.

South Raval housing market

In 2015, the real-estate market followed a global macroeconomic trend in recovering after the sharp fall in prices that accompanied the 2007–2008 economic crisis. In Barcelona, growing demand from international investors boosted the market, mainly because financial benefits from real estate seemed more secure and profitable than other products. This trend is not exclusive to Barcelona, and, according to the annual report of the Spanish Association of Property, Mercantile and Real Estate Registers (Fabra Garcés, 2015), foreign buyers are making a decisive contribution to this trend all over Spain, with upper-middle class Britons, French and Germans topping the ranking of international investors. They represent a demand focusing mainly on the Balearic and Canary Islands and the entire Mediterranean coast, corroborating the relevance of tourism-related issues. According to the report, this growing dynamic started in 2010 and reached its peak (13.2 per cent of foreign investors in overall property transactions) in 2015.

Within this Spanish scenario, Barcelona stands out as probably the strongest pole of attraction for international buyers interested in acquiring second homes and/or investing for financial purposes. Since 2013, in fact, the Spanish real-estate market has been a steady reference for global players and sovereign funds whose operations focus on hotels and resorts (ESADE, 2014). In Barcelona, recent years have seen the purchase of entire historic buildings or traditional hotels by major real-estate investors in order to turn them into high-end hospitality establishments. In these cases, there is a prominent presence of international capital from Russia, China and the Middle East (ESADE, 2014).

These operations prompt controversy both for disfiguring the morphological features of the architecture and for the abrupt economic effect they tend to have on the neighbourhoods involved. These controversies are regularly reflected in local newspapers and media, and feed a growing wave of protest on the part of residents.

Furthermore, the sector received a major boost from the growing rental market that sprang up after 2014 as a consequence of a significant increase in demand compared to a noticeable shortage in supply (Engel & Völkers, 2016). The spatial translation of these dynamics at ground level presents a complex situation, given the city's very varied geography. While, during the housing bubble, prices increased in all districts, the trend of the current recovery phase varies significantly throughout Barcelona. In order to underline these dynamics, a set of data referring to the Barcelona housing market at different scales was elaborated using the information included in the official statistical data base of the Municipality of Barcelona (Ajuntament de Barcelona, 2016).

At the beginning of the current decade, prices were still falling in Barcelona, yet clear signs of recovery started appearing in upper-class districts and in the main tourist areas as of late 2012. In this sense, it is significant to note that the old beachfront neighbourhood of Barceloneta (part of the central old town district) was the first area to show signs of this reversed trend. Despite its historical working-class profile, today Barceloneta is the city's most expensive neighbourhood for renting and among the most expensive on the real-estate market. Overall, the whole of Ciutat Vella was the first district in Barcelona to experience growth rates in the last quarter of 2012. In the course of 2013, this trend pushed the district's average price over the city average for the first time in history, thanks to the positive trend of its four central boroughs. In the areas of the Gòtic and the Raval, growth rates have been outstanding, being the second and the third highest in Barcelona (+46.8 per cent and +21.2 per cent respectively, for 2012–2015), only behind the upper- and middle-class beachfront area of Diagonal Mar and the seafront of Poblenou. Moreover, the real-estate vitality of the Raval can be seen in the city's biggest increase (+15.4 per cent) and the third highest number of registered sale contracts in the 2014–2015 period. This data also confirms the importance of the offer that ranks fourth in terms of property listings on Spain's main property portal, Idealista, mainly due to its southern sector, which includes 58.7 per cent of the overall whole Raval supply.[8]

This real-estate vitality of the Raval (specifically in its south-western sector) is indisputably linked to a noticeable improvement in the material quality of the supply. Traditionally, the rehabilitation of old buildings has not been the main option for Barcelona properties, and this is appreciable in a distinct lack of cultural and technical resources among local architectural professionals, who tend to prefer the comprehensive replacement of any building older than 50 years (Scarnato, 2016). Even today, the local school of architecture does not have a course in rehabilitation. It is the growing presence in recent years of a new, cosmopolitan middle class, prepared to invest in local property that is technologically upgraded (heating systems, efficient plumbing, soundproofing) but still possesses markedly vintage atmosphere (Múñoz, 2008), that explains the upswing in rehabilitation in the Raval. Whereas until recent years the main agents in these trends were mostly the international lower-middle classes, there is now a noticeable upgrade of the economic profile of investors. Sometimes they are upper class and sometimes they are major investment groups, foreseeing the lucrative benefits of rehabilitating buildings that local traditional sensibility (economically and aesthetically) considered good for nothing except demolition. This is the case of Can Seixanta, an early manufacturing complex dating from the late eighteenth century at the heart of the South Raval; languishing in a poor state of repair for decades, ready for demolition, it was eventually purchased in 2015 by the German group Jäger & Pachowiak to develop a high-end urban resort. The Robador area, too, for a long time an urban lost cause, is now showing signs of a possible future in the real-estate market, precisely due to the increasing interest in this kind of investment on the part of international actors.

All this is happening while the area and the entire historic centre is immersed in a negative demographic trend (−4.1 per cent since 2012) and the district's European population shows a clear increase (16.2 per cent). In this sense, the spatial correlation (R^2: 0.76) between the prices of the 2015 second-hand housing market and the percentage of EU-15 citizens of the residential foreign community at neighbourhood scale are significant.

The local demand for formal residential purposes was obviously not the only factor impinging on the definition of market price. Official population data and household statistics cannot in themselves describe the increasingly intense flows of populations temporarily or informally settling in the city, including undocumented migrants, seasonal workers, post-bohemians (Quaglieri Domínguez and Russo, 2010) or Erasmus students. The search for housing on the part of these groups can significantly influence the rental market and, therefore, the real-estate trend. In this respect, tourism mobility is also linked with growing residential tourism (Figure 5.5) and the boom of the short-term rental market around the offspring of the so-called collaborative economy, represented by webpages like Airbnb.

Figure 5.5 Experiencing 'cool Barcelona'. Tourist flat in the South Raval

Source: Authors

Barcelona has become one of the main destinations of this global network, and the Raval is the area, according to the data provided by the platform Inside Airbnb (2016), where Airbnb offers the largest number of places to stay (1,134 listings, 10.4 per cent of Barcelona supply), half of them (50.1 per cent) concentrated in South Raval.[9] Moreover, the proximity to the smaller and more expansive Gòtic neighbourhood could increase the perspective of profitability of real-estate investment aimed at the tourist market in the cheaper but increasingly appealing South Raval.

Even though Airbnb emphasizes its supposedly new model of decentralizing and decongesting the pressures of tourism, its offer in Barcelona demonstrates the same spatial patterns as the conventional hospitality sector (Arias Sans and Quaglieri Domínguez, 2016). Moreover, the number of short-term rental listings is strongly correlated (R^2: 0.89) with the size of the EU-15 communities as shown in Figure 5.6.

The fact that the so-called expat community seems more prone than residents to be involved in the short-term rental network (Arias Sans and Quaglieri Domínguez, 2016) explains this dynamic, but only as one factor among others. Despite these statistical correlations, it is not possible to establish a clear, linear causal relationship. The steadily growing touristic activity in Barcelona even

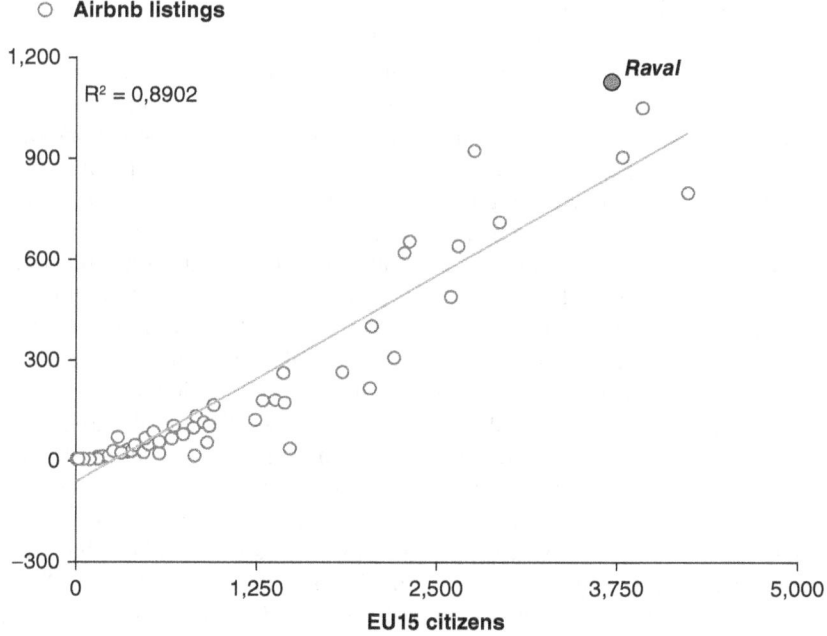

Figure 5.6 Spatial correlation between the number of Airbnb listings and foreign residents from EU-15 countries

Sources: Authors' own elaboration of data from insideairbnb.com and www.bcn.cat/estadistica

during the deepest economic crisis and, particularly, the increase of touristic pressure on the Raval had, in fact, not been able to prevent the price collapse of both the neighbourhood and the overall housing market in Barcelona. Rather, it seems more appropriate to suggest a circular relation between different mobile populations whose spatial strategies converge around and reciprocally feed on similar lifestyle considerations and expectations. In the framework of the current phase of macroeconomic recovery and the steady expansion of tourism activity in the city, increasing flows of visitors and lifestyle migrants boost the further development of the cultural and commercial landscape, which, in turn, enhances the appeal of the area's everyday scene. This, as well as providing the area with further contingents of lifestyle consumers, stimulates the related housing demand.

Conclusion

All over Europe, recent years have seen a growing number of central or historic areas affected by the significant penetration of tourist flows despite not being specifically monumental or, supposedly, related to the main tourist circuit. The rise of vibrant contemporary cultural and commercial scenes around iconic landmarks and/or picturesque urban fabric emerge as attractors for a wide range of urban users including tourist populations eager to venture off the beaten track and to experience appealing urban every-day lifestyle.

The case of the Raval is a clear example of these trends. Nevertheless, these trends have not been linear due to the simultaneity of deep-seated dynamics of change that actually boost a rapid process of socio-cultural transformation. Concretely, the area's touristic vocation has developed in the framework of a process of wider urban transformation intended to adapt the old popular neighbourhood into a new "landscape of pleasure" (Hannigan, 1998). In the meantime, the socio-cultural profile of its residents has undergone a major transformation due to the arrival of intense flows of international migrants during a phase of economic expansion. In addition, the complexity of this historical neighbourhood is linked to the diverging dynamics between its different areas. Overall, different patterns can be appreciated between the evolution of the North and the South Raval over the last 20 years, in both social and economic terms.

The municipal strategy of building a dense network of cultural and educational venues in the North Raval has been pivotal in channelling cultural lifestyle mobility towards the area and in opening it up to the rest of the city. Meanwhile, the South sector has seemed to be more affected by the persistence of social marginalization and petty crime. Despite interventions in its urban fabric that have provided unprecedented injections of urban glamour, the former Barrio Chino has not yet shaken off the social stigma still perceived in Barcelona's uptown neighbourhoods.

Reports about cases of mobbing or the underlying pressure to expel traditional residents have accompanied the rehabilitation of this area throughout the development of the main operations: the new Rambla del Raval, the Illa Robador complex and even the brand new Filmoteca de Catalunya. But even if there has been a degree of commercial development in keeping with the emerging circuit of cosmopolitan consumption, though firmly anchored in the northern sector, socio-demographic indicators seemed to ward off this risk. On the contrary, when urban reform reached its peak in the late 1990s, the growth in foreign population led to a demographic recovery thanks to more intense conventional (that is, economically motivated) migration flows.

It is true that a Western foreign community has significantly grown in the Raval, but this dynamic is more linked to a wider development of the local bigger "multiculturalist" Extra-European community that reinforced the working-class profile of the area. In other words, it is the balance of power among different urban groups converging in the area for a variety of reasons that curbed the gentrification process.

However, the accelerated penetration of the commercial "cosmopolitan wave" into until recently conflictive sectors, as well as the proliferation of large real-estate operations, reflects, visually too, a deeper-seated change. With the current macroeconomic recuperation, analysis confirms a trend of a reversal of pre-crisis dynamics as regards both the booming housing market and the reverse of the socio-demographic structure of residents.

This latter shift is prompted by an increase in the Western community, while the traditional Asian presence has begun a significant demographic downturn. If we make the comparison with other districts at municipal and provincial scale, rather than population "leaking out" towards more affordable urban peripheries, this phenomenon seems to have more to do with growing return migration along the North–South axis.

Therefore, the transformation of the socio-cultural landscape of the Raval is still driven in first place by macroeconomic factors defining Barcelona's differentiated appeal for international migration flows. In this respect, the recent economic crisis has acted as a watershed in the city's relation with these long-range flows, especially in the socio-economic profile of South Raval. Growing touristic pressure has been central in consolidating the local demand for cultural and leisure consumption, thereby enhancing perception of the area's appealing liveability. With economic recovery underway, then, the area seems even more suitable than before for the strategies of settlement of a cosmopolitan middle class. As we say above, the increase in real-estate activity and demographic trends confirms this interpretation.

Even the parallel boom in the short-term rental supply is suspiciously related with these dynamics. On the one hand, this new cosmopolitan scene increases the appeal for tourists; on the other, the growing tourist demand stimulates the upgrading of the consumption landscape and the local housing demand, in which short-term rentals represent a key point.

Therefore, converging needs and consumer interests in users with different mobility trajectories but similar lifestyle expectations are feeding a circular logic of growth in new tourist areas such as South Raval in the framework of the polarization of the geography of social destinations.

Notes

1 For the first time, real-estate agencies specializing in the Raval appeared. This was positive in itself, but it generated a huge, unchecked rise in prices leading to speculation
   with private property and the unexpected effect of momentary stagnation of the market.
   According to local press and going through the real estate market bulletins of the period
   (like the Barcelona Chamber of Real Estate Property Bulletin) the apartments overlook-
   ing the new Rambla del Raval were sold for 400,000 pesetas/m², and the average rent
   skyrocketed from 15,000 to over 75,000 pesetas/month.
2 At this point, Mayor Joan Clos personally called for the addition of a 4-star hotel with
   at least 110 rooms – clipping 7,000 m² from other uses – to benefit from Barcelona's
   booming tourist trade. It was also decided to include the new Filmoteca de Catalunya
   (a public film theatre and archive) to attract cultural consumers from the rest of the city
   and beyond Barcelona itself.
3 There is no available data about the place of origin of migrants at neighbourhood scale, but
   available data at district level could provide indicative insights. Unlike other city districts,
   the majority of the old town's new residents (51.9 per cent) came from abroad in 2014.
4 In 2014, despite representing 3.0 per cent of the population, the Raval registered
   6.0 per cent of all inbound flows at neighbourhood scale, taking into account both new
   residents from other Barcelona neighbourhoods and migrants from other Spanish municipalities or abroad. At the same time, it was also the area most affected by outbound
   flows (5.6 per cent), including changes of residence to other neighbourhoods or removal
   of names from the Municipal Register of Inhabitants.
5 According to the Dictionary of the *Real Academia de la Lengua Española* (RAE), the
   word 'guiri' refers to a supporter of Queen Cristina during the Carlist wars in the nineteenth century; nowadays it is colloquially used to mean a foreign tourist.
6 The 15 Member States of the European Union as of 31 December 2003, before the
   eastern enlargement: Austria, Belgium, Denmark, Finland, France, Germany, Greece,
   Ireland, Italy, Luxembourg, Netherlands, Portugal, Spain, Sweden, and the United
   Kingdom. For the purposes of analysis, the Spanish population was not considered.
7 The area obtained by grouping the new census sections coincide 98 per cent (1 ha smaller)
   with the surface obtained with the previous division. Because of the small difference,
   comparing trends referring to periods before and after, respectively, the new territorial
   division seems reasonable while no trends between the two periods were calculated.
8 Own elaboration of data produced by filtering property sales ads in http://www.idealista.
   com (retrieved 1 March 2016).
9 Data are based on information from http://insideairbnb.com/barcelona/ and refer to the
   situation registered in October 2015. Data for South Raval were obtained by extrapolat-
   ing the information for census sections by using GIS tools and summing the data for the
   concerned sections.

References

Abella, M. (2004) *Ciutat Vella. El centre històric reviscolat*, Barcelona: Aula Barcelona.

ADCV, *Arxiu del Districte de Ciutat Vella* (Municipal Archive of Barcelona Central
   District), Caixa 177, carpeta A161 'Acords'.

Aisa, F. and Vidal, M. (2006) *El Raval, Un espai al Marge*, Barcelona: Editorial Base

Arias Sans, A. and Quaglieri Domínguez, A.P. (2016) 'Unravelling Airbnb: Urban
   Perspectives from Barcelona', in Russo, A.P. and Richards, G. (eds), *New Localities in
   Tourism*, Bristol: Channel View.

Arias Sans, A. and Russo, A. (2017) 'The Right to Gaudí. What Can We Learn From the
   Commoning of Park Güell, Barcelona?', in Colomb, C. and Novy, J. (ed.) *Protest and
   Resistance in the Tourist City*, London: Routledge.

Ajuntament de Barcelona (2010) *Projecte d'Intervenció Integral del Raval Sud*, Barcelona: Ajuntament de Barcelona.

Ajuntament de Barcelona (2016) *Departament d'Estadística. Ajuntament de Barcelona*, [Online], Available at: www.bcn.cat/estadistica/angles/ [10 Sep 2016].

Ajuntament de Barcelona, Diputació de Barcelona and Turisme de Barcelona (2016) *Tourism statistics. Barcelona: city and surroundings.* Available: http://ajuntament. barcelona.cat/turisme/sites/default/files/documents/estadistiques_de_turisme_2015._ barcelona_ciutat_i_entorn.pdf [26 Aug 2016].

Bauman, Z. (2005) *Work, Consumerism and the New Poor*, 2nd edition, Maidenhead, Berkshire, England: Open University Press.

Bayona, J. and López-Gay, A. (2011) 'Concentración, segregación y movilidad residencial de los extranjeros en Barcelona', *Documents d'Anàlisi Geogràfica*, vol. 57, no. 3, pp. 381–412.

Bialski, P. (2012) *Becoming Intimately Mobile*, Frankfurt am Main: Peter Lang.

Bianchi, R. (2000) 'Migrant Tourist-Workers: Exploring the "Contact Zones" of Post-Industrial Tourism', *Current Issues in Tourism*, vol. 3, no. 2, pp. 107–137.

Bohigas, O. (1985) *Reconstrucció de Barcelona*, Barcelona: Edicions 62.

Cabré, A. and Domingo, A. (2002) 'Flujos migratorios hacia Europa: actualidad y perpectivas', *Arbor*, vol. 172, no. 678, pp. 325–344.

Cia, B. (2001) 'Oposición a que se levante un hotel de nueve plantas en la Rambla del Raval', in *El País Catalunya*, 25 Apr. 2001.

Cohen, E. and Cohen, S.A. (2015) 'A Mobilities Approach to Tourism from Emerging World Regions', *Current Issues in Tourism*, vol. 18, no. 1, pp. 11–43.

Cohen, S.A. (2010) 'Searching for Escape, Authenticity and Identity: Experiences of "Lifestyle Travellers"', in Morgan, M., Lugosi, P. and Ritchie, J.R.B. (eds), *The Tourism and Leisure Experience: Consumer and Managerial Perspectives,* Bristol: Channel View Publications.

Cohen, S., Duncan, T. and Thulemark, M. (2013) 'Lifestyle Mobilities: The Crossroads of Travel, Leisure and Migration', *Mobilities*, vol. 10, no. 1, pp. 155–172.

Coles, T. and Timothy, D. (2004) '"My Field is the World": Conceptualizing Diasporas, Travel and Tourism', in Coles, T. and Timothy, D. (eds), *Tourism, Diasporas, and Space*, London: Routledge.

Degen, M. (2003) 'Fighting for the Global Catwalk in Manchester and Barcelona', *International Journal of Urban and Regional Research*, vol. 27, no. 4, pp. 867–880.

Degen, M. and García, M. (2008) *La metaciudad: Barcelona, Transformación de una metrópolis*, Rubí: Anthropos.

Delgado, C. (2016) 'Spain's Tourism Sector Braces Itself For a Sixth Record Year', *El País*, 6 June. Available: http://elpais.com/elpais/2016/06/06/inenglish/1465206174_665153. html [27 Aug 2016].

Domingo, A., Sabater, A. and Ortega, E. (2014) '¿Migración neohispánica? El impacto de la crisis económica en la emigración española', *EMPIRIA. Revista de Metodología de Ciencias Sociales*, no. 29, pp. 39–66.

Edensor, T. (2001) 'Performing Tourism, Staging Tourism. (Re)producing Tourist Space and Practice', *Tourist Studies*, vol. 1, no. 1, pp. 59–81.

Edensor, T. (2006) 'Sensing Tourist Space', in Minca, C. and Oakes, T. (ed.), *Travels in Paradox: Remapping Tourism*, Lanham: Rowman & Littlefield.

Edensor, T. (2007) 'Mundane Mobilities, Performances and Spaces of Tourism', *Social & Cultural Geography*, vol. 8, no. 2, pp. 199–215.

Elliott, A. and Urry, J. (2010) *Mobile Lives*, London: Routledge.

Engel & Völkers (2016) *Market Report*, Barcelona: Engel & Völkers.

ESADE (2014) *Fondos soberanos 2014*, Barcelona: ESADE.

Fabra Garcés, L.A. (2015) *Estádistica Registral Inmobiliaria*, Madrid: Colegio de Registradores de la Propiedad, Bienes Muebles y Mercantiles de España.

Fainstein, S., Hoffman, L. and Judd, D. (2003) 'Making Theoretical Sense of Tourism', in Hoffman, L., Fainstein, S. and Judd, D. (eds) *Cities and Visitors: Regulating People, Markets and City Space*, Oxford: Blackwell.

Farbrother, C. (2010) *Non-Commercial Homestay: An Exploration of Encounters and Experiences of Guests Visiting the UK*, in CHME National Research Conference, 5–7 May 2010, University of Surrey, UK.

Fernández, M. (2012) 'Asaltar el Raval. Control de población y producción de plusvalías en el barrio barcelonés', *Revista de Estudios Urbanos y Ciencias Sociales*, vol. 2, no. 1, pp. 51–68.

Florida, R. (2002) *The Rise of the Creative Class*, New York: Basic Books.

Franklin, A. (2004) 'Tourism as an Ordering: Towards a New Ontology of Tourism', *Tourist Studies*, vol. 4, no. 3, pp. 277–301.

Franklin, A. and Crang, M. (2001) 'The Trouble with Tourism and Travel Theory?', *Tourist Studies*, vol. 1, no. 1, pp. 5–22.

Garay Tamajón, L. and Cànoves Valiente, G. (2010) 'Un análisis del desarrollo turístico en Cataluña a través del ciclo de evolución del destino turístico', *Boletín de la Asociación de Geógrafos Españoles*, no. 52, pp. 43–58.

García Ballesteros, A., Jiménez Blasco, B. and Mayoral Peñas, M. (2014) 'Emigración de retorno y crisis en España', *Scripta Nova. Revista Electrónica de Geografía y Ciencias Sociales*, vol. 18, no. 491. Available at: http://www.ub.edu/geocrit/sn/sn-491.htm [accessed 2 Feb 2017].

Germann Molz, J. (2011) 'CouchSurfing and Network Hospitality: "It's Not Just About the Furniture"', *Hospitality & Society*, vol. 1, no. 3, pp. 215–225.

Giddens, A. (1991) *The Consequences of Modernity*, Cambridge: Polity Press.

González, S. (2010) 'Bilbao and Barcelona "in Motion". How Urban Regeneration "Models" Travel and Mutate in the Global Flows of Policy Tourism', *Urban Studies*, [Online], pp. 1–22.

Hall, M. (2005) 'Reconsidering the Geography of Tourism and Contemporary Mobility', *Geographical Research*, vol. 43, no. 2, pp. 125–139.

Hannam, K. (2008) 'Tourism Geographies, Tourist Studies and the Turn towards Mobilities', *Geography Compass*, vol. 2, no. 1, pp. 127–139.

Hannam, K., Sheller, M. and Urry, J. (2006) 'Editorial: Mobilities, Immobilities and Moorings', *Mobilities*, vol. 1, no. 1, pp. 1–22.

Hannigan, J. (1998) *Fantasy City. Pleasure and Profit in the Postmodern Metropolis*, London: Routledge.

Harvey, D. (1989) 'From Managerialism to Entrepreneurialism: The Transformation in Urban Governance in Late Capitalism', *Geografiska Annaler. Series B, Human Geography*, vol. 71, no. 1, pp. 3–17.

Harvey, D. (2001) *Spaces of Capital: Towards a Critical Geography*, Edinburgh: Edinburgh University Press.

Harvey, D. (2014) *Seventeen Contradictions and The End of Capitalism*, London: Profile Books.

Haug, B., Dann, G. and Mehmetoglu, M. (2007) 'Little Norway in Spain. From Tourism to Migration', *Annals of Tourism Research*, vol. 34, no. 1, pp. 202–222.

Hayllar, B. and Griffin, T. (2005) 'The Precinct Experience: A Phenomenological Approach', *Tourism Management*, vol. 26, no. 4, pp. 517–528.

Hottola, P. (2004) 'Cultural Confusion. Intercultural Adaptation in Tourism', *Annals of Tourism Research*, vol. 31, no. 2, pp. 447–466.

Huete, R., Mantecón, A. and Estévez, J. (2013) 'Challenges in Lifestyle Migration Research: Reflections and Findings about the Spanish Crisis', *Mobilities*, vol. 8, no. 3, pp. 331–348.

Inside Airbnb (2016) *Inside Airbnb: Adding data to the debate*, [Online], Available: http://insideairbnb.com/index.html [15 Jan 2016].

Jacobs, J. and Fincher, R. (1998) 'Introduction', in Fincher, R. and Jacobs, J. (eds), *Cities of Difference*, New York: Guilford Press.

Jiménez, S. and Prats, L. (2006) 'El turismo en Cataluña: evolución histórica y retos de futuro', *PASOS. Revista de Turismo y Patrimonio Cultural*, vol. 4, no. 2, pp. 153–174.

Judd, D.R. (1999) 'Constructing the Tourist Bubble', in Judd, D.R. and Fainstein, S.S. (eds), *The Tourist City*, New Haven, CT: Yale University Press.

Larsen, J. (2008) 'De exoticizing Tourist Travel: Everyday Life and Sociality on the Move', *Leisure Studies*, vol. 27, no. 1, pp. 21–34.

Lynch, K. (1960) *The Image of the City*, Cambridge, Massachusetts: MIT Press.

McNeill, D. and While, A. (2001) 'The New Urban Economies', in Paddison, R. (ed.), *Handbook of Urban Studies*, London: SAGE.

Maitland, R. (2007) 'Culture, City Users and New Tourism Areas in Cities', in Smith, M. (ed.), *Tourism, Culture, and Regeneration*, Wallingford, Oxfordshire, UK: CABI Pub.

Maitland, R. (2008) 'Conviviality and Everyday Life: The Appeal of New Areas of London for Visitors', *International Journal of Tourism Research*, vol. 10, no. 1, pp. 15–25.

Martinotti, G. (1993) *Metropoli: la Nuova Morfologia Sociale della Città*, Bologna: Il Mulino.

Massey, D. (1991) 'A Global Sense of Place', *Marxism Today*, June, pp. 24–29.

Monnet, N. (2001) 'Moros, sudacas y guiris, una forma de contemplar la diversidad humana en Barcelona', *Scripta Nova. Revista Electrónica de Geografía y Ciencias Sociales*, vol. 94, no. 58. Available at: www.ub.edu/geocrit/sn-94-58.htm [10 Mar 2016].

Moreras, J. (2004) '¿Ravalistán? Islam y configuración comunitaria entre los paquistaníes en Barcelona', *Revista CIDOB d'Afers Internacionals*, no. 68, pp. 119–132.

Múñoz, F. (2008) *UrBanalización. Paisajes comunes, lugares globales*, Barcelona: Gustavo Gili.

Novy, J. and Colomb, C. (2017) 'Urban Tourism and its Discontents: An Introduction', in Colomb, C. and Novy, J. (eds), *Protest and Resistance in the Tourist City*, London: Routledge.

O'Reilly, K. (2003) 'When is a Tourist? The Articulation of Tourism and Migration in Spain's Costa del Sol', *Tourist Studies*, vol. 3, no. 3, pp. 301–317.

O'Reilly, K. and Benson, M. C. (2009) 'Lifestyle Migration: Escaping to the Good Life', in Benson, M. and O'Reilly, K. (eds), *Lifestyle Migration: Expectations, Aspirations and Experiences*, Farnham: Ashgate.

Paddison, R. (2001) 'Communities in the City', in Paddison, R. (ed.), *Handbook of Urban Studies*, London: SAGE.

Palou i Rubio, S. (2011) *Barcelona, destinació turística. Promoció pública, turismes, imatges i ciutat (1888–2010)*, Barcelona: Edicions Vitel·la.

Quaglieri Domínguez, A. and Russo, A.P. (2010) 'Paisajes urbanos en la época post-turística. Propuesta de un marco analítico', *Scripta Nova. Revista Electrónica de Geografía y Ciencias Sociales*, vol. 14, no. 323, Available at: www.ub.edu/geocrit/sn/sn-323.htm [29 March 2016].

Richards, G. and Wilson, J. (2007) 'The Creative Turn in Regeneration: Creative spaces, Spectacles and Tourism in Cities', in Smith, M. (ed.), *Tourism, Culture, and Regeneration*, Wallingford, Oxfordshire, UK: CABI Pub.

Ritzer, G. and Liska, A. (1997) '"McDisneyization" and "Post-Tourism": Complementary Perspectives on Contemporary Tourism', in Rojek, C. and Urry, J. (eds), *Touring Cultures: Transformations of Travel and Theory*, London: Routledge.

Rius Ulldemolins, J. and Zarlenga, M. (2014) 'Industrias, distritos, instituciones y escenas. Tipología de clústeres culturales en Barcelona', *Revista Española de Sociología*, no. 21, pp. 47–68.

Rojek, C. and Urry, J. (1997) 'Transformations of Travel and Theory', in Rojek, C. and Urry, J. (eds), *Touring Cultures: Transformations of Travel and Theory*, London: Routledge.

Russo, A.P. and Quaglieri Domínguez, A. (2013) 'From The Dual Tourist City to The Creative Melting Pot: The Liquid Geographies of Global Cultural Consumerism', in Smith, M. and Richards, G. (eds), *The Routledge Handbook of Cultural Tourism*, London: Routledge.

Russo, A.P. and Quaglieri Domínguez, A. (2016) 'The Global Geographies of Networked Hospitality', in Russo, A.P. and Richards, G. (eds), *New Localities in Tourism*, Bristol: Channel View.

Russo, A.P. and Scarnato, A. (2016) 'Barcelona in Common: Reclaiming the Right to the Tourist City', paper presented at the IV World Planning School Congress, Rio de Janeiro, Brazil, 4–8 July 2016, Available at: www.globaltur.org/files/MOVETUR/CONFERENCES/Russo_Scarnato_2016.pdf [26 Aug 2016].

Sargatal, M.A. (2001) 'Gentrificación e inmigración en los centros históricos: el caso del Raval de Barcelona', *Scripta Nova. Revista Electrónica de Geografía y Ciencias Sociales*, vol. 5, no. 94, Available at: www.ub.edu/geocrit/sn-94-66.htm [10 Sep 2016].

Sassen, S. (1991) *The Global City: New York, London, Tokyo*, Princeton, N.J.: Princeton University Press.

Scarnato, A. (2006) 'Multiple Exposures or New Cultural Values? European Historical Centers and Recent Immigration Fluxes', in Monclús, J. and Guàrdia, M. (eds), *Culture, Urbanism and Planning*, Burlington: Ashgate.

Scarnato, A. (2016) *Top Model Barcelona, 1979–2011*, Firenze: Altralinea.

Sheller, M. and Urry, J. (2004) 'Places to Play, Places in Play', in Sheller, M. and Urry, J. (eds), *Tourism Mobilities: Places to Play, Places in Play*, London: Routledge.

Soja, E. (2001) 'La mujer dominó las primeras ciudades', interviewed in *La Vanguardia*, 8 Aug. 2001, p. 60.

Steylaerts, V. and Dubhghaill, S.O. (2011) 'CouchSurfing and Authenticity: Notes Towards an Understanding of an Emerging Phenomenon', *Hospitality & Society*, vol. 1, no. 3, pp. 261–278.

Subirats, J. and Rius, J. (2006) *Del Chino al Raval: Cultura y transformación social en la Barcelona central*, Barcelona: Centro de Cultura Contemporanea de Barcelona.

Tatjer, M. (2006) 'La industria en Barcelona (1832–1992). Factores de localización y transformación en las áreas fabriles: del centro histórico a la región metropolitana', *Scripta Nova. Revista Electrónica de Geografía y Ciencias Sociales*, vol. 10, no. 218, Available at: www.ub.edu/geocrit/sn/sn-218-46.htm [15 Mar 2016].

Thorpe, H. (2012) 'Transnational Mobilities in Snowboarding Culture: Travel, Tourism and Lifestyle Migration', *Mobilities*, vol. 7, no. 2, pp. 317–345.

Urry, J. (1990) *The Tourist Gaze*, London: Sage.

Van der Duim, R. (2007) 'Tourismscapes: An Actor-Network Perspective', *Annals of Tourism Research*, vol. 34, no. 4, pp. 961–974.

Vertovec, S. and Cohen, R. (2002) 'Introduction: Conceiving Cosmopolitanism', in Vertovec, S. and Cohen, R. (eds), *Conceiving Cosmopolitanism: Theory, Context and Practice*, Oxford: Oxford University Press.

Williams, A. and Hall, M. (2002) 'Tourism, Migration, Circulation and Mobility: The Contingencies of Time and Place', in Hall, C. and Williams, A. (eds), *Tourism and Migration*, Dordrecht: Kluwer Academic.

6 Rome: a cultural capital with a poor working-class heritage

Strategies of touristification and artification

Sarah Lilia Baudry

Introduction

Contemporary tourists are willing to live new cultural, artistic and urban experiences (Aguas and Gouyette, 2011). Major urban tourism spots are no longer the only attractive places (Vivant, 2007b). "Off-the-beaten-track" exploration (Gravari-Barbas and Delaplace, 2015), alternative tourism, or tourism in the interstices of cities (Vivant, 2007a), are increasingly popular (Holbrook and Hirschman, 1982). These 'new' kinds of tourism may consist in discovering unfamiliar sites as well as everyday life areas (Füller and Michel, 2014; Maitland, 2008) or in living 'genuine' experiences in so-called trendy districts of both ethnic and creative character (Pappalepore *et al.*, 2010). These new experiences often take place in popular or formerly working-class neighbourhoods (Chapuis and Jacquot, 2014). Although characterised as alternative tourism destinations (Shaw *et al.*, 2004), these districts have undergone or are undergoing a gentrification phenomenon, through very diverse urban regeneration processes. Researchers have extensively documented the transformation of these abandoned areas and their 'refunctionalisation' in a context of both neo-liberalism and urban gentrification (Clerval and Fleury, 2009; Collet, 2015; Colomb, 2006; Derek, 2015; Hamnett, 2003; Harvey, 2001; Judd and Fainstein, 1999; Sassen, 2001; Smith, 2002; Zukin, 1995).

The reconversion of industrial wastelands and abandoned public spaces into cultural areas has become a genuine marketing instrument within a (neo)liberal context (Le Galès, 2016), where metropolises compete with each other (Plaza, 1999; Vivant, 2007a). These spaces, which may be of variable dimensions (from former working-class towns, neighbourhoods to isolated factory sites), may become either important tourist destinations or remain 'niche spaces' only known to 'insiders' (Djament-Tran and Guinand, 2014). In this respect, the artification, defined as the transformation of former residential/working urban landscapes into 'artscapes' addressed to global aesthetic consumption (Gravari-Barbas and Guinand, 2015) is a way to make such territories attractive. Dissanayake (2001) defines artification as "transforming things into art proper by making or producing art". According to Naukkarinen (2012: 1)

> the neologism 'artification' refers to situations and processes in which something that is not regarded as art in the traditional sense of the word is changed

into something art-like or into something that takes influences from artistic ways of thinking and acting. It refers to processes where art becomes mixed with something else that adopts some features of art.

Artification phenomena are observed in major tourist destinations like Rome, a capital city endowed with exceptional heritage and limited industrialised sites but looking to renew its 'classical' image. This chapter aims at addressing this paradox: the implementation of alternative tourism by public actors in a major historical and cultural capital, which is among the cities with the smallest industrial building legacy in Europe. May this political strategy of 'alternative' tourism be considered successful? Is this phenomenon concomitant with gentrification? How does the artification phenomenon manifest itself in Rome and what are its characteristics? Who are the main actors in this artification process and how do the various new tourism practices contribute to shaping it?

In order to answer these questions, we based our analysis on results of a survey we carried out with different stakeholders involved in artistic and urban projects in Rome (artists, gallery owners, urban architects, researchers, inhabitants, long term tourists – some of them may belong to several of these categories). We undertook semi-structured interviews and collected observations about Roman public spaces dedicated to 'alternative' culture.[1] In addition, we reviewed grey literature on cultural policies, which have been enforced since 1990.

The chapter stresses four main points. First, it presents the main features of city contexts in which the artification phenomenon takes place. This "cultural turn" (Ray and Sayer, 1999), which can be traced back to the last decades, occurs within the frame of urban regeneration policies implemented by institutional actors and is, in the scope of this chapter, analysed under the spectrum of gentrification. Second, we look at, and describe how, Rome has been affected by these trends. Through an alternative touristification, Rome's public administration wished to modernise the image of the eternal city and focused its strategy on the valuation of contemporary culture. We then show that this policy has not been a complete success but that it nevertheless created the opportunity for citizens to propose initiatives around wastelands or brownfields identified as 'countercurrent' actions. We show however that these 'bottom-up' practices are driven by people with a high cultural capital and that some of them become politically exploited by City authorities, who, while not intervening in their content, still 'City-label' them. We based our analysis on the Testaccio neighbourhood, a Roman district in socio-economical transition.

Art in the City and the 'cultural turn' of contemporary metropolises

The relationship between art and the city is not new (Bouchier, 2015). However, for thirty years, a 'cultural turn' has been observed in cities (Bouchier, 2015) through careful heritage restoration and redevelopment as well as the exploitation of culture. A double phenomenon occurs: while art is urbanised, the city is increasingly artified. The artification process follows up from a double condition. First, art becomes a process more than an object. In the city for instance, it is the constant changes in the urban landscape, the ephemeral interventions that are primordial. These actions are the conditions of the experiment in which the audiences can participate. Second, definition's frontier of art has become very porous as the number of prescribers has arisen. Today art is not uniquely being legitimised under canonical and classical principles. The media, festivals, *collectionneurs*, *amateurs*, hedge funds, etc. are increasingly taking part in what makes art today (Shapiro, 2004). Thus, the transformation of non-art into art generates changes in the definition and status of persons, objects, and activities. "It is the common result of all transactions, practical and symbolic, organisational and discursive, by which actors agree to consider an object or activity as Art"[2] (Heinich and Shapiro, 2015: 20). In this respect, the forms of artification (e.g. the heritagisation of old factories or the beautification of damaged walls), the duration of the process (permanent or temporary) and the operators, who concur in transforming these spaces, can vary considerably (from the street artist to the real estate investor).

The de-industrialisation of a number of European contemporary societies has been accompanied by the growth of the service sector, the development of highly skilled economic activities and the emergence of a creative class (Florida, 2004) and of artists (Smith, 1996) contributing to gentrification phenomena in inner-cities (Bidou-Zachariasen, 2003; Boltanski and Chiapello, 1999; Clerval and Fleury, 2009; Fleury and Van Criekingen, 2006; Honneth, 2006; Semi, 2015). Looking at the district of Belleville in Paris and opposing it to neighbourhoods in Anglo-Saxon territories, Vivant and Charmes (2008: 29) underline that "Artists are more indicators of gentrification than triggers of its development. They are part of a trend which upgrades the urban centrality and its resources (including cultural resources)". Artists do contribute to the symbolic revaluation of ancient abandoned sites or industrial wastelands. However, these transformations of public space or vacant land and buildings are not only the result of artists' production; it should be replaced in a broader context including urban policies, real estate developers, the cultural and tourism economy, etc., as these processes usually occur under the larger spectrum of urban regeneration.

Culture is indeed one of the pillars of urban regeneration and remains an economic strategy that is implemented by institutional actors in many megalopolises (Gravari-Barbas and Ripoll, 2010; Grésillon, 2010; Guinand, 2015; Pratt, 2011). The best known examples are, of course, the Guggenheim Museum in Bilbao (Plaza, 1999). The Tate Modern in London, the 104 or La Villette in Paris also deserve a mention. In some cities, the transformation of public spaces through art entails social and political dimensions. In Johannesburg, for instance, art is used as an instrument for both social appeasement and remembrance. By upgrading heritage assets in public areas, metropolitan authorities intend to recognise the exceptional past of the city but also to create a distance from a difficult history by conforming it to the standards of international tourism. This approach can however be challenged by artists and citizens offering alternative artistic forms in the city (Guinard, 2010). Temporary artification can also be a means to renew the image of a neighbourhood. For example, the 'Tower Paris 13', a social housing project in Paris's 13th district (Kullmann, 2015), was launched by the municipality with the objective to create an outdoor museum in an as yet non-touristic area of Paris.

As Bouchier (2015: 66) stresses:

> Cultural schemes of creative districts, European cities or Capitals of Culture, regions and cultural landscapes are implemented together with the establishment of cultural objects, festive events, where crowds are invited to celebrate Culture. They give a territorial dimension and international visibility to urban development. They also provide an institutional framework for the 'cultural turn' that connects art, city and territory.

Art can be an important marketing tool for urban regeneration projects. But it is also a factor creating strong metropolitan identity (Fagnoni, 2013) and by extension a tool to make territories socially more uniform and to control them (Capron and Haschar-Noé, 2007; Costa and Lopes, 2015; Gigot, 2012).

Finally, the re-functionalisation of degraded public spaces contributes to the renewal of cultural and tourism practices. They may not always be major international tourist attractions; nevertheless, these rehabilitated industrial wastelands are able to attract visitor profiles likely to be interested in so-called 'alternative' cultural spaces. The artification that can be observed there may be heterogeneous and of varying degrees of institutionalisation. These heterogeneous areas (Maitland, 2010) are visited and experienced by national and international tourists or by temporary city users (i.e. artists, resident researchers, long term visitors) but also by city residents like millenials, who are looking for unconventional areas. Thus the boundaries between the tourist, the visitor or the user are very difficult to tackle as artification goes along with a certain lifestyle and offers a break within the city. Such areas encourage the city user to have a different look upon off-the-beaten-track territories. Once a place gets popular, its success may go along with/or accelerate the already on-going gentrification process (Gotham, 2005). A marginal place can thus very quickly transform into a cultural district and become a touristic place.

Current Roman urban policies: a 'touristification' strategy in tune with the trends of European metropolises

Rome's exceptional cultural heritage, the cradle of western antiquity and Christianity, has been the subject of photographs, texts, poems but also studies and research in the field of social sciences, as well as art history or archaeology – so much so, that the Italian capital has become the stereotype of a museum and heritage city.[3] Internationally and universally famous, Rome's historical centre, home to important monuments of Antiquity, was listed as a UNESCO World Heritage Site in 1980, less than 10 years after the adoption of the World Heritage Convention in 1972. The inscribed perimeter was then extended in 1990 to the sixteenth-century Walls of Pope Urban VII. As opposed to Milan, the production centre and economical capital of Italy, Rome, home to prestigious national and international institutions, appears as the city of Culture. However, although no important industrial clusters nor big factories are settled in the Italian capital, industries are far from non-existent (Rivière, 1990; Seronde-Babonaux, 1980). Small industrial buildings, especially manufacturing ones, have been developed outside of the historical centre delimited by the Aurelian Walls in the eastern parts of the city (Tiburtina, Prenestina, San Lorenzo areas) and south of the historical centre (Testaccio and Ostiense areas).

As many European metropolises, Roman public policies under the mandates of the leftist mayors Francesco Rutelli (1993–2001) and Walter Veltroni (2001–2008) were in favour of enhancing contemporary urban heritage. They chose to implement a strong cultural policy as a means to foster growth and competitiveness based on negotiations with private actors. The Rome Master Plan (1993–2008) and the New Strategic Plan (2008) identified tourism and culture as the two key sectors affecting urban development. Researchers have extensively documented this "Modello Roma" (Annunziata and Violante, 2011; Delpirou, 2014; Insolera, 1993). This policy aimed at revaluating the working-class urban heritage while at the same time looking for added value to modern and contemporary art.

The 2000 Jubilee[4] helped to enhance Rome's urban image. As a matter of fact, the Holy Year event has been a good pretext not only for the restoration of religious buildings and façades, and urban embellishments in the historic centre – in particular the restoration of Rome's great patrimonial squares – but also of landmarks in the periphery (*Via Appia Antica*). The event brought together 35 million visitors:

> As in Paris, where the celebrations of the bicentenary of the Revolution have been an opportunity to endow the city with the *Opera Bastille* and the *Grande Arche*, likewise the 1996 *Olympic games* (OG) resulted in the regeneration of Barcelona and the 1998 OG led to the reorganisation of Eastern Lisbon, similarly, the *Jubilee* printed the seal of the XXI century architecture in the eternal city. (Vallat, 2004:140)

The mayors conceived their economic strategic lines focusing on the upgrading of the historic centre through ephemeral cultural events, inspired by examples from major European cities and interventions in public spaces.[5] The most important event was the so-called *Notte Bianca* which was initiated in 2003 and ended in 2008 when the right-wing politician Gianni Alemanno was elected as Mayor of Rome. The second edition of 2004 *Notte Bianca* brought together 1.5 million visitors in Rome.[6] It was organised by the Municipality, the Chamber of Commerce in collaboration with the Ministry of Culture, the Lazio Region and the Treasurers banks of the Municipality. The agency called Zetema Progetta Cultura was created in 1998 as the operational agency of the Cultural Department of the City of Rome. It took over the responsibility concerning design and management of all cultural and tourist events.

Another economic leverage that was used by the Municipality relied on tourism development, leading to the opening of several Art museums. The City of Rome, in close collaboration with the Zetema Agency, opened several museums on ancient industrial wastelands in the territory called Testaccio Ostiense. The aim was two-fold: first, to re-invest the space of the old slaughterhouse of Testaccio, the shops and its general market and, second, to upgrade the Ostiense Marconi area as an innovative and creative space (Djament-Tran, 2015). In 1999, the Central Montemartini power plant became an archaeological Museum. The *MACRO, (Museo di arte contemporanea di Roma),* was created in 2002 on the site of the old Peroni beer factory and became a contemporary Art Museum. Its annex building, the *MACRO Testaccio*, was created on the site of the city's ancient slaughterhouse. The opening of these two museums is linked to the willingness to pursue urban regeneration strategies on a former working class but deserted territory. The municipality of Rome carried out its strategy for the restoration of the historical centre together with major contemporary cultural projects. In this respect, the *Ara Pacis Museum*, located in the historical centre which was designed by Richard Meier and inaugurated in 2006, and became the home of the Altar of Peace of Augustus, was subject to many controversies. Cultural spaces emerged in the Flaminio district, which were designed by "starchitects" (Gravari-Barbas and Renard-Delautre, 2015). Among them, the *Auditorium Parco della Musica*, which was designed by Renzo Piano, can be compared to major European cultural institutions.[7] The *MAXXI Museum* of the twenty-first century was created under the leadership of Mayor Francesco Rutelli but did not open its doors until 2010. It was designed by the starchitect Zaha Hadid (Renard-Delautre, 2015).

An institutional strategy towards alternative touristification not completely successful

This strong public action aimed at renewing the image of Rome in order to make it a true cultural metropolis like Paris, London or Berlin. The creation of new museums of contemporary Art, the development of cultural interventions and events in public spaces, the revitalisation through Art of a number of territories, contributed, to some extent, to transforming Rome's image.[8] But, can this process be considered as a successful touristification or as a completed urban regeneration? The initiatives that seek to give higher value to industrial archaeology and promote contemporary Art usually take place within a specific context. Rome is a city that possesses an important cultural heritage, and therefore could be considered as not needing to upgrade industrial heritage or copy other European metropolises (interview with artist-architect of *All Nighter*, 2015) in order to be a touristic place. An artist-architect who worked for the so-called museums night and *All Nighter* criticised the city administration for its trend towards culture standardisation:

This year, I struggled to make people understand that the *Anno della luce di Roma*,[9] was an opportunity – through our cultural heritage – to be *the Anno*

della luce OF Rome, and to be unique in the world. But it seems the game is
over and we have lost in some way.

Moreover, Roman urban policies need better territorial consistency and have
difficulty creating coherent urban systems (Cremaschi, 2008; Delpirou, 2014;
Djament-Tran, 2015). It should be noted that there is a lack of a strategic policy.
There are considerable delays and paralysis, which greatly affect the implementation of urban regeneration projects. In the old industrial areas such as Ostiense or
Testaccio, urban projects do not have any territorial base. They are the sources of
conflict between different stakeholders (associations for the protection of heritage, squatters, and municipality and private actors). The lack of funds, the use of
private actors in a context of economic and political crisis,[10] the greater issue of
confidence between the inhabitants and public authorities may explain why the
City's strategy of artification and institutional touristification is not a complete
success. Tourist flows in these new touristic sights are still low compared to the
other well-known touristic sites.

These urban regeneration policies driven by cultural economy and supported
by institutions led to the upgrading of spaces and territories that might not have
been attractive places before. However, it is far from being a complete success.
This looks more like "random regeneration"[11] and urban embellishments than a
well-conceived thought process encompassing the entire metropolitan area. But
this regeneration phenomenon is happening together with other urban processes. It
is feeding alternative cultural and *off* (or 'underground') practices. Faced with an
uncompleted revitalisation process, Rome citizens had the opportunity to take initiatives and 'artify' their territories through diverse forms of non-institutionalised
artistic practices. As a matter of fact, the promotion of the city's heritage is not
only a political commitment but is also supported by associations and residents.
As the municipality failed to fully implement this strategy, the recovery and/or
the revalorisation of abandoned heritage is being carried out by inhabitants themselves. Thus, artification is a process which is also implemented by the citizens
of Rome, who want to offer tourists as well as residents unusual itineraries and
propose a 'counterculture' and a way to 'consume' the city differently (Maitland,
2008, 2010, 2013).

Miseries and splendours of Roman wastelands

The diversity of wastelands in Rome – wastelands initially subject to institutional reinvestment processes as well as completely abandoned wastelands – is
wide[12] (Cellamare *et al.*, 2011; Dazieri, 1996). As a consequence, a studio of
architects, *T-Spoon*,[13] in 2012 built a mapping system listing all the abandoned
spaces.[14] In order to give life to an idea or a project of artistic, cultural or social
nature, *T-Spoon* links the owners of these spaces (government or private actors)
and the people who need space. This way, in recent decades, initiatives and
re-appropriation driven by citizens, activists and architects have abounded

in Rome. The objective of this chapter is not to provide an exhaustive list of all such practices and initiatives but to highlight the fact that the opportunities offered by free spaces enable some Roman citizens to reinvent the city through bottom-up artistic practices.

The *self-managed social centres* may be the best illustration of this re-appropriation phenomenon. Historically, it is possible to identify the date of creation of these centres in Rome. They started to operate from 1980 to 1990 (*Centro Sociale Occupato Autogestito*). These autonomous social centres do not depend on any institution (unlike other 'social centres' which are not all autonomous) (Fominaya and Cox, 2013). Former garages, car parks, sheds, barracks or abandoned factories, in the majority located in the peripheries are spaces that were initially invested in by left-wing militant students who were affiliated to the leftist movement *Autonomia Operaia*. These activists installed the bases in squats that were used for housing or political and/or artistic practices; currently, they offer underground and alternative cultural and social activities. East of Rome, *Rampa Prenestina* and *Forte Prenestino* have become real cultural alternative venues. *Rampa Prenestina* is an urban art spot, a spontaneous Museum, offering courses and art workshops. *Forte Prenestino* is a former military fortress built in the late nineteenth century. Quickly abandoned, it was occupied by leftist young people in the mid-1980s. After restructuring work, the place became a cultural centre for alternative music, an 'outsider', and trendy place. We can also mention a less 'outsider' and more conventional place, *Pastificio Cerere*, an old pasta factory (in the area of San Lorenzo).

These spaces that were used for resistant actions in the first place are now included in various tourist itineraries. Hence, these places are listed in tourist guides, including Lonely Planet.[15] The book *Guida alla Roma ribelle*[16] also proposes to discover a different and resistant Rome, the 'off-the-beaten-track' Rome, far from stereotypes. These itineraries are becoming interesting not only for those tourists who want to visit alternative areas, but also for the Roman citizens themselves wishing to explore the city from another perspective. This brings us to questions about what is touristic and what is not. A French artist (Interview, 2015) who has been living in Rome for several years, noticed that these *off* itineraries have become attractive for a certain category of people, and eventually started to resemble the itineraries of ordinary tourists:

> You are like: 'I have to see this', 'I must see it', whether it is the Coliseum, the Roman Forum, the Pantheon or the occupied centre X, the occupied centre Y, [. . .] and as a matter of fact; your approach of the city becomes similar to this visiting attitude, but you are not looking for the same thing. This is the only difference.

By becoming alternative places with a touristic potential, these spaces are no longer such 'alternative' places, but are transformed into trendy places.

The street artification of Rome: between specific urban practices and politic 'exploitation' (the cases of two 'grassroots museums' MURo and MAAM)

This artification, even though a bottom-up phenomenon, is now part of the story-telling related to a creative metropolis. This is also apparent for *MURo, the Urban Museum of Rome*, a non-institutional Museum dedicated to open air street art (Figure 6.1). It is located in the Roman district of Quadraro, which is a middle-class residential area. The *MURo* is an association born under the leadership of a street artist named Diavu in 2010. He is also the project manager of the place. Diavu wanted to tell, through street-art, the memory of the district, which was formerly a resistant and partisan neighbourhood. He wanted to show that Rome could match London or Berlin as far as street-art is concerned. To this end, he invited internationally recognised street artists to participate in the project, such as Alice Pasquini, Ron English, etc. Local residents have the opportunity to have a say on what they want to see or not, on what can be shown on the walls of the neighbourhood. This museum does not benefit from any financial aid from the municipality even though the latter supports the project. Now, a touristic itinerary exists within this district, which has become 'open air museum'. Visits in Italian, Spanish and English languages for children and adults are organised. But this

Figure 6.1 Photograph of MURo Museum

Source: Author, 2015

project is not unanimously well-perceived among the neighbourhood residents.[17] Finally, the artification of the neighbourhood is not strictly speaking an urban regeneration. This is more an embellishment which is attracting people from outside. But it makes it more attractive. This ordinary neighbourhood, a residential place, is now a sightseeing place that city users like to visit.

The *MAAM, the Museum of the Other and the Elsewhere at Metropoliz*, was also created without any institutional support, but thanks to the initiative of people with high cultural capital and militancy experience (Figure 6.2). It is the first inhabited Museum of contemporary art in the world. Before this happened, the museum was a factory producing salami. It is located in a Roman suburb, *Tor Sapienza*, on Via Prenestina. This district has been said to have been socially abandoned by politicians. In 2009, migrants and Italians illegally occupied this former factory, *Metropoliz*, through the support of a group of activists fighting for housing rights. In 2011, a freelance anthropologist and documentary filmmaker, G, decided to transform common spaces of this squat into an artistic and experimental museum. Ironically, he called it the *MAAM*, echoing the *MAXXI* and *MACRO*, which are institutional Museums. He called for free support from artists from all over the world. He received the assistance of the *University Anthropology Laboratory Roma Tre* and of associations and contemporary art galleries (*Mondo Pop Gallery,*

Figure 6.2 Photograph of MAAM Museum

Source: Author, 2015

art gallery -1, Love Difference, Walls, Garage Zero etc.). While far from being the most touristic place in Rome, it is a place for so-called *interstitial tourism*.[18] This unusual venue is also taken as an area for research and as a topic for national and international media.[19] By becoming a common-good heritage and a contemporary Museum, the squat cannot be dismantled anymore. This was the objective of its curator, G. This 'success' does not prevent us from outlining the ambiguity of such a project. This informal Museum has its own limits. This once marginal space is becoming more and more mainstream due to its national and international recognition, so much that it now has its own page in the official Tourist Office guide.[20] An architect who worked with G underlines this fragility:

> The risk of the relation to institution is very fragile . . . I joke a lot with [him], but the risk of it becoming an institution is very high in such cases. [. . .] I give an example, it may be stupid, but relevant. I refer to an artist, Basquiat, who died when he ended up in museums, when he was working in the street, he was really an artist.

This site looks like an island known by the initiated, who come to see this unusual place; it is not really linked to the neighbourhood (Maitland, 2013). These Roman inhabitants have become exotic tourists in their own town. We observed that the Museum was not known by the inhabitants of the neighbourhood. The squatters, artists, visitors and residents do not communicate with each other. As an interviewee underlines:

> What I observe is: a residential community (squat) which is a kind of closed core group, a community of artists in the appendix, they are there for the touristic visit, which is linked with the museum aspect [. . .] Still, I see no connection between the two (the museum and squat), and I observe no special relationship with the context.

In both cases, the relationship with the territory is different. Their common point is that these initiatives come from the civil society. However, they have been 'exploited politically' by the city of Rome. For instance, both unofficial museums have their webpage on the Rome Tourism Office website. Thus, in 2015, the municipality of Rome, in partnership with the Zetema agency, decided to lead a campaign to promote all street-art works displayed throughout the city. A total of 50,000 flyers with maps locating the street-art works were distributed with the aim to let people discover the 'unfamiliar Rome' (Figure 6.3). The official tourism website and its thematic tourism tours offer a 'street-art tour', proposing to discover the contemporary Rome. The purpose is, without any doubt, to elevate Rome to the rank of major cities as far as this type of art is concerned and to come close to London and Paris on the podium of street-art capitals.[21] This tourist itinerary was placed in the same section as the archaeological tourist or religious itineraries. This action aims to show that Rome is not only the capital of archaeology but also a capital of contemporary and urban art, like any

Figure 6.3 Overview of street-art urban spots in Rome

Source: Tourist Office, 2015

other large metropolis. Playing on the specificity of Rome, well-known as a City Museum, the promotion campaign was baptised '*Change of perspective, the street is your new museum*'.[22] Just as in Paris with the *MyParisStreetArt* application, a similar application, *StreetArt Roma*, can be downloaded. However, originally, the majority of these street-art works were not driven by public administration.

Finally, abandoned open spaces in very different districts of Rome have been re-appropriated, and reinvested or 'artified' by a fraction of the civil society.

The latter usually has skills for planning and architecture. On top of that, it has an interest in art and in the cultural heritage or in major militant heritage. These alternative forms of collective management, which are more or less organised or autonomous, were born in very different contexts. They do not always have the same strategies and tools. However, they have in common the wish to stand away from so-called traditional culture. They want to 'tell' Rome differently. However, as their reputation grows, their status becomes more and more ambiguous. And the development of the *off* culture by the 'grass root initiative' may be mediated by institutions. These institutional bodies label these practices, once they are established, but they were originally grassroots initiatives.

Testaccio – Ostiense or the new trendy district

As we have seen, artification has affected Rome via complex processes. Through cultural policies, urban mobilisation and artistic practices (these are interrelated), there has been a symbolic revaluation of certain areas or neighbourhoods. The former Testaccio – Ostiense is an eloquent illustration of the transformation of an unattractive place into an underground and trendy neighbourhood. Formerly a working-class district, located on the outskirts of Rome, this old industrial area has become 'desirable' and has undergone a process of gentrification (Annunziata, 2011; Ranaldi 2014). Arousing little political interest until the 1990s, the municipality of Rome decided to take advantage of the international and national context to re-develop projects in this working-class district. In this perspective, the aim was to re-invest in the old slaughterhouse as well as the main shops and markets in Testaccio but also to upgrade further south the axis Ostiense Marconi with the idea of transforming it into a creative cultural and research cluster. Despite many achievements – the opening of *MACRO Museum*, the University of Architecture and a big market – its success is limited. In 2014 it received 145,670 visitors compared to 2 million for the Coliseum (Ministero dei Beni e delle Attività Culturali e del Turismo, 2014). This new urban area was not conceived and re-designed through a comprehensive and systemic approach (Djament-Tran, 2015). Urban regeneration – even partial – contributed to symbolically change the value of this formerly popular neighbourhood, which became a more central place enriched by a great number of galleries, cultural and leisure spaces (Ranaldi, 2014). This former working area is now inhabited, reinvested and visited by elite groups of the society (i.e. researchers, journalists, architects, artists, and freelance workers). It has become a 'cool place' to hang out (Maitland, 2008).

As in other areas of Rome, inhabitants' participatory practices have played a very important role in this neighbourhood's artification. Associations organise urban walkabouts and alternative tourism activities: for example, some associations offer walks around the industrial archaeology and the identity of the area. We can also mention the event *OpenTestaccio*, a cultural association of 'open doors', where from morning to evening shops, universities, theatres, museums and markets open for free in the neighbourhood. These residents' initiatives are driven by social groups possessing important cultural capital. 'Victims' of their success, they were subject to media coverage by the Municipality of Rome. For instance, the project *Ostiense District* – which was supported by the Municipality of Rome, in collaboration with *999CONTEMPORARY* art gallery, specialists in street-art and by the communication Agency *Pescerosso* – organised a visibility campaign to advertise various art works not well-known to the public, through mapping of the places.[23] It is worth noting that the majority of street-art works subject to publicity were not ordered by the Municipality. The latter took the opportunity of their existence to give more value to these initiatives by street artists.

Conclusion

Considering our results, it can be observed that artification or the staging of industrial art has reached Rome in the same way as it has taken place in other European capitals. Although Rome is a touristic city with a limited industrial heritage public authorities have astonishingly taken over the remaining industrial buildings. This strategy of alternative touristification is not completely successful. The political discourse is contradicted by the insufficient resources allocated to this strategy. The Coliseum still attracts more visitors than the *MACRO museum*. In addition, visitors who are interested in the revalorisation of cultural wastelands may prefer to experience so-called non-institutional art practices and wastelands which are rehabilitated and reinvested by civil society. These wastelands attract one category of the population who wants to discover an 'alternative' Rome but finally, very often, meets the criteria of a cosmopolitan and trendy metropolis. These visitors who are touristifying these territories may be sometimes the same who are 'artifying' others. Indeed, this (re)interpretation of the city is made by a specific population (i.e. artists, researchers, architects and the so-called creative class). This artification by civil society may be subject to 'political exploitation' by institutions. Public authorities may be keen to echo citizens' initiatives and give them high visibility. According to our initial investigations, what is being observed is not a unique strategy of artification but a multitude of artistic initiatives in the city. The porosity between the practices of inhabitants and the strategies of policy makers highlights the originality, the "palimpsest metropolis" of Rome (Mongin, 2015).

Notes

1 The survey was carried out among around 25 Italian and 2 French people in 2015. As a French researcher studying the Roman territory, my positionality was part of the research methodology. Am I as a French person, a tourist, a gentrifying person (social and cultural proximity with the participants of the survey, personal interest in urban areas studied), a temporary user of the city? How do I participate in the phenomenon that I am describing?

2 Translations by the author.

3 For instance, the French tourist guide *Guide du Routard* uses the expression "musée à ciel ouvert", Museum open to the sky to present the Italian capital www.routard.com/guide/code_dest/rome.htm (Accessed: 12 July 2016).

4   Celebration of the holy year, the year 2000 being 2000 years after the birth of Jesus.
5   The first edition of *All-Nighter* was launched in 2002 in Paris ('Nuit blanche').
6   These cultural events in the open public space are also a way to re-establish the links with Rome's history and the *Estati Romane*. These festive and cultural events were initiated in the open public spaces of Rome (centre and peripheries) at the end of the 1970s, under the impulse of the Mayor Giulio Carlo Argan and the architect Renato Nicolini.
7   For instance, the *Music city* in Brussels or *La cité de la musique* in Paris.
8   www.lonelyplanet.fr/destinations/europe/italie/rome/si-vous-aimez (Accessed: 12 July 2016).
9   "The International Year of Light is a global initiative which will highlight to the citizens of the world the importance of light and optical technologies in their lives, for their futures, and for the development of society. It is a unique opportunity to inspire, educate, and connect on a global scale." www.light2015.org/Home/About/Resources. html (Accessed: 13 July 2016).
10 A scandal was revealed in 2014 in the "*Mafia Capitale*". It involved the government in Rome (Gianni Alemanno, the former mayor) in which alleged crime syndicates misappropriated money normally allocated for municipality services. This generated a widespread crisis of confidence in the Municipality. After the mandate of the centre-leftist Ignazio Marino, who lost the confidence of this party and was not popular among Roman people, Roman voters elected Virginia Raggi (*5 stelle*) who opposed city corruption (it was the first time a politician from this party was elected in Rome).
11 The words of an interviewee, an urban sociologist who has funded an association that proposes to discover Roman territories off the beaten track.
12 In Rome, there are 2 million km² of abandoned and derelict spaces (source "censimento aree degradate in stato di abbandono, Settore Parchi e Giardini, Comune di Rome), quote by Giovanni Attili (2013).
13 www.tspoon.org/cityhound/?lang=en (Accessed: 12 July 2016).
14 The "Rome self-made urbanism" was the topic of an exhibition at the museum MAXXI in 2015. http://smu-research.net/ (Accessed: 12 July 2016)
15 www.lonelyplanet.fr/destinations/europe/italie/rome/si-vous-aimez (Accessed: 12 July 2016).
16 *Guida alla Roma ribelle*, Rosa Mordenti, Viola Mordenti, Lorenzo Sansonetti et Giuliano Santoro, Voland.
17 Graffiti was written on street-art walls of the museum ("*Fuck Muro*"). The author of the graffiti considered that the artists were people outside the district. He believes that this museum does not respect the essence of graffiti, which is to do things illegally and underground.
18 www.tripadvisor.fr/Attraction_Review-g187791-d9465609-Reviews-MAAM_ Museo_dell_Altro_e_dell_Altrove_di_Metropoliz-Rome_Lazio.html This attraction is ranked 905/1248 things to do in Rome (Accessed: 12 July 2016).
19 www.la-croix.com/Culture/Actualite/Dans-la-Rome-eternelle-un-musee-de-l-ailleurs-qui-revendique-la-marginalite-2014-11-28-1244481; www.economist.com/blogs/pros pero/2016/01/art-and-migration; www.brooklynstreetart.com/theblog/2016/01/07/nemos-mafia-capitale-on-a-pork-slaughterhouse-outside-rome/#.VpAen_nhBD9; www.rai.tv/ dl/RaiTV/programmi/media/ContentItem-6f50d290-61b6-4611-83dd-ea016de5c583. html#p=0 (Accessed: 10 July 2016).
20 www.turismoroma.it/cosa-fare/maam-museo-dellaltro-e-dellaltrove (Accessed: 10 July 2016).
21 http://roma.repubblica.it/cronaca/2015/04/26/news/la_guida_della_street_art_a_ roma-112917934/ (Accessed:10 July 2016).
22 www.turismoroma.it/news/roma-presenta-la-sua-prima-mappa-di-street-art (Accessed: 13 July 2016).

23 We can read on the website "OSTIENSE DISTRICT We are contemporary. The rest is history. [. . .] Rome is known all over the world for its ancient splendour. This is both an opportunity and a restraint for the eternal city. On one hand, it is always crowded with tourists, on the other tourists only occasionally cross the boundaries of the center looking for the most contemporary side of the city. How to change this dusty perception of a city, ready to offer much more?" http://www.pescerosso.com/works/ostiense-district/ (Accessed: 3 February 2017).

References

Aguas, J-C. and Gouyette, B. (2011) 'L'invention d'un tourisme de l'ordinaire. L'exemple des promenades urbaines', *Espaces*, vol. 292, pp. 8–13.

Annunziata, S. (2011) 'The Desire of Ethnically Diverse Neighbourhood in Rome. The Case of Pigneto: An Example of Integrated Planning Approach', *Future Urban Research Series*, no. 4, pp. 601–614.

Annunziata, S. and Violante, A. (2011) 'Rome-Model: Rising and Fall of an Hybrid Neo-liberal Paradigm in Southern Europe', Paper presented at the International RC21 conference.

Attili, G. (2013) 'Gli orti urbani come occasione di sviluppo di qualità ambientale e sociale. Il caso di Roma', in Scandurra E. and Attili G., *Pratiche di trasformazione dell'urbano*, Roma: Franco Angeli, pp. 47–67.

Bidou-Zachariasen, C. (2003) *Retours en ville*. Paris: Descartes & Cie.

Boltanski, L. and Chiapello, E. (1999) *Le nouvel esprit du capitalisme*. Paris: Gallimard.

Bouchier, M. (2015) 'Aestheticization of Public Space. Art as an Instrument or Instrumentalization of Art?', *Manzar,* vol. 30, no. 64, pp. 64–74.

Capron, G. and Haschar-Noé, N. (2007) *L'espace public urbain : de l'objet au processus de construction*, Toulouse : Presses Universitaires du Mirail, 2007, available at: https:// halshs.archives-ouvertes.fr/halshs-00176261 (Accessed : 10 July 2016).

Cellamare, C., Ferreti, A., Pisano, M. and Postiglione, M. (2011) *Progettualità dell'agire urbano: processi e pratiche urbane*. Rome: Carocci.

Chapuis, A. and Jacquot, S. (2014) 'Le touriste, le migrant et la fable cosmopolite: Mettre en tourisme les présences migratoires à Paris', *Hommes & Migrations*, no. 1308, pp. 75–84.

Clerval, A. and Fleury, A. (2009) 'Politiques urbaines et gentrification, une analyse critique à partir du cas de Paris', *L'Espace Politique*, no. 8, available at: http://doi.org/10.4000/ espacepolitique.1314.

Collet, A. (2015) *Rester bourgeois: les quartiers populaires, nouveaux chantiers de la distinction*. Paris: La Découverte.

Colomb, C. (2006) 'Le new labour et le discours de la "Renaissance urbaine" au Royaume-Uni: Vers une revitalisation durable ou une gentrification accélérée des centres-villes Britanniques?', *Sociétés Contemporaines*, vol. 63, no. 3, p. 15.

Costa, P. and Lopes, R. (2015) 'Is Street Art Institutionalizable? Challenges to an Alternative Urban Policy in Lisbon', *Métropoles*, no. 17, available at: https://metro poles.revues.org/5157 (Accessed: 10 July 2016).

Cremaschi, M. (2008) *Tracce di quartieri. Il legame sociale nella città che cambia*. Milan: Franco Angeli.

Dazieri, S. (1996) *Italia Overground: Mappe E Reti Della Cultura Alternativa*. Rome: Castelvecchi Editoria & Com.

Delpirou, A. (2014) 'La qualité environnementale, une question d'échelles?: Les dilemmes de l'aménagement urbain durable à Rome', *Méditerranée*, no. 123, pp. 23–29.

Derek, M. (2015) 'Le tourisme hors des sentiers battus à la conquête des friches industrielles', *Téoros. Revue de recherche en tourisme*, vol. 34, no. 1–2, available at: https:// teoros.revues.org/2754 (Accessed: 10 July 2016).

Dissanayake, E. (2001) 'An Ethological View of Music and its Relevance to Music Therapy', *Nordic Journal of Music Therapy*, vol. 10, no. 2, pp. 159–175.

Djament-Tran, G. and Guinand, S. (2014) 'La diffusion des grands équipements culturels, vecteur de métropolisation des quartiers populaires?: Une comparaison de trajectoires urbaines', *Belgeo*, no. 1, available at: http://doi.org/10.4000/belgeo.12737.

Djament-Tran, G. (2015) 'Quand les interactions entre nouveaux musées et nouvelles ères urbaines ne parviennent pas à produire de nouvelles mobilités touristiques pérennes: le cas de la Centrale Montemartini (Rome)', *Nouveaux Musées, Nouvelles Ères Urbaines, Nouvelles Mobilités Touristiques, Colloque Des*, vol. 20.

Fagnoni, E. (2013) 'Patrimoine versus mondialisation?', *Revue Géographique de l'Est*, vol. 53, no. 3–4, available at: https://rge.revues.org/5048 (Accessed: 10 July 2016).

Fleury, A. and Van Criekingen, M. (2006) 'La ville branchée : gentrification et dynamiques commerciales à Bruxelles et à Paris', *Belgeo*, no. 1–2, pp. 113–134.

Florida, R. (2004) *Cities and the Creative Class*. New York: Routledge.

Fominaya, C. F. and Cox, L (2013) *Understanding European Movements: New Social Movements, Global Justice Struggles, Anti-Austerity Protest*. London and New York: Routledge.

Füller, H. and Michel, B. (2014) '"Stop Being a Tourist!' New Dynamics of Urban Tourism in Berlin-Kreuzberg: New Dynamics of Urban Tourism in Berlin', *International Journal of Urban and Regional Research*, vol. 38, no. 4, pp. 1304–1318.

Gigot, M. (2012) 'Patrimoine en action (s), un regard sur les politiques publiques patrimoniales', *Internationale de L'imaginaire*, no. 27, pp. 401–421.

Gotham, K.F. (2005) 'Tourism Gentrification: The Case of New Orleans' Vieux Carre (French Quarter)', *Urban Studies*, vol. 42, no. 7, pp. 1099–1121.

Gravari-Barbas, M. and Ripoll, F. (2010) 'Introduction : De l'appropriation à la valorisation, et retour', *Norois. Environnement, aménagement, société*, no. 216, pp. 7–12.

Gravari-Barbas, M. and Delaplace, M. (2015) 'Le tourisme urbain hors des sentiers battus', *Téoros. Revue de recherche en tourisme*, vol. 34, no. 1–2, available at: https://teoros. revues.org/2790 (Accessed: 28 July 2016).

Gravari-Barbas, M. and Renard-Delautre, C. (2015) *Starchitecture(s): Celebrity Architects and Urban Space*, L'Harmattan Paris.

Gravari-Barbas M. and Guinand S., (2015) EIREST, Paris I Panthéon-Sorbonne, Association of the American Geographers Annual Meeting, Chicago, session Tourism & Gentrification, available at: https://www.univ-paris1.fr/fileadmin/Colloques_IREST/ AAG_Tourism_Gentrification.pdf (Accessed: 1 March 2017).

Grésillon, B. (2010) 'Villes, création et événements culturels en Méditerranée: un certain regard', *Méditerranée. Revue géographique des pays méditerranéens/Journal of Mediterranean geography*, no. 114, pp. 3–5.

Guinand, S. (2015) *Régénérer la ville: patrimoine et politiques d'image à Porto et Marseille*. Rennes: Presses universitaires de Rennes.

Guinard, P. (2010) 'Quand l'art public (dé)fait la ville?: La politique d'art public à Johannesburg', *EchoGéo*, no. 13, available at: http://doi.org/10.4000/echogeo.11855.

Hamnett, C. (2003) 'Gentrification and the Middle-Class Remaking of Inner London, 1961–2001', *Urban Studies*, vol. 40, no. 12, pp. 2401–2426.

Harvey, D. (2001) *Spaces of Capital: Towards a Critical Geography.* Edinburgh: Edinburgh University Press.

Heinich, N. and Shapiro, R. (2015) *De L'artification: Enquêtes Sur Le Passage À L'art*, vol. 20, Paris: Éditions de l'École des Hautes Études en Sciences Sociales.

Holbrook, M.B. and Hirschman, E.C. (1982) 'The Experiential Aspects of Consumption: Consumer Fantasies, Feelings, and Fun', *Journal of Consumer Research*, vol. 9 no. 2, p. 132.

Honneth, A. (2006) *La Société Du Mépris*, Paris: La Découverte, available at: www.edition sladecouverte.fr/Liens/ps/d04772_preface.pdf (Accessed: 10 July 2016).

Insolera, I. (1993) *Roma moderna. Un secolo di urbanistica romana 1870–1970.* Turin : Einaudi.

Kullmann, C. (2015) 'De l'exposition de la *Tour Paris 13* au concept de musée à ciel ouvert', in Gravari-Barbas M., Delaplace M. (ed.), *Teoros. Revue de recherche en tourisme*, special issue *Tourisme hors des sentiers battus* vol. 34, no. 1–2, available at: http://teoros.revues.org/2776 (Accessed: 10 July 2016).

Judd, D.R. and Fainstein, S.S. (1999) *The Tourist City*. New Haven and London: Yale University Press.

Le Galès, P. (2016) 'Performance Measurement as a Policy Instrument', *Policy Studies*, 2016, vol. 37, no.6, pp. 508–520.

Maitland, R. (2008) 'Conviviality and Everyday Life: The Appeal of New Areas of London for Visitors', *International Journal of Tourism Research*, vol. 10, no.1, pp. 15–25.

Maitland, R. (2010) 'Everyday Life as a Creative Experience in Cities', *International Journal of Culture, Tourism and Hospitality Research*, vol. 4, no. 3, pp. 176–185.

Maitland, R. (2013) 'Backstage Behaviour in the Global City: Tourists and the Search for the "Real London"', *Procedia - Social and Behavioral Sciences*, no. 105, pp. 12–19.

Mongin, O. (2015) *La condition urbaine. La ville à l'heure de la mondialisation.* Paris: Seuil.

Naukkarinen, O. (2012) 'Variations in Artification', *Contemporary Aesthetics*, Special Volume, Issue 4, available at: http://hdl.handle.net/2027/spo.7523862.spec.402 (Accessed 10 July 2016).

Pappalepore, I., Maitland, R. amd Smith, A. (2010) 'Exploring Urban Creativity: Visitor Experiences of Spitalfields, London', *Tourism Culture & Communication*, vol. 10, no. 3, pp. 217–230.

Plaza, B. (1999) 'The Guggenheim-Bilbao Museum Effect: A Reply to María V. Gomez' "Reflective Images: The Case of Urban Regeneration in Glasgow and Bilbao"', *International Journal of Urban and Regional Research*, vol. 23, no. 3, pp. 589–592.

Pratt, A.C. (2011) 'The Cultural Contradictions of the Creative City', *City, Culture and Society*, vol. 2, no. 3, pp. 123–130.

Ranaldi, I. (2014) *Gentrification in parallelo. Quartieri tra Roma e New York*, Rome: Aracne.

Ray, L. and Sayer, A. (1999) *Culture and Economy After the Cultural Turn*: London: Sage.

Renard-Delautre, C. (2015) 'La Vieille Europe et ses musées contemporains. Le cas de Rome: Renouveler le regard touristique sur la ville éternelle', in Fagnoni E. and Gravari-Barbas, M., *Nouveaux musées, nouvelles ères urbaines, nouvelles pratiques touristiques*, Laval : Presses de l'université de Laval, pp. 261–277.

Rivière, D. (1990) 'Stratégies territoriales et aménagement du territoire. Le cas de l'aire de développement industriel Roma-Latina", *Strates. Matériaux pour la recherche en sciences sociales*, no. 5, available at: https://strates.revues.org/1449 (Accessed: 10 July 2016).

Sassen, S. (2001) *The Global City: New York, London, Tokyo*, 2nd edition, Princeton: Princeton University Press.

Semi, G. (2015) *Gentrification. Tutte le città come Disneyland?* Bologna: Il Mulino.

Seronde-Babonaux, A.-M. (1980) *De l'Urbs à la ville. Rome, croissance d'une capitale.* Aix: Edisud.

Shapiro, R. (2004) 'Qu'est-ce que l'artification?' XVIIème Congrès de l'Association inter-nationationale de sociologie de langue française, *L'individu social*, Tours.

Shaw, S., Bagwell, S. and Karmowska, J. (2004) 'Ethnoscapes as Spectacle: Reimaging Multicultural Districts as New Destinations for Leisure and Tourism Consumption', *Urban Studies*, vol. 41, no. 10, pp. 1983–2000.

Smith, N. (1996) *The New Urban Frontier: Gentrification and the Revanchist City.* London: Routledge.

Smith, N. (2002) 'New Globalism, New Urbanism: Gentrification as Global Urban Strategy', *Antipode*, vol. 34, no. 3, pp. 427–450.

Vallat, C. (2004) *Autres vues d'Italie: lectures géographiques d'un territoire.* Paris: L'Harmattan.

Vivant, E. (2007a) 'L'instrumentalisation de la culture dans les politiques urbaines: un modèle d'action transposable?', *Espaces et sociétés*, vol. 131, no. 4, p. 49.

Vivant, E. (2007b) 'Les événements off: de la résistance à la mise en scène de la ville créa-tive', *Géocarrefour*, vol. 82 no. 3, pp. 131–140.

Vivant, E. and Charmes, E. (2008) 'La gentrification et ses pionniers: le rôle des artistes off en question', *Métropoles*, no. 3, available at: https://metropoles.revues.org/1972 (accessed: 1 July 2016).

Zukin, S. (1995) *The Culture of Cities.* Oxford: Blackwell.

7   Grunge authenticity

The tenement as upscale tourist destination

Elissa Sampson

Introduction

In New York's post-9/11 landscape, the Lower East Side Tenement Museum is touted as a Downtown economic engine. Its hip 'iconic neighborhood' draws visitors and buyers with galleries, upscale eateries, boutiques and new multi-million dollar condos. Nearby busy 'revitalized' streetscapes are moored to renovated tenements and shops that invoke collective memories of immigration in ways that help sell the past to new residents and visitors. The Lower East Side's built environment and ethnic communities are threatened, if not overwhelmed, by rapid gentrification.

This chapter discusses the Tenement Museum – and issues of ethnic representation, absence and presence – within a larger context of Downtown's post-9/11 landscape of rebirth and revival in which tourism, gentrification and the growth of museums are tied together, not least by governmental funding of cultural sites associated with commemorative national narratives. The Museum specifically participates in the commemoration and commodification of past and contemporary immigrant/migrant life, consistent with its goal of increasing funding, visitor numbers, gift shop purchases, and the real estate to accommodate future visitors. Far more people have passed through as tourists in any given year than ever actually lived in its tenement building.[1]

The Museum's role as an institutional actor in this drama sheds light on the processes redefining Lower Manhattan as the Museum increasingly becomes a surrogate for a past Lower East Side. New York's nineteenth-century tenements – multi-storey buildings in which three or more unrelated families lived under the same roof – were cheaply constructed to shoehorn poor and working-class tenants into tiny apartments. The ironies and virtues of the Tenement Museum's presence are evident at 97 Orchard Street, its poor landmarked tenement expressly reconstructed for upscale, heritage tourism. Its building serves as a useful lens through which to examine how and where the Museum's stories of local 'historic' and living immigrant communities get told in a popular but less known site often associated with other Lower Manhattan tourist destinations such as Ellis Island and the Statue of Liberty.

The Museum's founder Ruth Abram wanted visitors to see a past ethnic immigrant history that would 'speak' to them in a personal, experiential fashion, while engaging them in thinking about a new immigrant present. The Museum offers guided tours of ethnic tenement apartments staged to show the stories of selected immigrant families who once lived in its building. Thus, a tenement that is now presented as uniquely precious illustrates the link between tourism, real estate and ethnic representation. Analysis of the represented interplay of ethnic presences and absences helps illuminate the fate of collective memories in an immigrant neighbourhood reconfigured as a cosmopolitan destination. This analysis thus contributes to an ongoing conversation on various intersections of ethnicity, representation and the built environment in Lower East Side gentrification (Abu-Lughod, 1994; Smith, 1996b; Kirshenblatt-Gimblett, 1998; Mele, 2000; Kwong, 2009).

In 2001–2002, the Museum's local real estate dealings backfired because of community opposition to the proposed use of eminent domain to allow its nearby expansion. An ethnic politics of representation helped fuel this protest (Sampson, 2014; Gravari-Barbas, 2014). This controversy is presaged in early documents from 1987 to 1994, preserved in Museum archives, that discuss what sort of museum it should be, and how best to interpret its building and neighbourhood when connecting past and contemporary migration. By 2012, when the Museum opened up 103 Orchard Street as a new building at a time of intense gentrification, it needed to put into play an updated, modified version of its ethnic apartment template.

The Museum's interpretive politics of representation as seen at 97 Orchard Street offers a good vantage point on its future real estate dealings as gentrification and tourism rapidly accelerated. The Lower East Side changed dramatically since the Museum's 1988 founding; by 2002, Jason Hackworth had astutely defined gentrification as "the production of urban space for progressively more affluent users" (2002: 815). The rapidity of previously visible trends accelerated in the fluid aftermath of 9/11. As discussed below, Downtown's future was seen as tied to the growth of its museums as cultural visitor sites as well as to upscale housing. The Lower Manhattan Development Corporation's (LMDC) official funding and growth strategies that were put into place as a post-9/11 response reflect this approach as a basis for Downtown's 'renewal'.[2]

The tenement as authentic: 97 Orchard

How did a lowly, unadorned pre-Law tenement dating from 1863 come to epitomize the nostalgic and the hip in a remaking of the Lower East Side itself as an upscale tourist destination? The neighbourhood – flush with newly affluent visitors and residents – pays a high price for being a part of a reconfigured Downtown conveniently wedged between Midtown and Wall Street. While Museum visitors were initially likely to be interested in Jewish heritage tourism, today they increasingly include history aficionados and Euro-tourists. The Tenement Museum negotiates this particular tension in regard to the commemoration of East European Jews and Southern Italians, the 'New Immigrants' from a prior age of mass migration.

The Tenement Museum is a national landmark open almost every day of the year; its standardized tours are repeated up to seven times daily. Its main product is thematic guided tours, which provide the only way to enter the Museum's historic building. Interpretive constructions of past immigration are reiterated daily as guides uncover the building's physical layers and changes while presenting the personal stories of selected prior building residents. Today, 97 Orchard is deployed to represent its own story and that of the neighbourhood which housed many of the seventeen million immigrants who came through New York's port as part of a mass migration that ended when quotas were imposed shortly after the end of World War I. The Museum makes that past place come alive, for an hour or so, again and again, seven days a week, eight hours a day.

As one of the few remaining historic sites at the gentrifying intersection of a formerly Jewish Lower East Side and contemporary Chinatown, the Museum's footprint and public voice have become larger as its number of visitors grows. Its tours reflect its reliance on what is known as the power of place where performance and affect are greatly amplified as an audience connects through a setting's resonant appropriateness and authenticity.

Initial Museum visitors were most likely to be heritage tourists looking for their own past (Diner, 2000: 118). The Museum's founding Director, Ruth Abram, sought to harness that history to advance an ethical political agenda of 'tolerance' toward new immigrants through the portrayal of older immigrant life. Guided tours stressed connections between the sympathetic immigrant past shown in tenement apartments and the visible immigrant present surrounding 97 Orchard Street.

The stories of 97 Orchard Street are presented as inevitable, inherent and self-evident. When the Museum finally opened in late 1994, its stories of immigrant hardship and collective memory were constructed to stick in place. Sixteen years later, the Museum's (2010a) website proclaimed: "We tell the stories of 97 Orchard Street." The Museum's use of tours to impart various historic lessons highlights how narrative is joined to scripted movement to hold a building's and its residents' stories in place.

Tracing the Museum's earlier (1987–1994) stories of its building and tours exposes how contested narratives accrete and get reused, making them harder to later dislodge. The record shows an active reshaping of geographies of Lower East Side memory through narrative and material practices whose use and reception cannot be fully controlled (Certeau, 1980; Patraka, 1996; Benjamin, 2002). Most tellingly it shows that a multiplicity of ethnic representational possibilities had once been in play, before the interpretive scheme was irrevocably tied to the stories of building residents.

Whose heritage is it?

The appearance of inevitability surrounding the Museum's overall narratives of place reinforces 97 Orchard Street's current identity as a building whose one-time residents tell stories of Jewish, Italian, German, and Irish migration. But prior to renting (1988) and then acquiring 97 Orchard Street (1996), the fledgling Museum in 1987 had already decided upon an interpretive scheme in which six groups were, in the words of Ruth Abram (1987a), to form the Lower East Side's "constituencies whose history will be interpreted." By 1990, Abram's (1990c) Winter Trustee report touted that: "The Museum commissioned leading scholars to develop historically accurate composite profiles of members of six major immigrations to the Lower East Side, 1860–1935: Free African, Irish, Chinese, East European Jewish, German and Italian."

The Museum decided to use a historic composite family to represent each group (Abram, 1987b; 1990c). By 1990, the same year Ellis Island reopened for tourists, Abram (1990b) had negotiated the Museum's right to renovate its building prior to exercising its option to purchase it. At that point, historian and curator Richard Rabinowitz responded to Abram's request for a Conceptual Plan for 97 Orchard Street. Based on her write-up, he wrote: "One would have difficulty identifying a building more suitable than 97 Orchard Street to represent the ethnic diversity which has always characterized the Lower East Side" (Abram, 1990a: 5).

These memos suggest an institutional commitment to the unproven hypothesis that an 1863 Tenth Ward pre-Law tenement would provide proof positive of a past diversity reflecting standard chronologies of ethnic immigrant waves. This evocation of diversity was based on a fuzzy notion about the boundaries of the historic Lower East Side. That term clearly overlapped with "the Great New York Ghetto," a late nineteenth-century popularization that took into account not only the area's residents but its unsurpassed urban population density.

The 'Old East Side' was in fact a first New York destination for a number of groups. Broadly speaking, its past is marked by distinct areas with Jewish, Irish, German, Slavic, Hungarian, Bohemian, Italian, Chinese and Black settlement. This allows for an odd paradox: the highlighted personal stories of Irish and Italian families who are not representative of its building are produced by the Museum in apartments as evidence of 97 Orchard's authentic diversity. Early on, while doing research on commission for the Museum, historian Thomas Kessner had noticed that in 1870 "only one family [the Moores] was from Ireland, indicating how sharply the immigrants drew their ethnic turf boundaries" (1992: 18–19). By 1880, he noted:

> Russian Jewish families appear on the census [. . .] But no Italian families settled in the building, illustrating the pattern of succession that so struck Riis: eastern Europeans moving into German neighborhoods and southern Italians who were coming over at the same time moving into the Irish areas on the West Side. (Kessner, 1992: 21)

If Irish and Italian families were indeed scarce at 97 Orchard Street, Black and Chinese families apparently could not be found. The apartments of 97 Orchard first slowly evolved into a privileged space for telling the stories of ethnic groups – rather than of tenants – with proven building residency. While the ethnic remained central, adherence to its earlier heritage scheme started to drift further as historians looked for 'real' stories with contemporary resonance. By 1992–1993, the Museum decided to focus solely on interpreting the ethnic households of past building residents for its tours of upstairs apartments.

Practising stories

Once it committed to the scheme of showcasing actual residents, the Museum's interpretive schemes and funding were unavoidably affected by its building's own demographics. A Museum (2002) press release on its new Levine sweatshop-themed apartment stated:

> This new apartment will join [. . .] previously restored apartments: the Gumpertz family, Germans who made their home at 97 Orchard Street in the 1870s; the Sephardic Confino family from Turkey who came to America in 1914; the Baldizzi family, Italian Catholics who lived in the building in the 1930s; and the Orthodox Jewish Rogarshevsky family, who lived in the building for over 35 years.[3]

As seen here, the Museum's stories of its building's ethnic groups – and their attendant immigrant hardship and familial dilemmas – are the interpretive heart of its period room narratives.

The Gumpertz story is typically described elsewhere as German-Jewish, allowing that apartment to be retroactively positioned as the Museum's first Jewish apartment. However, in 1992 two ethnic apartments, one German (Lutheran) and the other Italian (Catholic) were planned for its opening. The Museum's architectural consultant Andrew Dolkart then wrote:

> I realize that the Tenement Museum is making a concerted effort to be seen as an organization that does not deal exclusively with Jewish issues. I think this is correct since the immigrant tenement experience is much more universal. However, for the longest period of occupation in this building it was largely inhabited by Eastern European Jews. In addition, as far as I know, funding and visitation have largely come from people of Jewish descent. It seems odd, even perhaps perverse, that the Eastern European Jewish experience (so closely connected with the lore of the Lower East Side and the history of 97 Orchard Street) is being overlooked. (Dolkart, 1992: 2)

Here Dolkart – unaware that the German-speaking Gumpertz family was Jewish – cited the building's own history as a corrective against the direction the Museum's ethnic narrative was taking, suggesting instead that it should lead with its most representative story. His memo was written just when new information was becoming available about the Gumpertz and Baldizzi families. A subsequent resistance to changing inappropriate parts of the already staged Gumpertz story reflects a prior determination that the Museum's sole Jewish apartment was envisioned to be East European. A building that it assumed would match its vision of the Lower East Side's past diversity was turning out to be increasingly Jewish throughout its tenancy, adding to its future attraction for Jewish heritage tourists.

Not residents

The Museum's representations of ethnicity were initially in flux for approximately six years and then fixed in place. Black, and then Chinese, representation in apartments was eventually jettisoned. But even as the Museum increasingly geared itself towards solely telling the personal stories of building residents, as late as 1992, Abram confirmed in an interview that Black New Yorkers were part of its story:

> living next door, we found a freed slave [Mr. Douglass]. So we've decided to use the building to tell the story of the Settlement of the Lower East Side by the major groups that settled here and not to confine our interpretive work only to the people who actually lived here [. . .]. The truth is that as early as 1820 the area is over twenty-percent freed slaves [. . .] we're determined to tell this story [. . .] blacks are very much associated with the tenement experience. (Kugelmass, 2000: 190–191)

Abram seemed sure that the Museum would still use its earlier heritage scheme to interpret the history of "the major groups that settled here." But Mr. Douglass merely merited a mention in the Museum's 'Urban Pioneers' (1994) slide-show that was shown downstairs when the Museum finally opened two years later: "According to the 1850 census, William Douglass, an African-American labourer, lived at 93 Orchard Street with his wife and three children, all of whom were born in New York" (p. 4). Abram's commitment to interpret the Black tenement experience at 97 Orchard was literally displaced into an overhead slide. The historic Chinese experience was similarly dislodged from earlier plans that had housed it in an upstairs apartment. All of this was in tension with the Museum's desire to tell a multicultural story of successive immigrant waves sharing a Lower East Side experience of hardship that eventually led toward successful Americanization.

As early as 1987, Ruth Abram explained how she envisioned a "Lower East Side Heritage Trail" to interpret the living yet historic "heritage" communities outside of its building.

> I had a grand time dreaming up the trails. We meet a pigtailed Chinese chef in a chop suey house in 1900. That tour starts over noodles. The Italian tour starts in a pasta shop. The Jewish one starts in a sweatshop. And so it goes. I think we could get various managers to let us use their restaurants, shops [. . .] this way. (Abram, 1987c: 1)

Abram's stillborn concept seems almost startling in its eagerness to use living places and people in ways that knowingly speak to the past exoticization of ethnic enclaves (ghettos) already viewed voyeuristically in Jacob Riis' (1890) time. Uptown tourists regularly came to the Lower East Side to gawk at its nursing women, bustling streets, and 'foreign' denizens. There is something more than a little audacious in a Lower East Side museum rendering the neighbourhood at large as a living set for ethno-tourism. What seems playful to a Museum director or entertaining to a visitor could offend community merchants and residents precisely because contemporary lives are exoticized and tied to past stereotypes. If heritage tourism intimates traveling to a place in order to temporarily travel back in time, it can imply that people who stay in place remain in an atavistic or timeless past, and thus are fair game to represent, display, and view.

Coveting thy neighbor: 99 Orchard

One Sunday in late 2001, I saw Museum visitors watching Fujianese migrants holding Mandarin and English signs on 99 Orchard's stoop, demanding an end to "Eminent Domain Abuse" (Sampson, 2014). A photo later showed a sign that said "Don't ERASE real history with artificial history." To accommodate twice the number of visitors, the Tenement Museum had attempted to expand beyond its 97 Orchard Street national historic landmark. The Empire State Development Corporation (ESDC), a quasi-state agency, had agreed to use eminent domain on behalf of the Museum after offers in the autumn of 2001 to buy 99 Orchard at below-market rates were refused. The 99 Orchard site, the Museum's one-time twin and neighbour, was a newly renovated, inhabited tenement whose street-level restaurant employed Fujianese workers.

According to the *Forward*, a Jewish weekly newspaper, the Museum's expansion was "part of a plan to secure a partnership with the National Parks Service-administered Statue of Liberty and Ellis Island, bringing thousands more people to the museum" (Keys, 2002). Abram (2002) tallied the numbers: "Ellis Island and the Statue of Liberty received 5 million visitors last year. The acquisition of 99 Orchard Street will enable the Museum to serve over 200,000 people [. . .]" To Abram's dismay, the controversy lent credence to the charge that the Museum's re-creation of historic immigration came at the expense of a contemporary immigrant neighbourhood. This fight had generated significant national news coverage just as its public visibility expanded in Lower Manhattan's post-9/11 landscape. The *New York Daily News* reported:

> Condemnation would put about 20 immigrant workers out of a job at a time when the Asian American Foundation estimates about 1,000 Chinese New Yorkers were left unemployed by the World Trade Center attack. (Kates, 2002: 2)

The Museum appeared to be placing its own interests – getting more people through the door, increasing revenue – over those for whom it purportedly advocated. The *New York Times* noted, under the headline "Your Tired, Your Poor,

Your Building?" that "The renters may not be the tired, poor and huddled masses of yesteryear. But there they are all the same" (Haberman, 2002, B1). The Los Angeles Times captured a local reaction voiced by Community Board 3's (CB3) manager: "They want to create a virtual tenement museum in a neighborhood that already has tenements" (Getlin, 2002: A.14) (see Figure 7.1).

At a raucous Community Board hearing, a 99 Orchard tenant asked, "What are they going to tell the tourists of 99 Orchard Street? This is the history of the people who lived here before we evicted them?" (Sayrafiezadeh, 2003). By mid-2002, the Museum's attempt had garnered enough opposition to prevent it from going forward. All of this took place on an Orchard Street otherwise eerily quiet, since most shoppers were scared away after 11 September 2001. Residents and merchants alike were just becoming aware that the City wanted Lower Manhattan to be restructured as a visitor arts and entertainment area that would also attract new residents (Appleseed in association with The Louis Berger Group, 2004). That Appleseed report, written for the LMDC (2004), described the LMDC's redevelopment strategy as strongly calling for "Supporting cultural activities that reinforce Lower Manhattan's identity as an attractive place to live and work, and at the same time attract visitors to the area" (p. 3). To support this strategy, the LMDC was funding marketing programmes for Lower Manhattan museums (p. 4).

Figure 7.1 W 99 Orchard Tenement Museum laundry

Source: Elissa Sampson. Lower East Side Tenement Museum SNAPSHOT Photo Event, July 20, 2010

Representations: free at last

Locally, the Museum's claim to represent the interests of newer immigrants and migrants seemed vitiated by its lack of direct representation for the area's earlier Chinese and Black residents, or for groups who had arrived in the area post-1935, that is after 97 Orchard's residential closing. In February 2002, the Museum announced:

> In 99 Orchard Street, the Museum will expand its interpretation to include stories of people in the neighborhood before and after those dates [1863–1935] including people from Africa, Latin America and Asia whose stories were not represented in 97 Orchard Street. (Abram, 2002)

Abram had previously hinted that she thought the ruckus was about historic representation. On January 3rd, a local reporter had quoted her as saying:

> If they do end up getting the building, Abrams [*sic*] said the museum would probably re-create tenement apartments inhabited by immigrants on the Lower East Side during the 19th century [. . .]. Since 99 Orchard St. has been nearly completely gutted and little of the original interior remains [. . .] any new exhibit would be interpretations based on the history of immigrant communities in the entire neighborhood. 'We would be free to interpret families who lived in the neighborhood but not necessarily in that building.' (Jensen, 2002)

That 99 Orchard, unlike 97 Orchard, was not a time capsule, 'freed' the Museum to interpret neighbourhood groups in its new building, thus at last achieving its goal of representing Black and Chinese 'heritage' groups. Yet, as seen in its land-marking filing and other documents, the scheme of limiting 97 Orchard to the interpretation of building residents was not inevitable. As discussed, after the Museum moved into 97 Orchard in 1988, it shed its original scheme focusing on six groups and eventually decided to interpret only the stories of residents found to have lived in its building. Its residential interpretive scheme, with its attendant dates and aura of inevitability, evolved into a set of rules that effectively bounded which groups' stories were told. Yet the Museum's agency in setting its own rules and then declaring them inviolable was invisible.

Gentrifying claims

The Museum's agreement to a broader portrayal of past Lower East Side life was made contingent upon agreement to its proposed expansion. Perhaps in response, the Fujianese community bolstered its opposition with the claim that the Museum's actions would take away housing in an area that didn't have enough. Chinatown's existence and growth had long been perceived as threatened by the impact of gentrification on affordable housing and business space (Lin, 1998, Kwong, 2009). It was thus plausible to depict the Museum's proposed expansion as a reduction of Chinatown's housing stock. Ultimately tenant rights became the issue politicians cited in withdrawing political support from the Museum. Here the need for state protection of tenants, and the need for the protection of small businesses from the state, were both invoked.

The Museum's actions triggered the rare involvement of documented and undocumented Fujianese workers with local organizations, and with business and ethnic associations that brought in communal support from Queens and elsewhere. The support of New York's relatively recent Fujianese community helped rally a local Community Board, Business Improvement District, and district politicians, all of whom desired the backing promised by the most unlikely of Lower East Side bedfellows: landlord and tenant groups.

9/11 as accelerator

Since 2001, a broader globalization of tourism and gentrification has intertwined with Lower Manhattan's museumification and urban 'revitalization' process, thus demonstrating how being in place becomes harder as place hollows out. Those tensions can be seen in the massive state and private investment transforming many Lower Manhattan museums into visitor sites that merge American-themed commemoration and memorialization, whether of immigrants or of 9/11 itself, into a larger destination mix of historic, cultural and entertainment venues. Lower Manhattan sites that touch on immigration include Liberty National Park (Ellis Island, the Statue of Liberty), the Lower East Side Tenement Museum, the Irish Hunger Memorial, the Italian American Museum, the Museum of Jewish Heritage and the rapidly evolving Museum of the Chinese in America. Dependent on state money as well as private/public partnerships and donors, some sites are nationally affiliated and/or run by the National Park Service. Their institutional origins can be seen as an offshoot of earlier ethnic pride movements that owe a largely unacknowledged debt to the Black Pride movement. While some sites place greater emphasis on earlier antecedents, all tell a tale of hard New York beginnings that shape what becomes American.

The City's post-9/11 re-envisioning of Lower Manhattan proposed it as a potentially attractive destination for upscale living and recreation that needed to be re-made on a far grander scale if its 'revitalization' was to attract new businesses, residents and visitors. One important player in this process was the Lower Manhattan Development Corporation (LMDC). As a subsidiary of the state-run Empire State Development Corporation (ESDC), the LMDC was put in charge of the Federal 9/11 monies given to New York State for remedying 9/11 impacts and attending to memorialization. LMDC related documents also show a deep commitment to remedying growth-inhibitors specifically identified as deficits in transit, parks and cultural amenities.[4] Boosting museums in Lower Manhattan's erstwhile harbour area was seen as key to transforming it into a cityscape that would speak to New York's continued centrality to American narratives of finance, immigration, and the arts. By 2011, the LMDC's new grants were seen as supporting a successful strategy.

'The revitalization of Lower Manhattan is a testament to the resilience of New Yorkers in the face of great tragedy,' said Empire State Development President, CEO & Commissioner Kenneth Adams. 'These grants will go a long way toward redeveloping Lower Manhattan into a premier center for culture, commerce, and tourism.' (Bloomberg, 2011)

New York promoted institutions such as museums as particularly good investments given the cultural and other damage wrought by the World Trade Center attacks. While governmental delays and controversy plagued the architectural and memorial plans for 'Ground Zero' and its environs, cultural and infrastructure investments attractive to visitors and new residents paved the way for real estate money, some of it from overseas investors. Out-of-scale construction of luxury condo buildings, hotels and residential conversions took off particularly after 2012, despite Hurricane Sandy's damage. Long-term residents increasingly find their neighbourhoods unrecognizable and unaffordable in a reconfigured Lower Manhattan.

What is designated as locally historic is increasingly marketed as an encapsulated space for visiting, dining, living or investing. A heady mix of new and old sells: "Downtown had the highest number of closed sales, highest median price, and highest average price per square foot of any neighborhood in Manhattan during First Quarter 2016" [$2,155 Average PPSF, +30%] (Corcoran Group Real Estate, 2016: 18). Meanwhile the continuing pressure on housing and small ethnic businesses has made it increasingly hard for those who are already there to remain. Although Chinatown received relatively little in the way of post-9/11 allocations, funds were allocated for a large visitor kiosk (Chan, 2009) as well as for a new Museum of the Chinese in America building. Census figures confirm that Chinatown – Lower East Side continues to lose its migrant population to outer borough neighbourhoods (AALDEF, 2013).

A museum landscape

History museums often yoke national memorialization to contemporary themes. New York's own history as a former national capitol long associated with finance capital is linked to its prominence as the U.S.'s biggest harbour and contemporary port of immigration. Thus the city's history is presented to both tourists and New Yorkers as an American story offering a presumptively assumed 'shared' past. Newer stories of shared American identity and immigrant acculturation are folded into older founding national narratives of place with differences elided as the travails of additional groups are recast to fit into an older acknowledged pantheon. This emphasis on past immigrant hardship and achievement ostensibly provides inspirational support for diverse depictions of present immigration.

As themed museums, they enshrine past immigrant adversity and arguably domesticate it through distilling bittersweet narratives of survival, resilience, and immigrant contributions. Their sites are now proximate to well-appointed streets and river promenades replete with stores, museum shops, street art, galleries, boutiques, food, drink and entertainment. Among other things, a channelled consumption of history is promoted through visitor tours of newly renovated or constructed spaces and those of kindred institutions.

Most strikingly, this commemoration of earlier immigrants is now offered in tandem with memorialization dedicated to the dead of 9/11. In a newly shining, rebuilt Lower Manhattan sites of mourning and national identity are presented in an insistent narrative of rebirth and revival that features America, New York, and Downtown (with its amorphous boundaries) as coming back stronger than ever. Not only has Downtown 'recovered' from 9/11, but as 2008's economic woes recede from public vision the anger against Wall Street is seemingly invisible as more and more designer architecture fills Downtown's remaining premium spaces.

Historically ethnic

Today what remains of urban centre-city ethnicity provides local colour and com-modified authenticity. Gentrification in the Lower East Side has its own particular aesthetic – it is not just a question of Smith's (1996c) revanchist city cleaning things up for the benefit of visitors. As Chris Mele (2000) and Sharon Zukin (2010) have shown, the neighbourhood's hip grunge ethos combined with the authenticity of its older tenement buildings and immigrant/migrant communities can form part of its attraction as it goes upscale. A small, renovated two-bedroom, tenement apartment rents for $3,000–$4,500 a month.

The gentrified tenements of the Lower East Side, Little Italy and what is now called NoLita (North of Little Italy) offer possibilities more akin to those found in certain historically poor but now chic European city centres. The grunge authen-ticity of the Tenement Museum and its environs gets produced in ways that make the now 'iconic' Lower East Side attractive even as there is less of it to be found. That the neighbourhood has shrunk is reflected in the relatively hip offerings found on pedestrian-friendly Orchard Street that seek to marry ironic nostalgia to perceived authenticity. That reduction of space has proportionately increased the ability of the Tenement Museum to represent a disappearing place called the historic Lower East Side.

As a local affordable housing provider noted, spatial reduction is reflected demographically. Studies of Community Board 3 census tract data by the Two Bridges Council (2011) divided it into four neighbourhood areas: the East Village, the Lower East Side, and Two Bridges (all part of the historic Lower East Side) and Chinatown (see Figure 7.2). During a period where there had been almost no change in the total population of CB3, the Council concluded that CB3 had changed regarding "Ethnicity and Foreign Born Residents." Based on a comparison of the 2000 Federal census to 2009 American Community Survey (ACS) data:

[T]he East Village contained the largest percentage of residents that identified as white at 65%, a 9% increase from 2000 [. . .] In all four neighborhoods, the Latino population decreased by double-digit percentages over the last decade, while NYC experienced an 8% rise in the total Latino population (5).

These particular shifts tie ethnicity to gentrification and a lack of affordable housing stock. This data is understated inasmuch as it stops at 2009. While the

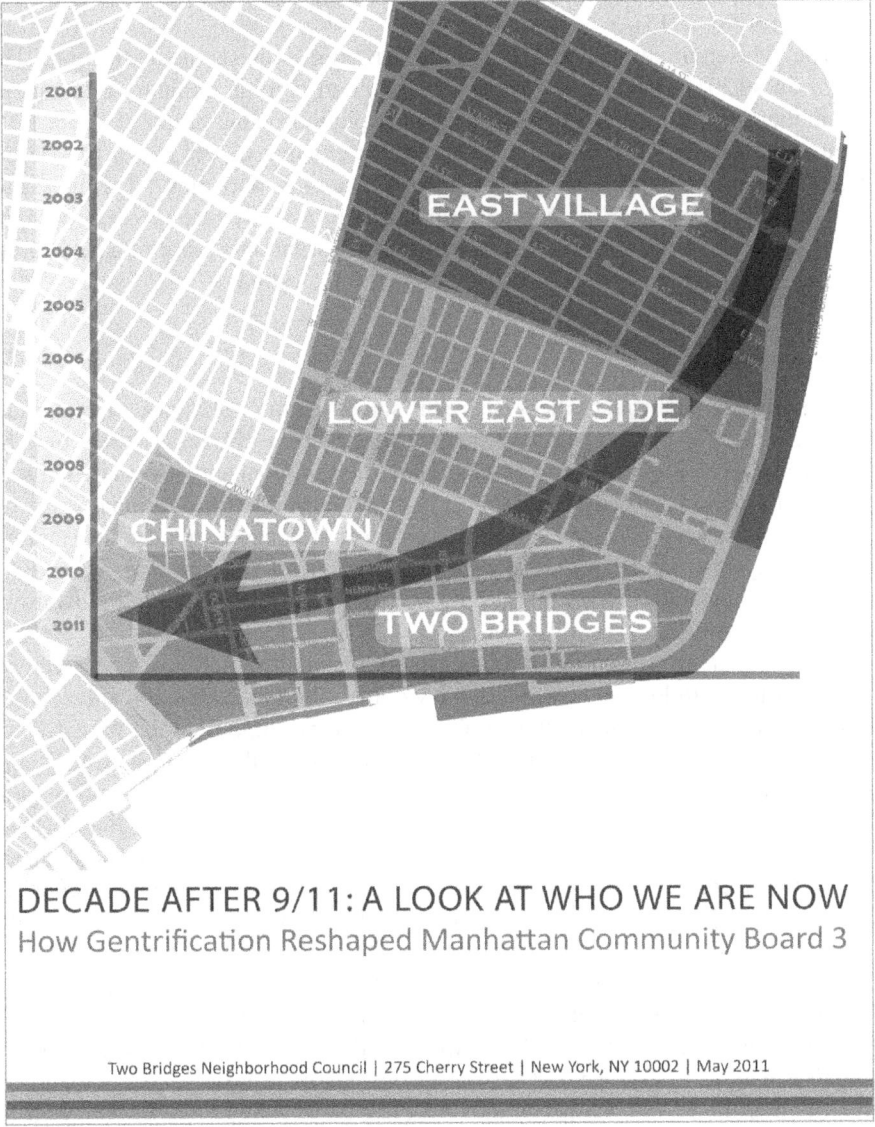

Figure 7.2 Two Bridges Neighborhood map

Source: Cover, Two Bridges Neighborhood Council Report

2009 ACS data showed that 56 per cent of Chinatown's residents were foreign-born, the boundaries of that once growing area have subsequently receded. New Fujianese migrants are far likelier to choose Brooklyn's less expensive Sunset Park area (AALDEF, 2013). Airbnb stats illustrate a related shift which favours visitors: "The East Village led the list with a whopping 28 per cent of its units going as illegal hotel rooms on the popular home-sharing site, according to an analysis from New York Communities for Change and Real Affordability For All" (Fermino, 2015).

The Museum attracts tourists while its blog protests gentrification. The Museum's (2010b) desire to convey a message of coming home to a revitalized neighbourhood can be seen in its website depiction of the area: "Explore [. . .] You can easily spend a morning, afternoon or evening on the Lower East Side. It's a thriving neighborhood, full of shops, restaurants and sites." This message, reassuring visitors that a historical destination offers newly safe access to past and present, is not discrete from a commodification reliant on a story of a new influx having once again tamed the urban frontier (Smith, 1996a).

State investment

Commodification is most salient where the stakes are highest: that is, in real estate dealings. Not-for-profit institutions need visitors' money both as a direct funding source and as an inducement to crucial donor, institutional and governmental support. Although in 2008 the Lower East Side was listed by the National Trust for Historic Preservation as one of the eleven most endangered places in America, real money has not come into the neighbourhood for preservation or even for affordable housing per se.[5] Instead, money has gone into remaking Downtown into a destination.

The Tenement Museum is at the periphery of the 9/11 funding zone. Even so, it is instructive to look at how that money helped fund its expansion and acquisition of new buildings. While the LMDC most notably supported the World Trade Center Memorial Foundation, LMDC allocations show that it has generally supported the development and expansion of Downtown cultural institutions so as to attract more visitors and residents (see Figure 7.3). For instance, the Museum received a $1,000,000 LMDC (2006) grant to fund its long-planned Irish apartment (the Moores), its downstairs (German) Schneider's Saloon, and a courtyard renovation that spurred the development of two new 97 Orchard Street tours (Kaysen, 2006).

The ESDC had previously supported the Tenement Museum's bid for 99 Orchard Street through its exercise of eminent domain on the Museum's behalf. As described on the LMDC's (2016) website, the "ESD mission is to provide the highest level of assistance and service to businesses in order to encourage economic investment and prosperity in New York State." This same LMDC website which has featured photos of new high-end luxury developments for a number of years also offers a description of its "Arts, Education and Tourism Advisory Council" formed in 2002.

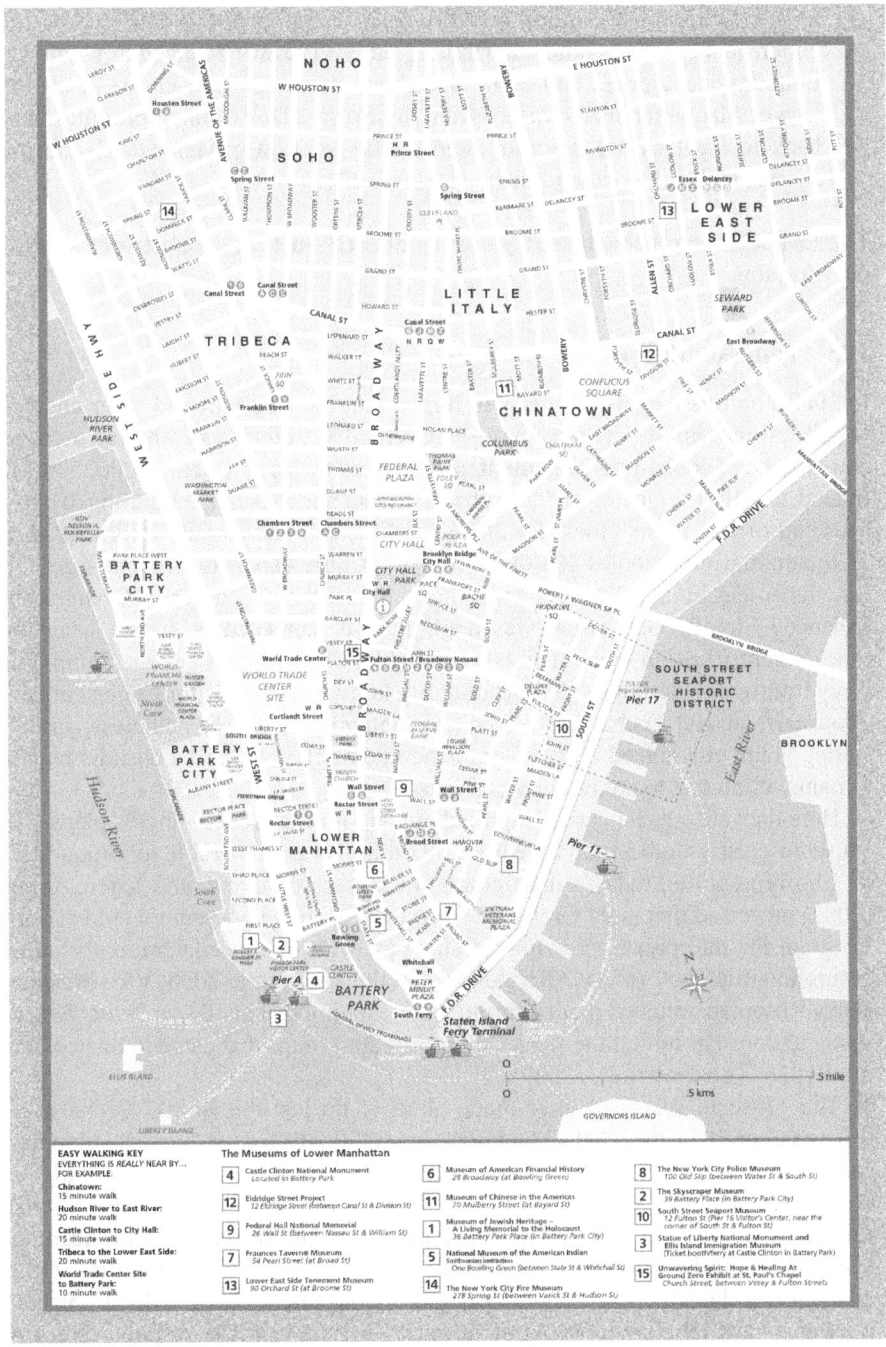

Figure 7.3 LMDC map

Source: Lower Manhattan Development Corporation. Map of Funded Post 9/11 Sites in Lower Manhattan 2004, p. 2, www.renewnyc.com/content/pdfs/History_Map_pg2.pdf

Arts, education and cultural institutions generate economic activity, create a better quality-of-life, and contribute greatly to Lower Manhattan's renewal. Encouraging tourists and cultural enthusiasts to take advantage of all that Lower Manhattan has to offer is a high priority. This Council provides feedback to [. . .] ensure the continued vitality of Lower Manhattan's cultural assets and [. . .] to expand them. (LMDC, 2016)

The LMDC's involvement thus brings us directly back to the post-9/11 story of the Museum.

103's representations

In 2002, the Los Angeles Times had hinted that the Museum had other choices: "While Abram insists she must have 99 Orchard St., others ask why the museum couldn't have looked for tenement property elsewhere" (Getlin, 2002: A.14). Indeed, shortly after losing on 99 Orchard, Abram bought a larger building to house Museum offices on its block. By 2007, the Museum had already fully embarked on the quiet acquisition of 103 Orchard – a far more expensive, much larger Old Law tenement building conveniently located at the busy corner of Orchard and Delancey Streets. Opposition was negligible: it happened at a time when the block's gentrification seemed all but inevitable. The Museum's Sadie Samuelson Levy Immigrant Heritage Center (popularly known as the Visitor Center) opened downstairs at 103 Orchard Street on 20 September 2011. Nonetheless, smooth sailing is not guaranteed *a priori* since the fate of 103 Orchard's upstairs tenants remains of interest to community activists.

Here then are some answers supplied by the Museum concerning its plans. The Museum clearly understood by early 2002 that it was vulnerable to the question of who represented today's immigrants. In a press interview and letter, Abram (2002) promised that at 99 Orchard Street, the Tenement Museum would "[f]ully integrate the interpretation of immigrants past with that of contemporary immigrants and migrants" (p. 2). After Abram left the Museum in 2008, Vice President Barry Roseman softened that commitment by stating that at 103 Orchard, "we are going to consider telling the stories of immigrant groups that came to this country after 1935" (Lo-Down TV, 2009).

The Museum seems to have made plans to do just this at 103 Orchard. By 7 September 2011 the LMDC had awarded $400,000 for "Exhibit development for three historic apartments; a Jewish Refugee Apartment, a Chinese Immigrant Apartment, and a Puerto Rican Migrant Apartment" (Bloomberg, 2011). A visitor orientation movie now comfortably juxtaposes more recent Dominican and Cantonese immigrant stories with those of Jewish and Italian families. Other changes implemented by new Museum management have helped improve relations with local groups. As it becomes officially more responsive to the neighbourhood's remaining ethnic communities, the Museum appears simultaneously to be more explicit about the area's Jewish past. Moreover, its immigrant theme if anything is becoming more universalized into an American narrative, all suggesting that tensions between these various claims for articulation remain contingent rather than absolute.

Yet this contingency contains distinct echoes of the Museum's past. The Museum's 2007 acquisition of 103 Orchard was also quietly supported by another ESDC entity that obligingly issued $15,000,000 in triple tax-exempt financing bonds to:

> Enhance the quality of cultural life in the City. Currently, the Museum has approximately 125,000 visitors per year. Due to space constraints, it also turns away 10,000–12,000 visitors per year. The availability of tax-exempt bonds will allow the Museum to admit more visitors, as well as to expand the range of immigrant experiences interpreted by the Museum to include Chinese, Dominican and African-American. (New York City Industrial Development Agency (NYCIDA), 2007: 4)

That New York City's cultural life would be enhanced by the addition of 12,000 visitors who would be able to see representations of "Chinese, Dominican and African-American life" suggests the continued interpretive centrality of ethnic households for a Museum that clearly envisioned different groups from those prioritized in its later 2011 LMDC grant. Both the Museum and state entities saw the value of directly tying economic support to a broader representation of immigration. The issue of representation has haunted the Museum since its very beginnings and its continued prioritization has remained both an issue and a focus.

The $15,000,000 raised from 2007 NYCIDA tax-free bonds was used primarily to pay for various Museum building mortgages; it was not used for upstairs new apartment exhibits (NYCIDA, 2007). To raise money for 103 Orchard Street's build-out, the Museum had to run a large, successful capital campaign that went ahead despite the economic downturn. According to the same NYCIDA document, New York City had agreed to directly contribute money much as it had with 99 Orchard, in this case $2,000,000 (p. 2). By 2011, the Museum had welcomed 173,000 visitors and was growing along the lines earlier envisioned by Abram, just in time for 103 Orchard's 2012 opening of its street-level Museum Shop.

By 2014, the Museum had 210,000 visitors. To accommodate even more visitors, additional private, city and state funding was made available to further 103 Orchard Street's 10,000-square-foot expansion at a projected cost of $8,000,000 (Iftikhar, 2014). Construction indeed started with a project kick-off held on September 15, 2016 for a $12,000,000 project (Litvak, 2016).

Updating a formula for a new tenement

Major cultural institutions are touted as economic engines for Lower Manhattan even if not-for-profit engines do not pay taxes. But engines need constant feeding and care. Unsurprisingly, Federal monies and recognition have been critical to the Museum's growth. On 13 December 2013, the House bill (HR1846) for 103 Orchard introduced by Congressperson Nydia M. Velázquez was approved; it was signed into law at the end of 2014. Velázquez's (2013) press release stated:

'Whether it is Chinatown, Little Italy, or émigrés from Germany, immigration has made New York a more vibrant place, offering invaluable economic and cultural contributions, while shaping the city's identity,' said Velázquez (D-NY). 'The LES Tenement museum [*sic*] honors these communities and pays tribute to the challenges they overcame when arriving in a new nation and city.'

[. . .] Morris Vogel, President of the Museum [stated] 'We'll use the site to tell the stories of real Puerto Rican and Chinese families – and Jewish Holocaust survivors – who lived and worked toward the American dream on the Lower East Side. Their stories are a critical chapter in our nation's vibrant melting pot history.'

An older melting pot formulation that ties overcoming adversity to achieving the American dream has returned with a celebratory as well as a commemorative twist. Validation of a past mass migration to the Lower East Side now incorporates a more current era of migration. In that reformulation, new groups are highlighted. Migrants, immigrants, émigrés, and survivors are now all eligible to celebrate a correctly calibrated amount of difference in being American.

Plentiful ironies abound, not the least of which is that it will have taken almost thirty years to open the Museum's apartment doors to a Chinese household. The Wong family apartment will be interpreted to represent post-1965 Chinese immigration, allowing the Museum to explain to visitors that Chinese immigration was banned by Congress in 1882. For the first time, 'internal' (rather than cross-border) migrants are not merely supplemental as the inclusion of Puerto Ricans shows, although Blacks and Dominicans did not make the final cut.

The Museum should have little trouble finding the remaining funding for its expansion. While its buildings have succeeded in outliving their erstwhile immigrant neighbourhood by telling 'their' stories to visitors, the Museum's assiduous fundraising efforts are integral to its story. The Museum's messages touch upon ethnic immigrants and American identity in ways that governmental, institutional, foundation and individual donors, and most recently, local real estate developers find congenial. Among other things, an acknowledgment of past 'New Immigrants' provides a model for acculturating contemporary ones.

Indeed, the introduction of "real Puerto Rican and Chinese families – and Jewish Holocaust survivors" at 103 Orchard represents a clever expansion of 97 Orchard's ethnic household formula into the post-War era, a sure sign that those years are now viewed as history by a new generation of anticipated visitors. At 103 Orchard residential apartments will remain the prioritized space for telling the personal 'real' stories of ethnic households associated with its building. Likewise, their American dream will require showing how ancestors surmounted initial hardship as a pre-requisite for the eventual success of descendants.

Upscale aesthetics

103 Orchard did mark an important change: its aesthetic is not preservationist. The bottom floors were 'reskinned' in glass to incorporate over 10,000 square feet for its large Museum Shop, Visitor Center, and conference and classroom space. Since it was expressly bought to be extensively renovated for those purposes, 103 Orchard was not landmark material despite Congress's post-hoc imprimatur. Nevertheless, architect Nicholas Leahy did in fact attempt to retain much of the building's past look and feel on its upper floors and basement. When asked in an interview about the choices made in refurbishing the building, Leahy replied that "It's the idea of creating a sort of a portal" (Hedlund, 2011).

> 'The reality is it generates revenue for the museum,' said David Eng, vice president of public affairs for the Museum, noting that the original building will be preserved as much as possible, including the tenants living on the top three floors (Hedlund, 2011).

Apparently without intending any irony, Leahy said in another interview: 'It's a great opportunity to keep the regeneration of the Lower East Side going, [. . .] 'You can turn the city into a museum or you can keep it lively' (Bortolot, 2011) (see Figure 7.4).

Gentrification in relation to tourism

As *Crain's* reported, the Museum is increasingly an actor in this escalating neighbourhood drama called gentrification. "Ironically, the museum helped foster the

Figure 7.4 103 Orchard

Source: Elissa Sampson. Exterior street photo of Tenement Museum Shop, 9 August 2015

growth of art galleries, trendy shops and chic bars in the neighborhood. 'It was a catalyst for change,' said Tim Laughlin, executive director of the Lower East Side Business Improvement District." (Agovino, 2014).

As its footprint grows, its presence attracts nearby upscale eateries, boutiques, art galleries, bars, hotels and chic condominiums that erode rent protections as well as a sense of belonging. The Museum's contemporary reconstructions invoke a historic Lower East Side popularly memorialized in novels, photography, and collective cultural memory. Its Museum Shop offers Lower East Side and New York classics alongside souvenir mugs and generic upscale gifts. The Museum's web site, and especially its blog, feature iconic photographs alongside newer ones accompanied by a constant stream of information about past and present.

Although it is far harder today to start a tour by instructing guides to "Introduce LES as immigrant neighborhood past and present," to do so is part of a balancing act for tours that depend on a semi-static built environment. While older tours are set in the context of a larger meta-story of immigrant absorption in the same place over time, newer tours now get to show the 'ravages' of gentrification and partake in food tastings at newer ethnic eateries. Questions of representation and the gaze of privilege are particularly visible in tour scripts that rely on a visible

Tenement Museum Will Hold Annual Gala at Historic Essex Street Market in Spaces Rarely Seen By Public in Over 50 Years
Sold-out event has raised nearly $1 million,
the Museum's most successful Gala ever

What: The Lower East Side Tenement Museum's Annual Gala Celebration

Date/Time: Wednesday April 23, 2014, 6:30 – 9:30 p.m.

Location: Essex Street Market Buildings B and D
               6:30 PM – Building D, 83 Essex Street
               8:00 PM – Building B, 130 Essex Street
Press RSVP: Elizabeth Tietjen, 212-431-0233 x235, etietjen@tenement.org

April 16, 2013, New York, NY — On April 23, the Lower East Side Tenement Museum will hold its annual gala event celebrating the rich history and vibrant future of New York City and the Lower East Side. This year's event has raised $1 million for the Museum, and is nearly sold-out. It is the most successful Gala in the Museum's 25-year history. Held at the Essex Street Market in buildings rarely seen by the general public over the last 60 years, the Tenement Museum Gala may be the last ever held in these historic spaces.

The Museum's 2014 Gala honors Taconic Investment Partners, L+M Development Partners, and BFC Partners; in September of 2013, these three firms were selected by the City of New York to develop the Seward Park Urban Renewal Area, which represents one of most significant urban renewal development opportunities remaining in Manhattan. Now known as Essex Crossing, the project will celebrate the neighborhood's past while promoting cultural energy and economic development for its future.

Figure 7.5 2014 media alert

Source: Press Release, 16 April 2013 [sic], Lower East Side Tenement Museum, "Tenement Museum Will Hold Annual Gala at Historic Essex Street Market"

backdrop of a Fujianese migrant community living in tenements in an area under tremendous housing pressure.

Salvaged landmarks such as 97 Orchard Street assume the unbearable weight of representing a past world in its fullness. The Museum gives every appearance of taking seriously the claim that what it is doing is History. But much like a building, history is not something that stands by itself. Questions of context bring us back not only to the question of how history is produced for visitors in ways that wield considerable ideological and affective heft, but to the related question of how tourism itself is tied to gentrification.

The Museum has grown to serve more visitors and provides those visitors with a compelling sense of access to an immigrant past. Yet that story is at the same time inseparable from the disappearance of local communities and their built environment, especially as the Museum turns to courting large scale real estate developers to enlarge its future. As the *Real Deal* noted, "Several big developers – including Taconic Investment Partners, L+M Development Partners and BFC Partners – [who] are helping to fund an $8 million expansion of the Lower East Side's Tenement Museum [. . .] are also being feted at the museum's fundraising gala" (Samtani, 2013). Indeed, those developers have been instrumental in raising more money.

97 Orchard Street thus appears as a celebratory bid to trump time by capturing a piece of a disappearing historic place. The hollowing out of the Lower East Side's identity is marked by a temporal confusion inherent in the mixed forces of nostalgia, tourist-oriented preservation, real-estate oriented gentrification, and larger capital flows. Nowhere, perhaps, is that confusion more evident than in Museum buildings whose ethnic household narratives are at once fixed in a past era and intended to describe as timeless the neighbourhood in which it is a visitor destination.

Notes

1 According to *Crain's*, over 210,000 visitors came to the Tenement Museum for the five years ended 30 June 2014, greatly adding to the wear and tear on the building (Agovino, 2014). By way of contrast, approximately 7,000 people lived at 97 Orchard Street over a 70-year period from 1864 to 1935. While Ellis Island itself served as a port of entry for over 12 million immigrants from 1892 to 1954, the Ellis Island Restoration receives two million visitors annually.

2 One small example can be seen in governmental documents such as the Lower Manhattan Development Corporation's (2006) "Projects and Programs" press release that explicitly touted "The LMDC has contributed $27 million in grants to 62 cultural institutions in order to enhance and restore the vital cultural life of every Lower Manhattan neighborhood."

3 A bit of a throwback, the press release just describes the Gumpertz family as "German" rather than as German-Jewish.

4 As cited in then Mayor Bloomberg's (2011) press release:

'Vibrant community and cultural institutions are a key ingredient in creating vibrant neighborhoods because they attract residents, visitors, jobs, and private investment,' Deputy Mayor for Economic Development Robert K. Steel said. 'These investments by the LMDC will help expand the depth and breadth of these vital institutions with deep roots in Lower Manhattan, one of the fastest-growing neighborhoods in New York City.'

Also see other LMDC-cited documents pertaining to housing, infrastructure as well as amenities and tourism.

5 New low-income housing, typically at a 10–20 per cent level, is provided as an offset by luxury developers in exchange for zoning bonuses and tax abatements. The neighborhood's biggest development, Essex Crossing, formerly known as the Seward Park Urban Renewal Area, is now being constructed. A Request for Proposal was awarded by the city to three large developers for six large very upscale sites; the hard-fought trade-off was obtaining a high percentage of permanently affordable housing. These developers were honored by the Museum as donors at its million dollars 2013 fund-raising gala. See Figure 7.5.

References

AALDEF 2013. *Chinatown Then and Now: Gentrification in Boston, New York, and Philadelphia*. New York: Asian American Legal Defense and Education Fund (AALDEF).

Abram, R. 1987a. 6/24/1987 Letter from Ruth J. Abram to Dr. Richard Rabinowitz, American History Workshop, THE TENEMENT is Funded. *The Lower East Side Historic Conservancy Files*. Memo on American History Workshop THE TENEMENT ed. New York: Lower East Side Tenement Museum.

Abram, R. 1987b. 9/17/1987 Letter from Ruth J. Abram to Dr. Richard Rabinowitz, American History Workshop. *The Lower East Side Historic Conservancy Files*. Memo on American History Workshop THE TENEMENT ed. New York: Lower East Side Tenement Museum.

Abram, R. 1987c. October 1987 Letter to Dr. Richard Rabinowitz, American History Workshop on NEH Heritage Trail Grant. *The Lower East Side Historic Conservancy Files*. Memo on American History Workshop on NEH Heritage Trail Grant ed. New York: Lower East Side Tenement Museum.

Abram, R. J. 1990a. 1/23/90 Abram Memo to Rabinowitz, Re: AHW Draft Report on the Tenement Museum. *The Lower East Side Tenement Museum Archives*. Memo Reacting to First Draft of Conceptual Plan from Abram and Kahl ed. New York: Lower East Side Tenement Museum.

Abram, R. J. 1990b. 6/26/1990 Memo from Ruth J. Abram to Richard Rabinowitz, American History Workshop. *The Lower East Side Tenement Museum Archives*. 97 Orchard Street ed. New York: Lower East Side Tenement Museum.

Abram, R. J. 1990c. Lower East Side Tenement Museum Report to Trustees, Winter 1990. *The Lower East Side Tenement Museum Archives*. Lower East Side Tenement Museum Report to Trustees, Winter 1990, Box of Strategic Plans and Quarterly Notes to Trustees from 1988; folders 4.1 Annual Reports and Strategic Plans; 4.2 Organizational Schedules and Charts ed. New York: Lower East Side Tenement Museum.

Abram, R. J. 2002. *Statement Regarding the Planned Acquisition of 99 Orchard, Tenement Museum* [Online]. New York. Available: http://web.archive.org/web/20090106043110/http://tenement.org/statement.html [Accessed 20 February 2010].

Abu-Lughod, J. L. (ed.) 1994. *From Urban Village To East Village: The Battle for New York's Lower East Side,* Cambridge: Blackwell.

Agovino, T. 2014. 'Tenement Museum Keeps Memories Alive: The Lower East Side Museum is Boosting its Fundraising', *Crain's New York Business*, 18 September 2014.

Benjamin, W. (ed.) 2002. *The Arcades Project*, New York: Belknap Press, Harvard.

Bloomberg, M. 2011. *Mayor Bloomberg And Empire State Development Corporation Announce $17 Million In Funding To Community And Cultural Non-Profits In Lower Manhattan* [Online]. New York City: NYC.GOV. Available: http://www1.nyc.gov/office-of-the-mayor/news/320-11/mayor-bloomberg-empire-state-development-corporation-17-million-funding-to#/3 [Accessed 10 November 2014].

Bortolot, L. 2011. 'At Tenement Museum, New Look Reflects Lower East Side's Evolution', *Wall Street Journal*, 2 March 2011.

Certeau, M. D. 1980. *The Practice of Everyday Life*, Berkeley, M.A.: University of California Press.

Chan, D. 2009. *Déjà Vu Gentrification: Prospects for Community Mobilization in New York City's Lower East Side and Chinatown in the Post-9/11 Era*. Berkeley, M.A.: University of California Berkeley.

Corcoran Group Real Estate 2016. 'The Corcoran Report, 1Q16 Manhattan'. New York: Corcoran.

Diner, H. 2000. *Lower East Side Memories: A Jewish Place in America*, Princeton: Princeton University Press.

Dolkart, A.S. 1992. 11/17/1992 Memo to Li-Saltzman Concerning the Museum Plan for the Tenement Museum; Faxed Copy From Roz Li to Ruth Abram 11/24/92. *The Lower East Side Tenement Museum Archives*. Memo on Museum Plan ed. New York: Lower East Side Tenement Museum.

Fermino, J. 2015. 'Airbnb Taking Up 1 out of 5 Vacant Apartments in Popular New York City Zip Codes: Study', *The New York Daily News*, 29 July 2015.

Getlin, J. 2002. 'Museum Plan Hits Too Close to Home', *The Los Angeles Times*, 18 April 2002.

Gravari-Barbas, M. 2014. 'Patrimoine, culture, tourisme et transformation urbaine; le Lower East Side Tenements Museum, NY', in Géraldine Djament-Tran, P.S.M.C. (ed.), *La métropolisation de la culture et du patrimoine*, Paris: Editions Le Manuscrit.

Haberman, C. 2002. 'Your Tired, Your Poor, Your Building?' *The New York Times*, 13 February 2002, p.B1.

Hackworth, J. 2002. 'Postrecession Gentrification in New York City', *Urban Affairs Review*, no 37, 815–843.

Hedlund, P. 2011. *Lower East Side Tenement Museum to Open New Visitors' Center* [Online]. New York: DNAinfo.com. Available at: www.dnainfo.com/20110302/lower-east-side-east-village/lower-east-side-tenement-museum-open-new-visitors-center [Accessed 6 June 2011].

Iftikhar, B. 2014. 'Tenement Museum Nails Down $8M Expansion', *Crain's New York Business*, 13 January 2014.

Jensen, J. 2002. 'State Seeks Building To Expand Tenement Museum', *The Villager*, 3 January 2002.

Kates, B. 2002. 'Immigrants Museum vs. Locals, Lower East Side Divided', *The New York Daily News*, 28 April 2002.

Kaysen, R. 2006. 'Arts Groups Cash in on L.M.D.C. Cash', *Downtown Express*, 10–16 March 2006.

Kessner, T. 1992. 'The Immigrant Ghetto as Symbol and Community', *The Lower East Side Tenement Museum Archives*. Revised Draft ed. New York: Lower East Side Tenement Museum.

Keys, L. 2002. 'Immigration Museum Called Bad Neighbor In Expansion Battle'. *The Jewish Daily Forward*, February 6 2002.

Kirshenblatt-Gimblett, B. 1998. 'Ellis Island', *Destination Culture: Tourism, Museums, and Heritage.* Berkeley, London England: University of California Press.

Kugelmass, J. 2000. 'Turfing the Slum: New York City's Tenement Museum and the Politics of Heritage', in Hasia, R. Diner, Jeffrey Shandler, Beth S. Wenger (eds), *Remembering the Lower East Side: American Jewish Reflections.* Bloomington: Indiana Press.

Kwong, P. 2009. 'Ask About the Gentrification of Chinatown', *The New York Times*, 14 September 2009.

Lin, J. 1998. *Reconstructing Chinatown: Ethnic Enclave, Global Change*, Minneapolis, London, University of Minnesota Press.

Litvak, E. 2016. 'Tenement Museum Officially Kicks Off Expansion Project This Morning', *The Lo-Down*, 16 September 2016.

Lo-Down TV 2009. 'Tenement Museum's Visitor Center Renovation, Interview with Barry Roseman', in Lower East Side Tenement Museum 103 Orchard Street Video (ed.). New York: www.thelodownny.com.

Lower Manhattan Development Corporation (LMDC). 2004. (Appleseed In Association With The Louis Berger Group, I.) 'The Lower Manhattan Development Corporation's HUD-Funded Projects and Programs: Leading Economic Revitalization in Lower Manhattan'. New York City: Lower Manhattan Development Corporation.

Lower Manhattan Development Corporation (LMDC). 2006. *Projects and Programs* [Online]. New York City: NY.GOV, New York State. Available at: www.renewnyc.com/ProjectsAndPrograms/Cultural_Enhancement_Funds.asp [Accessed 10 November 2014].

Lower Manhattan Development Corporation (LMDC). 2016. *Renew New York: The Plan for Lower Manhattan* [Online]. New York City: NY.GOV, New York State. Available at: www.renewnyc.com/overlay/ThePlan [Accessed 9 June 2016].

Mele, C. 2000. *Selling the Lower East Side: Culture, Real Estate and Resistance in New York City*, Minnesota: University of Minnesota Press.

Museum: The Lower East Side Tenement Museum 1994. 'Urban Pioneers Slide Show – Script (attributed to Suzanne Wasserman) Slide Show – Urban Pioneers', Written 4/12/94; Revised 4/26/94. *LESTM Programs: Urban Pioneers Slide Show.* Slideshow script ed. New York: Lower East Side Tenement Museum.

Museum: The Lower East Side Tenement Museum. 2002. 'First Permanent Sweatshop Exhibit in the U.S. Opens at the Tenement Museum: The Levine Family Home & Garment Shop', in Museum, L. E. S. T. (ed.) *Press Release.* New York: Lower East Side Tenement Museum.

Museum: The Lower East Side Tenement Museum. 2010a. *A Landmarked Building, a Groundbreaking Museum* [Online]. New York: Lower East Side Tenement Museum. Available: http://web.archive.org/web/20100915120952/http://www.tenement.org/about.html [Accessed 19 November 2010].

Museum: The Lower East Side Tenement Museum. 2010b. *Visit: Explore the Lower East Side* [Online]. New York: Lower East Side Tenement Museum. Available: http://web.archive.org/web/20101128165026/http://tenement.org/guider.html [Accessed 28 October 2010].

New York City Industrial Development Agency (NYCIDA). 2007. *New York City Industrial Development Agency Project Financing Proposal Meeting of September 11, 2007 Lower East Side Tenement Museum* [Online]. New York City: New York City Industrial Development Agency (IDA). Available: www.goodjobsny.org/sites/default/files/docs/idaboardbook_part_1_of_3_september2007.pdf [Accessed 10 November 2014].

Patraka, V. M. 1996. 'Spectacles of Suffering: Performing Presence, Absence, and Historical Memory at U.S. Holocaust Museums', in Diamond, E. (ed.) *Performance and Cultural Politics*, London and New York: Routledge.

Sampson, E. J. 2014. 'Moral Lessons from a Storied Past in New York City', in Hannam, M.M.A.K. (ed.), *Moral Encounters in Tourism*. First ed. Farnham, Surrey, England ` and Burlington VT, USA: Ashgate in association with the Geographies of Leisure and Tourism Research Group of the Royal Geographical Society with the Institute of British Geographers.

Samtani, H. 2013. 'Industry Bigwigs Bankroll LES Museum Expansion', *The Real Deal*, 16 December 2013.

Sayrafiezadeh, S. 2003. 'The Fight Over 99 Orchard Street', in Beller, T. (ed.) *Mr. Beller's Neighborhood*. Available at: http://mrbellersneighborhood.com/2003/09/the-fight-over-99-orchard-street [accessed 01 Feb 2017]

Smith, N. 1996a. 'Introduction: Class Struggle on Avenue B: The Lower East Side as Wild Wild West', *The New Urban Frontier: Gentrification and the Revanchist City*, London: Routledge.

Smith, N. 1996b. 'Part I: Towards a Theory of Gentfication: Local Arguments', *The New Urban Frontier: Gentrification and the Revanchist City*, London: Routledge.

Smith, N. 1996c. 'Part III: The Revanchist City', *The New Urban Frontier: Gentrification and the Revanchist City*, London: Routledge.

Velázquez, N. M. 2013. *House Passes LES Tenement Museum Bill* [Online]. Washington, D.C.: House of Representatives. Available at: https://velazquez.house.gov/media-center/press-releases/house-passes-les-tenement-museum-bill [Accessed 20 November 2014].

Zukin, S. 2010. *Naked City: The Death and Life of Authentic Urban Places*, Oxford, New York: Oxford University Press.

Part III

Who are the tourism gentrifiers?

8 The sharing economy and its role in metropolitan tourism

Natalie Stors and Andreas Kagermeier

Introduction

The 'sharing economy' has been evolving rapidly in recent years, with Airbnb becoming one of the largest actors within this business segment. However, in addition to pure market share, the Californian online platform for peer-to-peer room and apartment rentals has also sparked a large controversy on the nature of sharing and how it affects the metropolis. The latter line of discourse in particular will be the central element of the following chapter, focusing on the way in which Airbnb influences the development of residential neighbourhoods in Berlin.

Within the wider framework of this book, which intends to illuminate the links between gentrification and tourism, this chapter focuses on Airbnb as a new stakeholder in the hospitality industry and its implications on urban transformation processes in Berlin. As gentrification in general is a multi-dimensional and highly complex field of research, the results presented here focus on stakeholders' perspectives. We aim to explore what collaborative consumption in tourism means to "explorer tourists" (Griffin, Hayllar and Edwards, 2008: 55) seeking authentic experiences "off the beaten track" and outside the "tourist bubble" (Judd, 1999; Maitland and Newman, 2009). The empirical basis consists of quantitative online and face-to-face questionnaires among the target group of Airbnb guests, as well as in-depth qualitative interviews with Airbnb hosts and guests in Berlin.

This research approach makes it clear that we have concentrated our empirical work primarily on the reasons and motivations of hosts and guests who participate in sharing economy businesses. As a result, the following paragraphs illuminate the motives of participants in online sharing platforms as well as the experiences of both the demand and supply side made by using Airbnb. The empirical results provided below are lastly discussed against the background of current urban developments in Berlin, such as gentrification, and the government's initiatives intended to tame the rapid rise of such room and apartment 'sharing' platforms.

The sharing economy in gentrified neighbourhoods and the role of Airbnb

Ever since Rachel Botsman and Roo Rogers's (2011) *What's Mine is Yours – How Collaborative Consumption is Changing the Way We Live* became a bestseller, the 'sharing economy' has become a buzzword in current debates in society. Originally regarded as a result of economic decline following the financial crisis in 2008–2009 (Heinrichs and Grunenberg, 2012: 2), today's connotation has shifted so that the term is used in many contexts. These range from discussions about collaborative consumption supporting environmentally friendly practices – in line with the sustainability paradigm – to criticism of capitalist consumption patterns and self-expression as a post-materialistic lifestyle. Different factors drive this development. Above all, the internet and its function as an enabler and facilitator of the matchmaking process between the demand and supply side of goods and services represents the heart of the sharing economy (Linne, 2014: 9). For a long time, high transaction costs and a lack of critical mass inhibited the resale and reuse of second-hand products or products that are used only temporarily. Constant access to the mobile internet, together with the emergence of large commerce platforms such as eBay, provided the basic conditions required to make the sharing economy and its sub-industries accessible and manageable for large parts of society (Behrendt, Blättel-Mink and Clausen, 2011).

This boom was not only supported by technological transformations, but also in participants' value systems – particularly in trend-sensitive and trend-responsive environments. Changing values towards post-materialistic positions play a similar role here, as people's awareness of sustainability issues increase. The blurring of a previously clear differentiation between the producer and the consumer and the resulting hybrid form of the "prosumer" (Surhone, Timpledon and Marseken, 2010) was not a new phenomenon of the sharing economy. This has been discussed in depth, particularly in tourism, mainly with regard to the role played by consumers in co-creating the tourist experience (Günther, 2006: 57; Kagermeier, 2015: 57 *et seq.*; Pappalepore, Maitland and Smith, 2014). Along these lines, Nora Stampfl has asserted that, "Sharing is nothing new; it has always been part of human co-existence" (2014: 13; author's translation) and the same applies for sharing activities in tourism.

Consequently, in this chapter the sharing economy is not considered to be a fundamental paradigm shift. Instead, it is understood as an evolutionary development of existing societal and behavioural transformations, which is certainly being accelerated by the aforementioned multi-dimensional shift in values. Due to the leading role played by the internet and the wide range of social media options available, these transformations have gained a previously unknown dynamism with unforeseeable ultimate consequences. Considering the central driving forces behind sharing offers in tourism, it can be assumed that the search for "authentic" visitor experiences (Gilmore and Pine, 2007) may play a major role. For a long time, visitors have been yearning for off-the-beaten-track experiences outside the confines of the tourist bubble, particularly in city tourism (Judd, 1999; Freytag, 2008; Maitland and Newman, 2009; Stors and Kagermeier, 2013; Stors, 2014).

Airbnb seems to fulfil these needs. The internet platform is one of the major actors in the sharing business and probably the one with the highest impact on the tourism industry. The company is a 2007 San Francisco start-up with an innovative internet-based business model and disruptive potential (Guttentag, 2015). The online platform offers the possibility for ordinary people to rent out their homes as accommodation for visitors. As a result, this new option to leave the tourist bubble and to stay at a private person's place opens up spaces for tourists that were previously mainly used by the local population. In addition, locations with the highest Airbnb density, such as Prenzlauer Berg, Kreuzberg and Neukölln in Berlin (Skowronnek, Vogel and Parnow, 2015, see Figure 8.1 as well) are often hit by large-scale urban transformation processes such as gentrification in general (Holm, 2013: 175; Krajewski, 2013: 26).

So-called 'explorer-tourists' searching for contact and interaction with the local population appreciate this specific spatial pattern. According to Pappalepore *et al.* (2014), people visit these creative urban areas in order to accumulate or display their cultural capital. They have similar preferences as the group of pioneers that live in such locations (Krajewski, 2004) and they frequent the same cafés, restaurants, second-hand stores, and local art markets as well as the nightlife infrastructure. All in all, the tourists' way of producing and consuming the urban space is very similar to the one observed by user groups that are generally characterised as pioneers and gentrifiers, so that the behaviour of both groups can be described as 'conspicuous consumption' (Beauregard, 1986). Consequently, the various interests of the illustrated actors supplement each other. The city users, regardless of being a long-term inhabitant, a temporary migrant or a weekend visitor, likewise co-produce the urban experience (Edensor, 2001; Pappalepore *et al.*, 2014). With a particular regard to the urban infrastructure, we argue that the additional demand created by the visitors – either from within or outside the city – makes the supply of certain goods and services in the neighbourhood just profitable enough. It also may happen that local businesses change their product range towards the visitors' desires due to these persons' elevated purchasing power. This means the loss of a local daily supply of goods and services for the neighbourhood's long-term inhabitants, which has been currently described as retail gentrification (Cócola Gant, 2015; Zukin *et al.*, 2009). In some specific cases, the perceived social carrying capacity has already been exceeded by large influxes of tourists to certain areas. Residents feel disturbed by party and nightlife noise, close to clubs but sometimes also within Airbnb apartments, and rubbish lying on the streets. They then no longer perceive the visitors as a positive element within their living environment. This situation has occurred in some parts of Berlin, such as Warschauer Brücke and Simon-Dach-Straße (Bezirksverordnetenversammlung Friedrichshain-Kreuzberg 2014), where political initiatives have already taken place to steer the immense tourism and nightlife economy demand in a more socially responsible direction (*Der Tagesspiegel*, 2015; *Berliner Morgenpost*, 2016); a later section of this chapter looks at these concerns.

Airbnb in Berlin and its implications

Based on the outstanding position of Airbnb in Berlin in comparison to other German cities (Kagermeier, Köller and Stors, 2016: 72 *et seq.*), it seems reasonable to primarily focus on Berlin to analyse Airbnb's implications for the city's housing market in particular and for the interplay between tourism and gentrification in general.

The scope of Airbnb in Berlin

At the end of 2014, official figures about the volume of the Airbnb business in Berlin were released for the first time. According to a study that Airbnb commissioned and that the GEWOS institute conducted, the figures illustrated that within 12 months a total of 13,802 apartments were booked via Airbnb; 69 per cent (9,267) of these were entire apartments and 30 per cent (4,193) single rooms. GEWOS (2014) then calculated the ratio between the 9,467 Airbnb flats booked in Berlin within the one-year timeframe and the city's total housing stock of 1,883,161 flats in 2013 (Amt für Statistik Berlin Brandenburg, 2016b). The outcome was that only 0.5 per cent of all apartments in Berlin have ever appeared on the Airbnb webpage, regardless of being let regularly or just one single time. They further noted that entire homes rented for more than 120 days a year – a number that indicates a professional type of renting and the dedication of the apartment to a holiday home – accounted for only 0.06 per cent of all apartments in Berlin (GEWOS, 2014).

Besides this Airbnb study, another completely independent research project was conducted by the University of Applied Sciences Potsdam. Students at the Faculty of Design documented all available Airbnb offers in Berlin that were online on one single day in January and again in February 2015. The data was accessed via Airbnb's API on 11 January 2015 and 25 February 2015 (Skowronnek *et al.*, 2015). They found about 11,700 offers, which include fully let apartments, private and shared rooms. If only entire flats are taken into account, the number amounts to 7,714 flats that were online on one single day in Berlin, which equals about 0.4 per cent of all apartments (Skowronnek *et al.*, 2015). The students further found a total of 34,418 offered beds, which is on average 2.9 beds per Airbnb flat.

These figures provide an initial quantitative approach to understanding the extent of Airbnb use in Berlin. They show that about the same number of apartments are listed on the Airbnb webpage as the number of flats built annually since 2013 (Investitionsbank Berlin, 2015a: 36). Particularly against the background of current urban developments in Berlin, including the increasing housing shortage (Holm, 2016c) fuelled by a net influx of about 40,000 people annually – not counting refugees (Investitionsbank Berlin, 2015b: 12) – and the concomitantly rising rents (Investitionsbank Berlin, 2015a: 59 et seq.) these figures must have been an alarming signal for the city's government, which decided to pursue an uncompromising course against apartment letting via Airbnb. This was done because the Berlin Senate Department for Urban Development and the Environment feared a large-scale transformation of regular flats into holiday homes, leading to a further decrease of affordable housing. Their efforts to protect the Berlin housing market against short-term rental resulted in the Misappropriation Ban Act (*Zweckentfremdungsverbot-Gesetz*; Senatsverwaltung für Justiz und Verbraucherschutz Berlin, 2013), which basically prohibited the conversion of regular apartments into holiday homes (Senatsverwaltung für Stadtentwicklung und Umwelt Berlin, 2016a). The law came into effect in May 2014 (Senatsverwaltung für Justiz und Verbraucherschutz Berlin, 2014) with a transition period of two years. At the end of April 2016, all holiday apartments had to be registered and licensed at the respective district administration. Otherwise, the host may be punished by a fine of up to €100,000. The Berlin Senate Department for Urban Development and the Environment recently stated that about 6,300 holiday apartments had been reported at district administration offices; of these, only 87 received permission (Blume, 2016). Despite this immense effort, the actual number of full apartments let out via Airbnb decreased between 26 April and 25 May 2016 from 6,760 to 5,860, according to another data analysis conducted by the students from the University of Applied Sciences Potsdam (Blume, 2016: 6). All in all, this illustrates that Berlin's government has made a great effort to prohibit the transformation of residential units into holiday homes, but with only marginal results to show for it. In addition, it appears necessary to ask if the government's fight against Airbnb is truly an intervention against rising rents and displacement, or if the online sharing platform simply became a scapegoat for the complex and expensive challenge to provide enough affordable housing for the city's rising number of inhabitants.

After having discussed the actual number of Airbnb flats, the following section deals with the spatial distribution of Airbnb listings in Berlin. Figure 8.1 shows that most of the urban neighbourhoods – which are called LOR (Lebensweltlich Orientierte Räume) and represent the lowest administrative level in Berlin (Senatsverwaltung für Stadtentwicklung und Umwelt 2016b) – are only very slightly affected by Airbnb phenomena. Out of the 447 LOR units, in about 100, not a single Airbnb offer could be identified in spring 2015. Another 169 urban neighbourhoods saw one Airbnb listing or fewer per 1,000 inhabitants. At the same time, the concentration of the phenomena to a few LOR units can clearly be identified on the map. Nine LOR neighbourhoods exhibit a density of more than 20 Airbnb offers per 1,000 inhabitants. Concentration in the districts of Mitte, Friedrichshain-Kreuzberg and Pankow (especially the southern part around Prenzlauer Berg) is clearly visible. Thus the density of Airbnb beds is especially high in the central parts of former East Berlin. These are also the neighbourhoods which have been affected the most by the gentrification process since reunification (Bernt, Grell and Holm, 2013; Krajewski, 2013).

In a nutshell, although the share of Airbnb apartments measured as a portion of the total housing stock in Berlin is comparatively low (Skowronnek, Vogel and Parnow, 2015), in certain areas of the city that have been greatly affected by different types of gentrification (Holm, 2013) professional apartment letting via Airbnb and other short-term rental platforms (mainly since the Misappropriation Ban Act came into effect) has especially increased this housing shortage even further (Blume, 2016, Holm, 2016b).

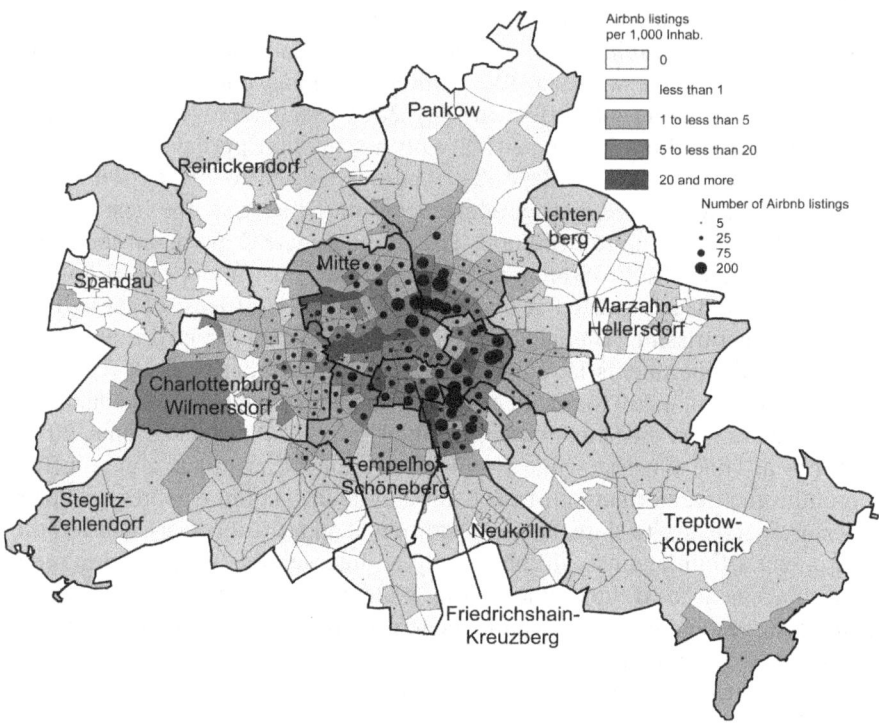

Figure 8.1 Airbnb in the urban neighbourhoods of Berlin

Sources: Skowronnek *et al.*, 2015; Land Berlin 2015 and Senatsverwaltung für Stadtentwicklung und Umwelt Berlin, 2016b

Implications of Airbnb on Berlin's hotel industry

From the perspective of Berlin's hotel industry, measures against the competitor Airbnb could be interpreted as an expression of structural problems within the hospitality industry. Hotel options in Berlin have expanded significantly in the 25 years since reunification. An overcapacity of hotel rooms and the accompanying decline of incomes could explain the industry's fears about facing this new competitor. However, statistical figures rebut this presumption. Since 1992, the number of beds has increased by 222 per cent (see Figure 8.2; Amt für Statistik Berlin-Brandenburg 2016d). Simultaneously, Berlin is among the most dynamically growing destinations in Europe. While the number of guests and overnight stays rose by 130 per cent to 140 per cent in German cities overall in this time (Statistisches Bundesamt 2015; see also Kagermeier 2015: 209), overnight stays in Berlin rose by 265 per cent and arrivals by even more, 274 per cent. Within the last five years, the demand for hotel rooms has grown faster than supply, so that the occupancy rates of Berlin hotels have continued to improve (Deloitte 2013: 6).

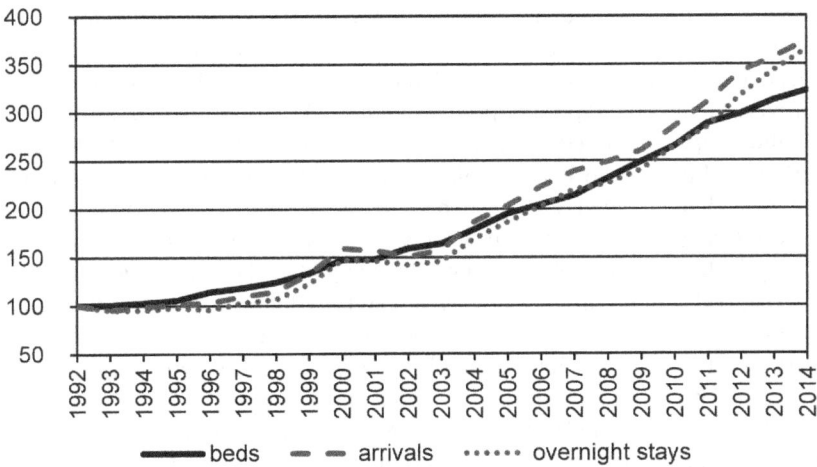

Figure 8.2 Development of the indexed accommodation capacity in Berlin and arrivals and overnight stays of tourists between 1992 and 2014 Index 1992 = 100

Source: Amt für Statistik Berlin-Brandenburg 2016c

In comparison to the aforementioned 30.25 million officially registered guest nights in Berlin (Amt für Statistik Berlin-Brandenburg, 2016c), the number of Airbnb nights of about 2.6 million between 1 January 2015 and 1 January 2016 seemed acceptable (Airbnb, 2016). The platform itself states that 20,200 hosts offered their rooms and apartments to about 568,000 guests who stayed for approximately 4.6 nights on average (Airbnb, 2016). Taking into account private overnight stays at friends or relatives (the VFR segment) as well, which amounts to about 32.5 million overnight stays in Berlin (Berlin Tourismus & Kongress GmbH, 2015: 4), the number of Airbnb nights appears even more modest. Compared to the combined number of private and officially registered overnights in Germany's capital, which is more than 60 million, the proportion of Airbnb nights is around 4 per cent of all generated overnight stays by visitors, despite the high media coverage and enormous criticism of the platform regarding its contribution to housing shortages, rising rents, and even evictions.

Airbnb and the touristification process in Berlin

The figure above illustrates that no general inefficiencies in the Berlin hotel market can be identified. Moreover, continuously rising guest arrivals keep hotel occupancy rates at a high level of 60.5 per cent (Amt für Statistik Berlin-Brandenburg, 2016d), while hotel rooms cost on average €86. This is relatively cheap in comparison to other German cities, such as Munich (€110) or Hamburg (€101) and considerably less expensive than hotel rooms in other European metropolises, such as London (€165) and Paris (€139) (HRS, 2015).

The total number of Airbnb listings available in Berlin, which amounts to about 15,000, according to data collected by the internet platform Inside Airbnb (Cox, 2015), is also significantly less than in other leading tourism destinations in Europe, such as Paris (45,000) and London (42,000) (Cox, 2016a and b).

However, media coverage and protests against Airbnb – ranging from occupying holiday apartments (Beitzer, 2016; Holm, 2016a; Jacobs, 2016) to large anti-Airbnb billboards in affected neighbourhoods (Mai, 2016) – seem to be much more intense than in other cities, such as Paris (Gravari-Barbas and Jacquot, 2016), even though these other cities have significantly more Airbnb listings.

But why is there such a debate around it? One part of the explanation is surely Berlin's relatively immature mass tourism history, which started after the reunification but which really took off with the rapid increase of tourism overnight stays and guest arrivals in 2003 (see Figure 8.2). This development coincides with or has been co-initiated through a boom for low cost carriers in Germany (DLR, 2015) and has made Berlin a techno and party tourism destination for the so-called "easyjetset," (Rapp, 2009). Since the majority of clubs are historically located in the less developed and much cheaper areas of the Eastern part of Berlin, the nightlife infrastructure and gastronomic options in these areas encouraged an influx of specific tourism demographic groups.

Since the early 2000s, the city's tourism marketing has also used the underground images of these emerging districts as being ethnically diverse, tolerant, creative and hip in order to promote them as appealing destinations for "authentic" urban tourism (Füller and Michel, 2014: 5). At the same time, gentrification processes catalysed "[r]ising rents and a growing perception of neighbourhood change [that] dominated local media reports and the public discourse" (Füller and Michel, 2014: 5) in these neighbourhoods. In particular, the south-eastern part of Kreuzberg and northern parts of the district of Neukölln – both currently experiencing relatively high densities of Airbnb apartments – experienced an increase in rents that was above the city average, while unemployment rates remained high (Füller and Michel, 2014). Holm (2013: 179) found that despite rent increases from 23 per cent to 30 per cent between 1999 and 2008 in the southern neighbourhoods of Kreuzberg, the transformation has not yet led to a displacement of poor and lower-class residents. Nevertheless, the area faced a large influx of households with higher incomes (Holm, 2013: 179) as well as an economic transformation: "[N]ew bars, restaurants, cafés, bicycle shops, small art galleries and independent fashion labels opened weekly" (Füller and Michel, 2014: 5f) due to initially lower rents compared to districts like Prenzlauer Berg or Mitte. Finally, the image of Kreuzberg and the northern part of Neukölln has shifted from being known as a "ghetto" (Best and Gebhardt, 2001, in Füller and Michel, 2014) to "the epicentre of cool" (Dyckhoff, 2011) within ten years.

The year 2011 also saw 'Berlin does not love you' stickers appear for the first time in Berlin (Novy, 2013) and the Green Party of the district of Friedrichshain-Kreuzberg invited residents for an initial roundtable discussion on the topic "Help, tourists are coming" (n-tv, 2011). This was also the year in which protests against tourism in residential neighbourhoods of Berlin started (see also Colomb and Novy, 2017).

The brief section above illustrates that tourism and gentrification processes were heavily intermingled even before Airbnb entered the market. However, the financial crisis of 2008–2009 and the resulting investment in the real estate market have further exacerbated the situation in Berlin's housing market. In areas where immigrants and the creative industries were able to find cheap apartments for the past 20 years, rising investment has led to rents that are no longer affordable for many of the city's residents. Real estate prices have risen by more than 65 per cent since 2007. This is a relatively high figure, but similar magnitudes of rent increases have occurred in other comparable cities. In Munich, for example, real estate prices have risen by about 80 per cent (Jung 2015) in the same time span. Comparatively, real estate prices in Berlin (at approximately 3,500 €/m²) are only about half the prices in Munich (at approximately 6,500 €/m²) and still much lower than in Hamburg or Cologne (Wohnungsbörse, 2015). Nevertheless, if we focus on rent increases in recent years, Berlin does have a leading position. Since 2007, rents have risen by almost 50 per cent, whereas in Munich or Hamburg, they have risen by only one-third (Jung, 2015). Nevertheless, rents in Berlin are still well below the level of those of many western German cities. Due to its new function as the German capital and the attractiveness of Berlin for young people of the creative class in particular, the city has witnessed a net migration gain of about 40,000 inhabitants per year (Amt für Statistik Berlin-Brandenburg, 2016a, not taking into account the additional number of refugees in 2015). In other words, the expected effect of banning Airbnb totally would have been offset by the migration gain within a couple of months. Moreover, when looking at the type and price of professionally let apartments on the Airbnb webpage – which is officially forbidden now in terms of short-term rental according to the Misappropriation Ban Act – it becomes clear that returning these apartments back into the traditional rental market would not solve the problems illustrated by Holm (2016b) and Investitionsbank Berlin (2016) that major deficiencies in affordable housing continue to exist, particularly for households with lower incomes.

The rapid, significant rise in rental rates is certainly one important reason for the inhabitants' sceptical positions towards Airbnb. The city government also quickly identified the platform as contributing to the aggravation of the housing shortage and rising rents. Nevertheless, as the above section describes, its effect is much too small to be held responsible for the general housing problem in Berlin.

Methodological approach

The empirical research leading to the results discussed below did not focus primarily on the discussion about Airbnb in Berlin. Interest has been more oriented to basic research on the motivations of participants in the sharing economy on the one side, and the experiences of Airbnb hosts and guests in Berlin on the other. Nevertheless, these findings might contribute valuable insights to the discussion on gentrification and touristification processes in urban centres.

An overview of sharing economy participants based on online surveys

A digital questionnaire was created to obtain an initial outline of the socio-demographic and motivational structure of sharing economy participants. The main objective of this online survey was to identify people's reasons for participating in the sharing economy (Kagermeier, Köller and Stors, 2015). In order to collect this data, convenience sampling was conducted involving students, employees and mainly young Tourism Studies graduates from Trier University (Germany). Sampling resulted in 271 completed questionnaires. Due to this specific selection, it cannot be claimed that the results are statistically representative of the German population as a whole. As Heinrichs and Grunenberg (2012: 13) illustrated, there is a high positive correlation between the age, level of education and income of sharing economy participants. By selectively addressing mainly young academics, our sample contains a disproportionately large number of "social-innovative collaborative consumers" (Heinrichs and Grunenberg, 2012: 14; also similar to Nielsen, 2014: 9). Compared to the German population, one-quarter can be assigned to this group of social-innovative collaborative consumers (Heinrichs and Grunenberg, 2012: 14). Regarding the awareness of internet platforms that offer overnight stays, the bias becomes even more striking. According to a GfK survey representative of the general population, two-thirds of the population are unaware of offers such as Airbnb (Marquart and Braun, 2014), whereas in our sample, only 1.5 per cent did not know of such possibilities. However, focusing on such a target group enabled more precise statements to be made on their motivations for taking part in sharing activities, which was the main reason for conducting the study. Regarding methods, in order to explore the initial results generated by the online survey in greater depth, the use of structured interviews with Airbnb hosts and guests in Berlin took place, leading to a qualitative study.

The motivations of Airbnb hosts and guests

A specific segment of the large number of collaborative consumption offers was identified and analysed in order to gain a clear picture of sharing economy participants. For the purposes of this chapter, it makes sense to focus on the segment of private accommodation within the sharing economy relevant to tourism, which has gained considerable media interest in recent years. While this small section of the sharing economy is characterised by multiple suppliers, our analysis focuses solely on the market leader, Airbnb.

Since the number of Airbnb hosts in Berlin is considerably large – there are more than 11,700 (Skowronek *et al.*, 2015) units available – it was not possible to contact all of them. Instead, the number of requests was based on the number of listings in Berlin's districts. The most important districts were those with the largest numbers of listings, which were Prenzlauer Berg, Friedrichshain, Kreuzberg, Neukölln and Berlin Mitte. More than 1,000 Airbnb units are available in each of these districts. A total of 46 interview requests were sent in these areas, resulting in 13 interviews. In the districts with listings between 250 and 1,000 (Schöneberg, Wilmersdorf, Charlottenburg, Moabit and Wedding), at least one interview was conducted. Fewer Airbnb hosts were contacted and interviewed in other neighbourhoods. After a one-week interview pre-test in March 2014, interviews were conducted over the space of four weeks in August and September 2014. Despite the relatively short data collection period, more than 100 requests were sent to Airbnb hosts, resulting in 25 personal interviews. This extensive data provides a solid basis for conducting an in-depth analysis of motivational structures and interaction between Airbnb hosts and guests.

To further illuminate the Airbnb guest perspective, it would have been desirable to conduct extensive qualitative interviews with a group of guests as well. However, since they are extremely difficult to approach, data about this group was gathered using a multi-method approach. Socio-demographic characteristics were determined from general studies on participants in the sharing economy and carefully applied to Airbnb users. In addition, the online questionnaire provided information on socio-demographics and motivations of Airbnb users, although the number of Airbnb guests in our convenience sample is clearly relatively small. We also distributed a primarily quantitative questionnaire in German and English to our interviewed Airbnb hosts in Berlin to pass on to their guests. As might be expected, the response rate of these questionnaires was quite low. Finally, we also intended to directly interview Airbnb guests when they were with their hosts. This approach resulted in five guest interviews, the results of which are also presented in the next section.

Motives for participating in Airbnb

As the quantitative oriented survey showed (Kagermeier, Köller and Stors, 2015), the main target groups of the share economy in tourism are younger tourists. These findings regarding the respondents' age structure are similar to those generated by the quantitative questionnaires distributed to Airbnb hosts. This group also has experience in other segments of the sharing economy, such as online and offline swapping, buying and selling goods, and hiring people for services; additionally, the group has a medium income level.

In order to conduct a more detailed characterisation of the respondents beyond simply socio-demographic figures, we created a profile of their personalities using a five-point Likert Scale (Figure 8.3).

Other than age there are no significant differences within the sample, for example between students and professionals. Also with regard to the use of sharing offers, there are only marginal differences in personality between users and non-users. One reason for this is likely to be that the sample was drawn from a sharing-prone population, which also means that these results cannot be translated easily to the German population. However, it should be noted that sharing economy participants are slightly more risk-tolerant and open to new things than their non-user counterparts. One comparatively strong feature that most sharing

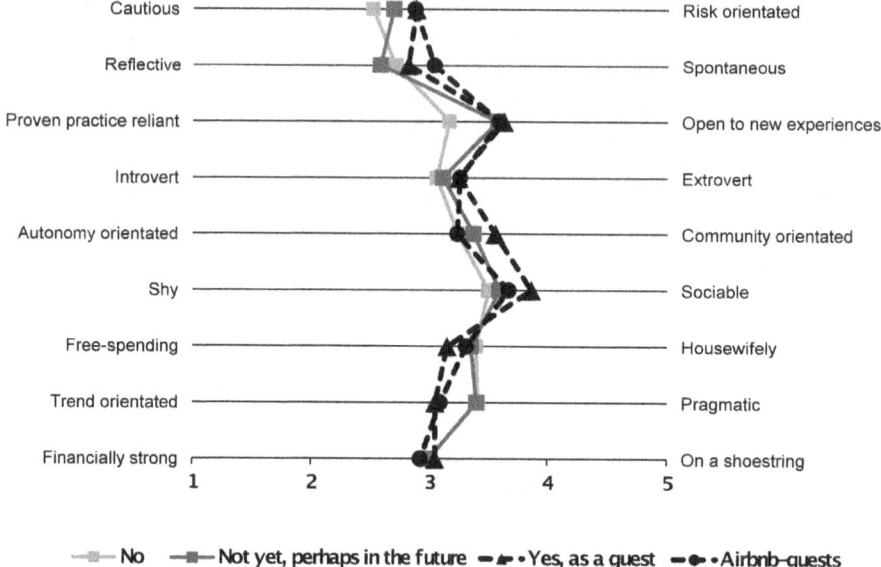

Figure 8.3 Personality profile of respondents by level of participation

Notes: N = 271 in online survey and N = 61 Airbnb guests in Berlin

economy users have in common is their openness to new things and their sociable personality. This is also mirrored in the descriptions Airbnb hosts gave of their guests. The hosts describe their guests as open, sociable and communicative.

In contrast, financial motivations are less relevant than expected. Actual and potential users of sharing-economy accommodation are no more frugal or thrifty than non-users. At the least, their reason for participating in the tourism sector of the sharing economy is not that they are unable to afford anything else. Their internal driving force must be another kind of motivation.

Figure 8.4 depicts various potential motives for using private sharing-economy accommodation and how the respondents evaluated them. As expected, the economic dimension within the motivational structure is relevant, but it is not the only driving force. Similar results can also be found in Liedtke's study, which focuses solely on Couchsurfing: in this study, too, financial aspects were less important than other motives, such as meeting new people, cultural exchange and establishing new friendships (Liedtke, 2011: 34 *et seq.*). Visitors' expectations concerning specific experiences at the destination – such as having direct contact with the local population, gaining insider information from the host about bars, restaurants or the neighbourhood in general, and experiencing the destination from the locals' perspective – are at least as relevant as the monetary factor. These are the most important motives in the leisure segment in particular. More general aspects, such as 'expanding one's horizon' or 'trying new things' together with

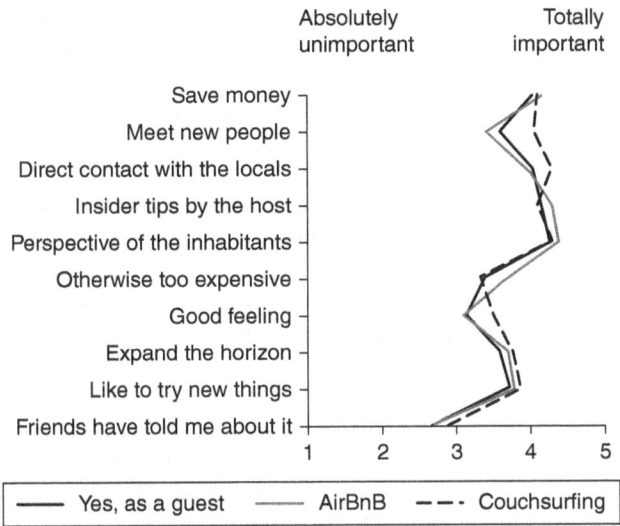

Figure 8.4 Motivations of sharing-economy accommodation users – differentiated by Airbnb and Couchsurfing users

Note: N = 112 in online survey

recommendations from friends are also relevant, but they are much less important than those dedicated to the on-site visitor experience.

Comparing the three lines in Figure 8.4, it becomes obvious that no significant differences exist between Airbnb users and Couchsurfers. The only noticeable deviation can be found in the social contact items 'meeting new people' and 'direct contact with the local population'. Couchsurfers seem to attach greater importance to these very specific social objectives, while differences decrease in the next item – gaining insider tips from the host.

In a nutshell, the online survey revealed two leading motivational dimensions that were supported by the quantitative questionnaires distributed to and the qualitative interviews conducted with Airbnb hosts: monetary aspects and the social interaction between the guest and the host.

Monetary dimension and the housing market

The role of financial motivation became a key aspect during the analysis of the quantitative offline questionnaires. This survey revealed that one-third of leisure guests and half of business tourists booked private accommodation via websites such as Airbnb to save money. Leisure visitors also stated that these sharing platforms enable them to visit destinations that they would otherwise be unable to afford.

The price aspect came up for discussion in the qualitative interviews as well. From the guest perspective and from a tourism point of view, the visiting of destinations that would otherwise be too expensive was often mentioned. Moreover, business travellers and people who intended to spend a longer period of time in Berlin (so-called 'temporary migrants') appreciated the better value for money of Airbnb apartments. The apartments or rooms are often fully furnished and they offer amenities, particularly for extended stays, that most hotels cannot provide.

> I got kind of frustrated by staying in hotels, they are really nice [. . .] but you are paying €150 a night and there is no kitchen, and internet is an extra €10 a day. And from the budget I have for work travel, I could rent the whole [Airbnb] place. I can get a two-bedroom place, with a kitchen and a balcony [. . .]. And there is also the appeal, instead of just staying somewhere you feel as if you are living somewhere for a couple of days (Guest_Berlin_3).

The high relevance of the monetary dimension likewise applies for the host side, although their motivations seemed to be even more diverse. Reasons for offering the whole apartment on Airbnb or renting out a private room within a shared flat are wide-ranging, as the following examples show. Of course, there are professional suppliers renting out several rooms and apartments to make a business out of it. Skowronnek *et al.* (2015) identified about 1,200 persons and thus about 10 per cent of all hosts in Berlin that offer more than one room or apartment.

However, in many interviews, a distinct connection between the tight housing market and the private supply of accommodation was drawn, as the following statements illustrate:

> It just worked out like that. I went back to Germany and I took over her shared apartment [. . .]. The flat is really large, four rooms, and I didn't think of anything else than just continuing it as a shared apartment. I can't use it only for myself, I don't have enough money. And in the case I leave, it'll be transformed into a luxury apartment [. . .]. Before Airbnb, I advertised the apartment as a student shared apartment at WG-Gesucht [www.wg-gesucht.de is a Germany-based internet platform where one can advertise and search for shared apartments]. But that is always such a hassle. You always have certain problems [. . .]. And now, I am with the holiday guests, which is fantastic. It works out even better than the shared apartment (Host_Mitte_8).
>
> We have a three-room apartment, my husband, the son of my husband, he is 12 years old, and me [. . .]. And now, we have a baby. That means we are lacking one room and we need one additional room. And we don't want to leave our house in which we are so rooted [. . .]. That's why we rented the other apartment. We plan to combine the two units so that our flat is simply enlarged in the end. It is a one-room apartment with a kitchenette and a bathroom and that is going to be our bedroom in about 2 years, when our little son needs his own room. And until then, we've planned to finance the room via Airbnb (Host_Rummelsburg_2).

Due to the specific situation in Berlin after reunification and until the 2000s, living space, even in central areas, was comparatively cheap, and low-income to medium-income people could afford to rent quite spacious apartments in these areas. Due to major investments in the real estate market, particularly after the financial crisis in 2008–2009 and in the course of general re-urbanisation processes, real estate prices have risen by more than 65 per cent since 2007 and rental rates increased by almost 50 per cent (see above). This context must be taken into account when interpreting the motivations of people who rent out their private homes to total strangers. We do not want to deny the certain existence of a considerable group of suppliers who own or rent apartments and sublet them to tourists for solely professional reasons. And of course it falls within the government's responsibility to ensure that this happens within the legal framework. However, even if an estimate of 30 per cent of Airbnb hosts are professionals, there is a very large majority of 70 per cent of hosts made up of small-scale businesses of people who rent out their rooms or apartments on an occasional basis to co-finance their living in Berlin, to pay for small luxuries, such as a cleaning person or a holiday trip, or even to just be able to keep the apartment they are living in.

Interaction between hosts and guests as an important element of the visitor experience

Besides the financial aspect, personal interaction between hosts and guests plays a major role for the majority of the tourists interviewed. In particular, visitors from the leisure segment consider it very important to get to know new people (a significant difference from business travellers) and to receive personal information and recommendations from the host (also a significant difference). This element is also reproduced in the contact intensity between hosts and guests. Based on 58 questionnaires completed, one in seven stated that contact was limited to formalities, e.g. receiving keys or brief information about the room/ apartment. In some cases, a third party dealt with these formalities (Figure 8.5). In one in four cases, the host had also prepared written information for the guest. Almost half of the visitors said that the host provided personal information about the city; another 12 per cent undertook activities with the host. In all of the latter cases, personal information and joint activities were supplemented by written information about the city.

Qualitative interviews with the hosts confirmed that most had personal contact with their guests:

> I hand over the key, I show them around, show them the bathroom, their room of course and I give them the Wi-Fi password. Mostly, I ask them why they are in the city. Sometimes, I leave the keys at a kiosk close by, when I am not at home. Sometimes, it happened that people arrived when I was not in the city at all. I just want them to find everything. But generally, I am always interested in the reasons why they are here, ask if everything is ok, and make some small talk (Host_Moabit_18).

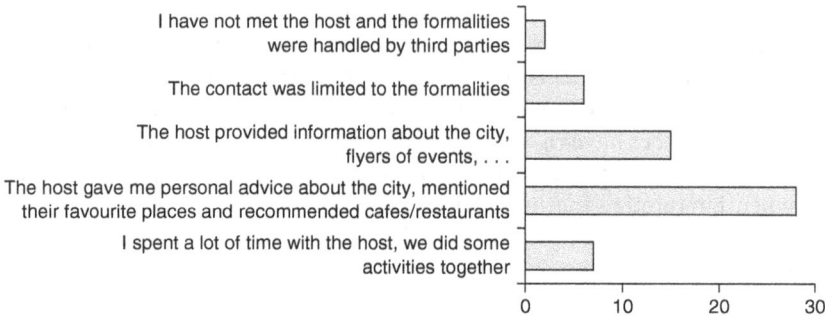

Figure 8.5 Interaction between hosts and guests

Note: N = 58; offline survey of Airbnb guests in Berlin

The intensity of contact between the host and the guest primarily depends on the type of rental. People who actually share an apartment and just rent out one of their rooms have more intense contact with their guests than hosts who rent out one or more apartments.

One last important aspect to consider at this point is the type of information on the city that is transferred between hosts and guests. It has already been pointed out that many hosts provide written tourist information material for their guests, also in order to support their personal advice. But personal, 'insider' tips (see also Figure 8.4), seemed to be of high relevance. Examples of this type of advice include nearby restaurants, cafés, and bars. This is done with the help of flyers or other information material for the guests, but also with the Airbnb website itself, where a map of the area can be supplemented by icons indicating a favourite café, club, grocery store and so forth.

This is also an important dimension focusing on the touristification process and the much more general questions of how tourist spaces emerge. In these cases, urban residents designate certain localities to be of interest for their guests. They bundle such places in their Airbnb neighbourhood maps and by doing so confer a certain, 'touristy' meaning on that place. In doing so, neither the city's government nor the destination management organisation influence these processes directly. Instead, the residents themselves, who often claim to suffer from a tourist glut, actually co-produce such areas and promote them on their Airbnb websites. At this point, it appears promising to go into greater detail about the early and resident-led phase of tourist space production. This is particularly relevant when considering that many Airbnb localities in Berlin are also claimed to be creative quarters and immigrant neighbourhoods, such as large parts of Kreuzberg and Neukölln. It would be worth analysing how other residential groups confer meanings on the same place and see to what extent they form a symbiosis, or if they conflict.

In addition to the written information, hosts often give tips about certain locations in their neighbourhood when talking. In doing so, they likewise direct visitor flows towards certain sights and neighbourhoods.

> I told him about things that I find are interesting. I told him about the war memorial in this area. That's always something that annoys me, when it comes to tourists. When they come to Berlin and only visit the East Side Gallery and think that you actually could have jumped down the wall. And they don't have any idea of what a divided city actually means. They want to experience the divided city and then they went visiting a gallery and not the wall. That's something I like to show people. And strangely enough, there are no tourists, there are the people themselves. That's not sufficiently promoted in the city. And it is something that I think is a bit different. I have the impression that thus they can also see other neighbourhoods (Host_Moabit_18).

In contrast to the aforementioned motivational sphere of using Airbnb, the monetary aspect and the personal interaction, the last section deals with an aspect that was only identified in the course of the on-site personal interviews.

Individuality of the facilities and design of the accommodation

The qualitative interviews conducted with the hosts revealed an element that was underestimated in the previous quantitative surveys. Due to their relatively intense guest contact, Airbnb hosts were able to observe that visitors greatly appreciate the ambience of private accommodation. As a result, not only does direct contact with the host and the creation of an inside perspective contribute to the specific visitor experience of Airbnb and the like, but the design and amenities of the accommodation are important too.

> And those who participate in something like that [Airbnb], and say, I don't want to go to a hotel, don't head for a standardised 70s-style flat, but prefer the charm of an old Berlin building [. . .]. But I think – for a relatively low price – they want this 'That's Berlin!' feeling. A hostel, in contrast, is of course completely interchangeable; it always looks the same everywhere. I think that's the first thing they want. (Author's translation) (Host_Berlin_15).

Finally, further aspects are also relevant when it comes to choosing private accommodation in the sharing economy. Some visitors stated that these online platforms are easy to use, offering a comparison of different accommodation and prices, and fast access to relevant information. For others, the straightforward and instant contact and communication with the host is the greatest advantage. In addition, both the quantitative surveys and the qualitative interviews revealed that the specific location of the rooms and apartments within a city or even a neighbourhood may be highly relevant to visitors, and may be the decisive factor for choosing private accommodation over a hotel.

At the same time, the guests showed a specific interest in accommodation options that are perceived as authentic and that provide an opportunity for personal contact with the locals. In return, this leads to the conclusion that even reducing the rates of commercial accommodation offers of hotels and hostels would not cause the disappearance of the Airbnb phenomenon. Airbnb is only the tip of the iceberg of a tendency in tourism demand where more and more experiences perceived as unstaged outside the traditional tourist bubble are sought out. In other words, gentrifying neighbourhoods are shaped by different actors, of which urban tourists and Airbnb users are just two groups among others. Disturbances caused by noise, waste or the mere presence of large visitor flows cannot be attributed to the growth of Airbnb in Berlin. It is much more the atmosphere, the urban setting and the people in these areas that will continue to attract residents who behave like tourists, visitors from other neighbourhoods of the city, traditional tourists staying in hotels, etc., even if Airbnb were completely banned from Berlin.

Challenges of the Airbnb phenomenon for urban governance reactions

The Airbnb phenomenon as a new aspect of the (urban) tourism market has evoked sharp reaction from the public sector. In order to restrict the fast-growing holiday apartment rental market and to gain some degree of oversight of this grey market segment, Berlin's Senate enacted a new law. The law prohibiting the misappropriation of living space (Zweckentfremdungsverbot-Gesetz; Senatsverwaltung für Justiz und Verbraucherschutz Berlin, 2013) came into effect in May 2014. The regulation states that each supplier of holiday apartments and rooms has to report the business to the responsible district office for request of approval. Meanwhile, about 6,000 holiday apartments have been reported to the districts' offices, with the majority of them located in Berlin Mitte with 1,728 registrations. According to these figures, the share of professional hosts is about 30 per cent, with 'professional hosts' defined as people offering two apartments or more (Holm, 2016b). At this point, one has to take into account that these figures also contain professional holiday apartment suppliers and even some hostels or hotels which used the Airbnb website as an additional distribution channel for their rooms and apartments. Within the discussion of Airbnb as a whole, it becomes particularly obvious that the hospitality industry in Germany lacks clear distinction between private and professional holiday home and apartment rentals (the statistical threshold from which renting is regarded as professional is ten beds).

However, what has further aggravated the situation in Berlin and what finally triggered the enactment of the law prohibiting the misappropriation of living space was not just the hotel industry defending itself against a new competitor (IHA, 2015). It was rather the highly strained housing market as well as the greatly increased rental rates in recent years that drove Berlin's officials to take legal action against private rentals to tourists. What the government, at this point, has surely not taken into account is the multi-dimensional background of the privately renting Airbnb hosts for offering their rooms or apartments to strangers.

The 2014 law and its long-lasting consequences can be characterised as an ill-conceived reaction of the local government. To a certain extent, their defensive position is not unusual for managing innovative but unconventional business models, which are in this case fostered by the proliferation of mobile internet and social media. Instead of trying to analyse the background of the development and the underlying driving forces that lead local residents to rent out their private space, banning private rentals seems to be rather short-sighted, trying to eliminate the symptoms without really touching on the causes of this new phenomenon. There is no doubt that financial regulations and measures in particular need to be adjusted to new business models introduced by the internet in general and by the sharing economy and its subsectors in particular. Mainly, semi-professionals and professional suppliers need to be identified and taxed correspondingly. Moreover, they have to fulfil the general regulations that apply to the professional hospitality industry. On the other hand, private rentals also have to adapt to the existing legal framework. This applies mainly for rules regarding the taxation of rental properties or small-scale economic activities (including local tourism taxes). However, such adjustments need a certain amount of time and are by no means exclusively connected to Airbnb or the sharing economy. On the contrary, they are typical for innovations that existing rules and laws cannot adequately address. Yet hindering these developments to preserve the status quo is surely not the most tolerant and liberal coping strategy that one could have expected from Berlin's government.

Suggesting that Airbnb is a major factor in the upward spiral of real estate and rental prices is, from our perspective, overestimating the platform's influence on the housing market. A very large majority of users do not rent units to simply sublet them to visitors. Many hosts have idle space available and sublet it to simply maintain their current standards of living in times of rising rental rates. However, these figures do not sufficiently take into account the large majority of Airbnb hosts who rent out their whole apartments only from time to time, for example when they are on holiday themselves. As GEWOS (2014) has illustrated, 88 per cent of the apartments were only let 120 days a year or less.

In order to understand the full scope of Airbnb's impact on Berlin's housing market and on the related gentrification discourse, it is indispensably necessary to further differentiate between different types of hosts and hosting on Airbnb. At least two different groups need to be addressed in this context: first, those professional hosts who have bought or rented one or several apartments in order to transform them into holiday homes and make a business out of it. In these cases, the flats were actually withdrawn from the classic housing market and they were let on Airbnb for short-term use for prices far higher than regular tenancy agreements. However, the share of such professional hosts is comparably small. If the frequency or the volume of renting accommodation serves as an indicator of professionalization, it is noteworthy that only 12 per cent of listings are rented out more than 120 days a year (GEWOS, 2014). And those who offer more than one unit cover 10 per cent of the hosts (Skowronnek *et al.*, 2015). In order to gain a full understanding of this host group, it would be necessary to analyse it in greater detail. It is also obvious that hotel chains and hostels use the Airbnb platform as a distribution channel as well (Köller, 2015) and in these cases, of course, one host has several listings.

Airbnb (2016) themselves published figures saying that between 1 January 2015 and 1 January 2016 a typical listing in Berlin was booked on average for 34 days, indicating that the vast majority of the hosts rented properties out only occasionally. This second group of Airbnb hosts consists of people renting out their whole apartment for less than 120 days a year. In other words, with such an occupancy rate they could not survive as professional hosts and the obvious assumption must be that these people actually inhabit their apartments and just rent them out when they are away.

A comparison between all available Airbnb beds and the amount of actual overnight stays indicates that most Airbnb hosts do not do so as professionals. This is an important point, since it is particularly the professionalisation of this business that is regarded as illegal in terms of lacking taxation regulations and the conversion from housing into business properties. This aspect of sharing flats is one of the main reasons for its decreasing social acceptance, since it supposedly leads to a reduction of affordable living space, particularly in highly desired, central areas of the city.

Concerning the gentrification process, it must be stated that sharing accommodation – especially via the Airbnb platform – cannot be regarded as the origin of the gentrification process. The existence of pioneers and gentrifiers in a neighbourhood who offer accommodation in their flats is as much a precondition of gentrification as the development of shops and gastronomy services focusing on a gentrified target group. New urban tourists staying in shared accommodation and frequenting the shops and restaurants in gentrifying/gentrified neighbourhoods can at the same time be seen as intensifiers of the gentrification process. But as the figures have shown in the case of Berlin – one of the metropolises with a very visible Airbnb presence – the sharing economy is far from being the most important driving force of the gentrification process and only to a limited extent contributes to its development, even though it is often seen in public discussions as the prime factor.

At the same time, sharing accommodation is not a totally new phenomenon, but has been facilitated by internet platforms like Airbnb such that it has become much more common than in former times. As a result, regulation schemes still have to adapt to this new form of market supply. Questions of taxation, local tourism fees and security certainly have to be discussed, these systems have to be adapted to this new type of accommodation supply alongside traditional providers, while also distinguishing between professional and non-professional providers. At the same time, existing legal instruments concerning the opening hours of restaurants and bars in specific neighbourhoods must also be consistently applied to reduce the negative impacts on inhabitants – always bearing in mind that a totally undisturbed environment cannot be achieved in central urban quarters.

As the empirical findings illustrate, new ways of experiencing urban tourism are on the advance, fostered by the accommodation opportunities initiated by

Airbnb and similar platforms. They do not seem to represent a mere temporary fashion or hype, but instead are the manifestation of a fundamental tendency in urban tourism. As is often the case with innovations, established ways of governing have difficulties adapting to new phenomena, such as the sharing economy. Using the example of Berlin, it is clear that the way the public sector has handled the challenge is barely adequate. A much more comprehensive and less hasty approach appears to be necessary.

References

Airbnb (2016) *Overview of the Airbnb Community in Berlin*, [Online] Available at: http://1zxiw0vqx0oryvpz3ikczauf-wpengine.netdna-ssl.com/wp-content/uploads/2016/04/airbnb-community-berlin-en.pdf [accessed 17 August 2016].

Amt für Statistik Berlin-Brandenburg (2016a) *Bevölkerung. Wanderungen. Zeitreihen*, [Online] Available at: www.statistik-berlin-brandenburg.de/basiszeitreihegrafik/Zeit-Wanderungen. asp?Ptyp=400&Sageb=12035&creg=BBB&anzwer=10 [accessed 1 February 2016].

Amt für Statistik Berlin-Brandenburg (2016b) *Statistiken. Gebäude und Wohnen. Lange Reihen*, [Online] Available at: www.statistik-berlin-brandenburg.de/statistiken/lange reihen1.asp?Ptyp=450&Sageb=31000&creg=BBB&anzwer=9 [accessed 17 August 2016].

Amt für Statistik Berlin-Brandenburg (2016c) *Statistiken. Handel, Gastgewerbe, Tourismus. Tourismus. Basisdaten*, [Online] Available at: www.statistik-berlin-brandenburg.de/statistiken/langereihen.asp?Ptyp=450&Sageb=45005&creg=BBB&anzwer=6 [accessed 17 August 2016].

Amt für Statistik Berlin-Brandenburg (2016d) *Statistiken. Handel, Gastgewerbe, Tourismus. Tourismus. Lange Reihen*, [Online] Available at: www.statistik-berlin-brandenburg.de/statistiken/langereihen.asp?Ptyp=450&Sageb=45005&creg=BBB&anzwer=6 [accessed 4 January 2016].

Beauregard, R. A. (1986) 'The Chaos and Complexity of Gentrification', in Smith, N. and William, P. (eds), *Gentrification of the City*, London and New York: Routledge.

Behrendt, S., Blättel-Mink, B. and Clausen, J. (2011) *Wiederverkaufskultur im Internet. Chancen für nachhaltigen Konsum am Beispiel von eBay*, Heidelberg: Springer.

Beitzer, H. (2016) 'Hartz-IV-Empfänger gegen Airbnb', *Süddeutsche Zeitung*, 19 January 2016, [Online] Available at: http://www.sueddeutsche.de/panorama/protestaktion-hartz-iv-empfaenger-gegen-airbnb-1.2824833 [accessed 17 August 2016].

Belk, R. (2010) 'Sharing', *Journal of Consumer Research*, vol. 36, no. 5, pp. 715–734.

Berlin Tourismus & Kongress GmbH (2015) *Wirtschaftsfaktor für Berlin: Tourismus- und Kongressindustrie*, [Online] Available at: www.visitberlin.de/sites/default/files/visitber lin_wirtschaftsfaktor_tourismus_und_kongressindustrie.pdf [accessed 18 August 2016].

Berliner Morgenpost (2016) 'Simon-Dach-Kiez verschwindet aus Berliner Touristen-App', [Online] Available at: www.morgenpost.de/berlin/article206961705/Simon-Dach-Kiez-verschwindet-aus-Berliner-Touristen-App.html [accessed 17 August 2016].

Berner, L. and Wickert, J. (2011) 'Keine Gespensterdebatte. Die Zweckentfremdung durch Ferienwohnungsnutzung nimmt weiter zu – Studie zeigt Umfang und Verteilung von Ferienwohnungen in Berlin', *MieterEcho. Zeitung der Berliner MieterGemeinschaft e.V.*, no. 350, pp. 4–9.

Bernt, M., Grell, B. and Holm, A. (eds) (2013) *The Berlin Reader: A Compendium on Urban Change and Activism*, Bielefeld: Transcript Verlag.

Bezirksverordnetenversammlung Friedrichshain-Kreuzberg (2014) *Aktionsplan Warschauer Brücke/Revaler Straße*, [Online] Available at: http://gruene-xhain.de/media/filer_public/99/37/993730e3-2ce0-47af-b82f-1001b699a59c/ds1448_aktionsplan_warschauer_brucke.pdf [accessed 17 August 2016].

Blume, J. (2016) 'Dunkelziffer bleibt hoch. Ferienwohnungen im Zeichen des Wohnungsmangels und neue Strategien von Vermietern', *MieterEcho, Zeitung der Berliner MieterGemeinschaft e.V.*, no. 382, pp. 4–7.

Botsman, R. and Rogers, R. (2011) *What's Mine is Yours – How Collaborative Consumption is Changing the Way We Live*, London: Harper Collins.

Cócola Gant, A. (2015) 'Tourism and Commercial Gentrification', Conference paper for the RC21 International Conference on The Ideal City: Between Myth and Reality. Representations, Policies, Contradictions and Challenges for Tomorrow's Urban Life, [Online] Available at: www.rc21.org/en/wp-content/uploads/2014/12/E4-C%C3%B3cola-Gant.pdf [accessed 18 August 2016].

Colomb, C. and Novy, J. (2017) *Protest and Resistance in the Tourist City*, London and New York: Routledge.

Cox, M. (2015) *Inside Airbnb: Adding Data to the Debate. Berlin*, [Online] Available at: http://data.insideairbnb.com/germany/be/berlin/2015-10-03/visualisations/listings.csv [accessed 18 August 2016].

Cox, M. (2016a) *Inside Airbnb: Adding Data to the Debate. London*, [Online] Available at: http://data.insideairbnb.com/united-kingdom/england/london/2016-06-02/visualisations/listings.csv [accessed 18 August 2016].

Cox, M. (2016b) *Inside Airbnb: Adding Data to the Debate. Paris*, [Online] Available at: http://data.insideairbnb.com/france/ile-de-france/paris/2016-07-03/visualisations/listings.csv [accessed 18 August 2016].

Deloitte (2013) *Hotelmarkt Berlin. Eine Hauptstadt für sich*, [Online], Available at: https://www2.deloitte.com/content/dam/Deloitte/de/Documents/consumer-business/Hotelmarkt%20Berlin%202013%20Deutsch.pdf [2 February 2017]

Der Tagesspiegel (2015) 'Pantomime gegen Ballermannisierung von Berlin', [Online] Available at: www.tagesspiegel.de/berlin/party-kieze-und-touristen-pantomime-gegen-ballermannisierung-von-berlin/11502698.html [accessed 18 August 2016].

DLR Deutsches Zentrum für Luft- und Raumfahrt (2015) 'Low-Cost-Carrier – Ein aktueller Trend?' [Online] Available at: www.dlr.de/fw/Portaldata/42/Resources/dokumente/pdf_dokumente/4AB_01_LF_LCC.pdf [accessed 18 August 2016].

Dyckhoff, T. (2011) 'Let's Move to Kreuzkölln', *The Guardian*, 19 March 2011, [Online] Available at: www.theguardian.com/money/2011/mar/19/move-to-kreuzkolln-berlin [accessed 17 August 2016].

Edensor, T. (2001) 'Performing Tourism, Staging Tourism: (Re)producing Tourist Space and Practice', *Tourist Studies*, vol. 59, no. 1, pp. 59–80.

Freytag, T. (2008) 'Making a Difference: Tourist Practices of Repeat Visitors in the City of Paris', *Social Geography Discussions*, vol. 4, pp. 1–25.

Füller, H. and Michel, B. (2014) '"Stop Being a Tourist!" New Dynamics of Urban Tourism in Berlin-Kreuzberg', *International Journal of Urban and Regional Research*, vol. 38, no. 4, pp. 1304–1318.

Germann Molz, J. (2014) 'Toward a Network Hospitality', *First Monday*, vol. 19, no. 3, pp. 1–17.

Germann Molz, J. and Gibson, S. (eds) (2007) *Mobilizing Hospitality: The Ethics of Social Relations in a Mobile World*, Hampshire, Burlington, VT: Ashgate Publishing.

GEWOS Institut Für Stadt-, Regional und Wohnforschung GmbH. (2014) Airbnb and the Berlin Housing Market: The Impact of Airbnb Properties on the Housing

Supply in Berlin, [Online] Available at: http://publicpolicy.airbnb.com/wp-content/uploads/2014/12/AirbnbandtheBerlinhousingmarket.pdf [accessed 4 January 2016].

Gilmore J. H. and Pine, B. J. (2007) *Authenticity: What Consumers Really Want*, Boston: Harvard Business Review Press.

Gravari-Barbas, M. and Jacquot, S. (2016) 'No Conflict? Discourses and Management of Tourism-Related Tensions in Paris', in Colomb, C. and Novy, J. (eds), *Protest and Resistance in the Tourist City*, London and New York: Routledge, pp. 31–51.

Griffin, T., Hayllar, B. and Edwards, D. (2008) 'Places and People: A Precinct Typology', in Hayllar, B., Griffin, T. and Edwards, D. (eds), *City Spaces – Tourist Places: Urban Tourism Precincts*, Amsterdam *et al.*: Elsevier.

Günther, A. (2006) '20 Jahre Erlebnisgesellschaft – und mehr Fragen als Antworten. Zwischenbilanz oder Abgesang auf die Erlebniswelten-Diskussion', in Reuber, P. and Schnell, P. (eds), *Postmoderne Freizeitstile und Freizeiträume. Neue Angebote im Tourismus*, Berlin: Erich Schmidt Verlag.

Guttentag, D. (2015) 'Airbnb: Disruptive Innovation and the Rise of an Informal Tourism Accommodation Sector', *Current Issues in Tourism*, vol. 18, no. 12, pp. 1192–1217.

Heinrichs, H. and Grunenberg, H. (2012) 'Sharing Economy. Auf dem Weg in eine neue Konsumkultur?' *SSOAR. Open Access Repository,* [Online] Available at: http://nbn-resolving.de/urn:nbn:de:0168-ssoar-427486 [accessed 24 January 2016].

Holm, A. (2013) 'Berlin's Gentrification Mainstream', in Holm, A., Grell, B. and Bernt, M. (eds), *The Berlin Reader: A Compendium on Urban Change and Activism*, Bielefeld: Transcript-Verlag, pp. 171–187.

Holm, A. (2016a) 'Berlin: Erwerbsloseninitiative fordert Beschlagnahme von Ferienwohnungen', 20 January 2016, [Online] Available at: http://gentrificationblog.word press.com [accessed 24 January 2016].

Holm, A (2016 b) 'Berlin: Wie verändert Airbnb den Wohnungsmarkt? Eine politische Ökonomie der Ferienwohnungen', [Online] Available at: http://gentrificationblog.wordpress.com/2016/07/05/berliin-wie-veraendert-airbnb-den-wohnungsmarkt-eine-politische-oekonomie-der-ferienwohnungen/ [accessed 17 August 2016].

Holm, A. (2016c) *Sozialer Wohnraumversorgungsbedarf in Berlin*, [Online] Available at: www.sowi.hu-berlin.de/de/lehrbereiche/stadtsoz/forschung/projekte/bericht-wohnraumversorgungsbedarf-berlin-holm-2016.pdf [accessed 17 August 2016].

Hotel Reservation Service HRS (2015) *Hotelpreisentwicklung 2014: Stabile Preise in Deutschland, starke Schwankungen in Europa und weltweit*, [Online] Available at: http://hrs.de/presse/wp-content/uploads/2015/01/2015-01-02-PM-HRS-Hotelpreisradar-2014_final.pdf [accessed 17 August 2016].

IHA Hotelverband Deutschland e.V. (2015) 'Hotelverbände fordern von Regierungen und EU-Institutionen gleiche Wettbewerbsbedingungen und Verbraucherschutz in der so genannten "Sharing" Economy', [Online] Available at: www.hotellerie.de/de/hotelverbaende-fordern-von-regierungen-und-eu-institutionen-gleiche-wettbewerbs-bedingungen-und-verbraucherschutz-in-der-so-genannten-sharing-economy [accessed 6 February 2016].

Investitionsbank Berlin (2015a) *IBB Wohnungsmarktbericht 2015*, [Online] Available at: https://www.ibb.de/media/dokumente/publikationen/berliner-wohnungsmarkt/wohnungsmarktbericht/ibb_wohnungsmarktbericht_2015.pdf [accessed 7 Feb. 2017].

Investitionsbank Berlin (2015b) *IBB Wohnungsmarktbarometer 2015*, [Online] Available at: https://www.ibb.de/media/dokumente/publikationen/berliner-wohnungsmarkt/woh nungsbarometer/ibb_wohnungsmarktbarometer_2015.pdf [accessed 7 Feb. 2017].

Investitionsbank Berlin (2016) IBB Wohnungsmarktbarometer 2016, [Online], Available at: https://www.ibb.de/media/dokumente/publikationen/berliner-wohnungsmarkt/woh nungsbarometer/ibb_wohnungsmarktbarometer_2016.pdf [06 February 2017].

Jacobs, L. (2016) 'Erst belegen, dann besetzen', *Zeit Online*, 4 Mai 2016, [Online] Available at: www.zeit.de/gesellschaft/2016-05/berlin-hausbesetzung-ferienwohnun gen-airbnb-polizei [accessed 17 August 2016].

Judd, D. R. (1999) 'Constructing the Tourist Bubble', in Judd, D. R. and Fainstein, S. S. (eds), *The Tourist City*, New Haven: Yale University Press.

Jung, A. (2015) 'Spekulationsblasen am Immobilienmarkt: Hier laufen die Kaufpreise den Mieten davon', *Spiegel Online*, 6 May 2015, [Online] Available at: www.spiegel. de/wirtschaft/unternehmen/immobilien-wohnungspreise-steigen-schneller-als-die-mieten-a-1032645.html [accessed 24 January 2016].

Kagermeier, A. (2015) *Tourismusgeographie. Einführung*, München/Konstanz: UVK/ Lucius.

Kagermeier, A., Köller, J. and Stors, N. (2015) 'Sharing economy im Tourismus. Zwischen pragmatischen Motiven und der Suche nach authentischen Erlebnissen', *Zeitschrift für Tourismuswissenschaft*, vol. 7, no. 2, pp. 117–145.

Kagermeier, A., Köller, J. and Stors, N. (2016), 'Airbnb als Sharing economy-Heraus-forderung für Berlin und die Reaktionen der Hotelbranche', in Bauhuber, F. and Hopfinger, H. (ed.) Mit Auto, Brille, Fon und Drohne. Aspekte neuen Reisens im 21. Jahrhundert, Mannheim: MetaGis (=Studien zur Freizeit- und Tourismusforschung, vol. 11), pp. 67–94.

Köller, J. (2015) *Sharing Economy im Tourismus. Reaktionen der Berliner Hotelbranche auf den neuen Wettbewerber Airbnb*, unpublished Master Thesis, Trier University.

Krajewski, C. (2004) 'Gentrification in Zentrumsnähe. Das Beispiel Spandauer Vorstadt in Berlin-Mitte', *Praxis Geographie*, vol. 34, no. 10, pp. 12–17.

Krajewski, C. (2013) 'Gentrification in Berlin. Innenstadtaufwertung zwischen etablieren "In-Quartieren" und neuen "Kult-Kiezen"', *Geographische Rundschau*, vol. 65, no. 2, pp. 20–27.

Land Berlin (2015) *Einwohnerinnen und Einwohner in Berlin in LOR Planungsräumen am 31.12.2014*. [Online] Available at: http://daten.berlin.de/datensaetze/einwohnerinnen-und-einwohner-berlin-lor-planungsräumen-am-31122014 [accessed 24 January 2016].

Liedtke, A. (2011) *One Couch at a Time – Analysing the Travel Behaviour and Target Group of Couchsurfer*, unpublished Bachelor Thesis University of Applied Sciences Stralsund.

Linne, M. (2014) 'Sharing economy im Tourismus', in Linne, M. (ed.) *Smart Tourism – Sharing Economy im Tourismus. Produkte, Grenzen, Folgen*, Elmshorn: IDT-Verlag.

Mai, D. (2016) '#boycottairbnb Plakate in Berlin rufen zum Boycott von Airbnb auf'. *Berliner Zeitung*, 14 July 2016, [Online] Available at: www.berliner-zeitung.de/ berlin/-boycottairbnb-plakate-in-berlin-rufen-zum-boykott-von-airbnb-auf-24394240 [accessed 17 August 2016].

Maitland, R. and Newman, P. (eds) (2009) *World Tourism Cities*, Oxon: Routledge.

Marquart, M. and Braun, K. (2014) 'Share Economy. Deutsche teilen nicht', *Spiegel-Online*, 17 October 2014, [Online] Available at: www.spiegel.de/wirtschaft/ service/share-economy-uber-und-airbnb-deutsche-wollen-nicht-teilen-a-997502.html [accessed 18 August 2016].

Nielsen (2014) *Is Sharing the New Buying? Reputation and Trust are Emerging as New Currencies*, [Online] Available at: www.nielsen.com/content/dam/nielsenglobal/apac/ docs/reports/2014/Nielsen-Global-Share-Community-Report.pdf [accessed 4 January 2016].

Novy, J. (2013) 'Berlin Does Not Love You'. Notes On Berlin's "Tourism Controversy" and its Discontents', in Bernt, M., Grell, B. and Holm, A. (eds), *The Berlin Reader. A Compendium on Urban Change and Activism*, Bielefeld: Transcript Verlag, pp. 224–237.

n-tv (2011) 'Hilfe, die Touris kommen', 1 March 2011, [Online] Available at: www.n-tv. de/reise/Hilfe-die-Touris-kommen-article2732231.html [accessed 17 August 2016].

Pappalepore, I., Maitland, R. and Smith, A. (2010) 'Exploring Urban Creativity: Visitor Experiences of Spitalfields, London', *Tourism, Culture & Communication*, vol. 10, no. 3, pp. 217–230.

Pappalepore, I., Maitland, R. and Smith, A. (2014) 'Prosuming Creative Urban Areas. Evidence from East London', *Annals of Tourism Research*, vol. 44, pp. 227–240.

Picard, D. and Buchberger, S. (eds) (2013) *Couchsurfing Cosmopolitanisms: Can Tourism Make a Better World?* Bielefeld: Transcript-Verlag.

Rapp, T. (2009) *Lost and Sound: Berlin, Techno und der Easyjetset*, Frankfurt am Main: Suhrkamp Verlag.

Senatsverwaltung für Justiz und Verbraucherschutz Berlin (2013) *Gesetz über das Verbot der Zweckentfremdung von Wohnraum (Zweckentfremdungsverbot-Gesetz – ZwVbG) vom 29. November 2013*, [Online] Available at: http://gesetze.berlin.de/jportal/?quelle= jlink&query=WoZwEntfrG+BE&psml=bsbeprod.psml&max=true [accessed 4 January 2016].

Senatsverwaltung für Justiz und Verbraucherschutz Berlin (2014*) Verordnung über das Verbot der Zweckentfremdung von Wohnraum* (Zweckentfremdungsverbot-Verordnung – WwVbVO) vom 4. März 2014, [Online] Available at: http://gesetze.berlin.de/jportal/ portal/t/hhn/page/bsbeprod.psml/action/portlets.jw.MainAction?p1=7&eventSubmit_do Navigate=searchInSubtreeTOC&showdoccase=1&doc.hl=0&doc.id=jlr-WoZw EntfrVBEpP6&doc.part=S&toc.poskey=#focuspoint [accessed 18 August 2016].

Senatsverwaltung für Stadtentwicklung und Umwelt Berlin (2016a) *Wohnungsbestand. Zweckentfremdungsverbot von Wohnraum*, [Online] Available at: www.stadtentwick lung.berlin.de/wohnen/zweckentfremdung_wohnraum/ [accessed 18 August 2016].

Senatsverwaltung für Stadtentwicklung und Umwelt Berlin (2016b) *Lebensweltlich ori-entierte Räume (LOR) in Berlin. Planungsgrundlagen*, [Online] Available at: www. stadtentwicklung.berlin.de/planen/basisdaten_stadtentwicklung/lor/ [accessed 1 February 2016].

Skowronnek, A., Vogel, L. and Parnow, J. (2015) *Airbnb vs. Berlin. Was sagen die Daten?* Fachhochschule Potsdam, [Online] Available at: www.airbnbvsberlin.de [accessed 4 January 2016].

Stampfl, N. S. (2014) 'Sharing Economy – Neue Konsumeinstellungen und verändertes Konsumverhalten', in Linne, M. (ed.) *Smart Tourism – Sharing economy im Tourismus. Produkte, Grenzen, Folgen*, Elmshorn: IDT-Verlag.

Statistisches Bundesamt (2014). *Gebiet und Bevölkerung – Haushalte*, [Online] Available at: www.statistik-portal.de/Statistik-Portal/de_jb01_jahrtab4.asp [accessed 4 January 2016].

Statistisches Bundesamt (2015) *Ankünfte und Übernachtungen in Beherbergungsbetrieben (Städtetourismus): Deutschland, Jahre*, [Online] Available at: https://www.destatis. de/DE/ZahlenFakten/Wirtschaftsbereiche/BinnenhandelGastgewerbeTourismus/ Tourismus/Tabellen/AnkuenfteUebernachtungenBeherbergung.html [accessed 7 Feb. 2017].

Stors, N. (2014) 'Explorer-Touristen im Städtetourismus. Ein Charakterisierungsversuch unterschiedlicher Besuchergruppen in Kopenhagen', *Zeitschrift für Tourismuswissen-schaft*, vol. 6, no. 1, pp. 97–105.

Stors, N. and Kagermeier, A. (2013) 'Crossing the Border of the Tourist Bubble: Touristification in Copenhagen', in Thimm, T. (ed.) *Tourismus und Grenzen*, Mannheim: MetaGis (=Studien zur Freizeit- und Tourismusforschung, vol. 9), pp. 115–131.

Surhone, L. M., Timpledon, M. T. and Marseken, S. F. (2010) *Prosumer*, VDM Verlag, Saarbrücken.

Wohnungsbörse (2015) *Immobilienpreisentwicklung in Deutschland*, [Online] Available at: www.wohnungsboerse.net/immobilienpreise [accessed 4 January 2016].

Zukin, S., Trujillo, V., Frase, P., Jackson, D., Recuber, T. and Walker, A. (2009) 'New Retail Capital and Neighborhood Change: Boutiques and Gentrification in New York City', *City & Community*, vol. 8, no. 1, pp. 47–64.

9 Post-tourism on the waterfront

Bringing back locals and residents at the Seaport

Sandra Guinand

Introduction

The first two weeks of May 2016 were designated Design week festival in New York City. On this occasion, South Street Seaport, historic district formerly devoted to port and merchant activity, located on the southern tip of Manhattan, was hosting various events on its space attracting audiences linked to or interested in fashion, design, culture and art.

This place used to be Benjamin Thompson and James Rouse's festival marketplace built between 1983 and 1985. Located on the waterfront's derelict land, devoted to recreation and consumption based on the traditional concept of European markets, one would come to look at the old ships or enjoy a walk by the water, have an ice-cream and a drink, shop in the retail pavilion and eat a bite at the food court market. The place would be animated with concerts or performances and the historical setting was part of the cultural and (re)creative enjoyment. In the 1980s cities were fun (Demarest, 1981) and following the acclaimed success of Faneuil Hall in Boston (1978) and Harbor Place in Baltimore (1980), festival marketplaces became a successful economic model for redevelopment of US inner cities. National tourism played an important role in this redevelopment process as these places quickly became flooded with tourists and suburbanites (Bloom, 2004). For instance, Harbor Place was said to have received 22 million annual visitors (during the years which followed its inauguration) among whom 6 million tourists (Hall, 2002: 386) and South Street Seaport 12 million annual visitors from 1983 to 1988 (Yarrow, 1988). By the mid-1980s local retail sellers had given space to popular shops and festival marketplaces translated into large commercial and entertainment ventures.

Today, South Street Seaport is undergoing major redevelopment. The developer, Howard Hughes, a major Texan developer firm, supported by the economic development corporation agency of the city (EDC), is imposing its own agenda of the future urban and social landscape development for this area. After much turpitude, the marketplace was pulled down and any references to its tourism era reduced to silence. Howard Hughes actually stresses the fact that this area should go beyond tourism and tourists as it has much more to offer, especially to New Yorkers. The developer operates a "reclaiming" strategy over the site for the sake of bringing this piece of land into the twenty-first century and repositioning it as the heart of Manhattan and the rest of the metropolis (http://www.southstreetsea port.com/LIFE-CULTURE/seaport-of-tomorrow.html).

Seen in the context of the (neo)liberal city (Le Galès, 2016), waterfronts in New York have since the Bloomberg's administration (2002) been identified as strategic sites for twenty-first-century accumulations (Schaller and Novy, 2010). Alongside federal policies (1972) to improve the outdoor water quality and the Lower Manhattan (redevelopment) plan (1966), the Seaport was very early on identified as a place to live and play by the water. Consequent to a new urban regime (Hydra, 2012) led by urban renewal policies and post-industrial growth, these redevelopments were adding new features and social profiles and contributing to gentrification on the water's edge (Schaller and Novy, 2010). Today, the developer's use of qualitative urban furniture, architectural design, careful place-making, creation of amenities and organization of events, all contribute to capture and capitalize on future public, residents, customers and consumers of the site inducing a new stage in the gentrification process. Although this strategy is set in a post-tourism perspective, it is in fact not so remote from the original and initial idea first proposed by Rouse's festival market place project, the latter being replaced by high-end facilities and products meant to make sure it remains a 'top destination' for Lower Manhattan residents and visitors.

In this chapter I examine the socio-economic trajectory of a prime tourist destination, a festival marketplace whose notoriety has faded away, and look at the means the principal actor in charge of this piece of land is implementing in order to bring residents and users back. I show that if the strategies sought by the developer are playing on upscale facilities and uses, creating a super-gentrification process, the rhetoric and story-telling behind it is rooted in mundane sense of time and community-led values proper to the new shift operated in the tourism economy (Edensor, 2007). First, I expose the concept of post-tourism or tourism of everyday life and explain the super-gentrification process at work in New York City, especially on its waterfronts. I then turn to South Street Seaport's socio-economic development leading to its preservation and the festival marketplace project. I examine the main features and ideologies at the heart of the project that led to what Peter Hall coined the "rousification" of America (2002: 383). As festival marketplaces were widely acclaimed as epicentres of US urban renaissance, they also displaced activities, if not people, as a means to upgrade the urban landscape. New York City is in a constant process of reinventing itself and Rouse's project was no exception. I then look at how it went from an emblematic model of retail and (re)creative tourism to a post-tourism place anchored in the reinvention of mundane routine. I will finally show how these relationships with daily life induce super-gentrification and how this place is becoming a high-end destination.

This work is based on a 12-month field work research project using qualitative data based on semi-directive interviews (including photographic ones), observations, official plans and reports, as well as web-based marketing or advocacy materials. South Street Seaport is an interesting case as this project takes place in a global and highly visited city where (neo)liberal recipes and private actors have been predominantly active in the conduct of urban affairs.

Post-tourism and super-gentrification on the waterfront

Scholars (Novy and Huning, 2008; Novy, 2011; Gravari-Barbas and Fagnoni, 2013) in tourism studies have witnessed a shift in tourism features and geographies. Tourism mobility is establishing new links with time, place and landscape. It has integrated the domain of the ordinary, the pace and rhythm of an idealized mundane life focusing on the experiential economy. As some tourism Meccas might be losing popularity and becoming places of the ordinary (Stock and Lucas, 2012), the ordinary is in turn shaping more and more new tourism itineraries, places and formulating new demand.

Post-tourism, a term coined by post-modern scholars (Feifer, 1985; Urry 1990; Rojek, 1997), appears as a specific process, which redefines tourism's issues without necessarily and explicitly following the tourism line (Viard, 2000, 2006). Indeed, strong changes experienced within the shift of the economy and technology have democratized tourism while at the same time contracting space and time (Harvey, 1989) and blurring the lines between proximity and distance, the exotic and the ordinary (Bourdeau, 2012). Structural changes have also induced a series of adjustments within tourism referents of action. For instance, tourism practices and policies have integrated notions of sustainability, strengthened the importance of social interactions, preservation and integrity while at the same time embraced the vocabulary of globalization, of acute competition and entrepreneurship.

This new geography of tourism consequently implies the harnessing of the tourism economy by a number of varied new actors proposing new products, as the well-known Airbnb phenomenon illustrates. But it also implies a voluntary confusion between the role of the tourist and the non-tourist as visitors' and ordinary behaviours are "mixing within a fun-based urban world" (Stock and Lucas, 2012). The duality between tourist and non-tourist travel is increasingly giving way to a continuum of mobility practices (Knafou, 2000; Maitland, 2010). People increasingly have multi-residences, endorse multiple lifestyles as they have multicultural points of reference. This form of tourism can be described as 'collage tourism' (Rojek, 1997: 62).

The experiential economy seems to better respond and capitalize on this phenomenon as it allows the penetration of everyday life. As Girard stresses, the tourist can no longer be defined as: "the one who made the transition between the periodical and temporary of ordinary industrial work space and time towards the extraordinary time of holiday departures" (2013: 45). In the broader context of entrepreneurship and fierce competition among territories, the experiential economy gets imbedded within the commodification of culture (Zukin, 1995) within the scope of tourism, entertainment and marketing. The "alternative", "authentic", "exceptional", the "off the beaten track" or the "underground" then can all become part of redevelopment project and place-marketing strategies (Colomb, 2012).

Gentrification has usually been the end result of such policies (Colomb, 2012) but in the context of New York City, and more specifically Manhattan (already highly gentrified), this gentrification process takes the aspects of super-gentrification (Lees, 2003). As Lees and Butler explain:

> The term super-gentrification describes a further process of gentrification that has already been occurring, a process that includes a significant step change in social class composition and evidence of social replacement (rather than displacement) with a significant transformation in community relations. (2006: 469)

This super-gentrification trend on the waterfront in New York City can be traced back to the Bloomberg administration (2002) and shift in the management of urban affairs. Before Bloomberg, the shorelines had not been seen as major assets for economic, housing and tourism strategy. Except for the Seaport, Battery and the Hudson Park that had been led under private initiatives, no major project had been City driven until then (Schaller and Novy, 2010). The shift in urban policies consisted in a more proactive role of public authorities in planning processes and urban design (by re-zoning, place-making, etc.) aiming at attracting major high-end businesses and upper-class residents. Bloomberg never hid his vision of New York's position as a high-end and luxury product with tremendous value for businesses able to capitalize on it (Schaller and Novy, 2010).

The waterfront then quickly became the prime location to implement these new precepts. If Howard Hughes' current project is not directly linked to Bloomberg's redevelopment schemes, it follows the general trend of ambitious high-class redevelopment projects that have since then spurred on New York's waterfront.

A festival market in New York City

Unfolding the urban renewal processes

The Seaport district has always been depicted as the core of New York's history, the origin of the merchant city (Figure 9.1). A place of exchanges of people and goods, this Lower area of Manhattan - not the Seaport per se – has always been under the scrutiny of businesses and economic actors.

By 1965 Lower Manhattan's working waterfront had largely faded out: of the fifty-one piers only eighteen were still active (Willis, Willen and Rossant, 2002). Strong economic shifts, movement of capital, dilapidated semi-vacant structures (Boyer, 1992) and depression in inner cities called for a shift in paradigm and action. Lower Manhattan was suffering tremendous losses of capital and forces towards Midtown. While under the Downtown Lower Manhattan Association (DLMA) David Rockefeller pushed to keep the Chase bank headquarters there by encouraging the creation of a new financial and business district, New York City officials backed by the Port Authority agreed that Manhattan should become white and not blue collar. The end result was a major shift in land uses. The port industry

Figure 9.1 Location of South Street Seaport in Lower Manhattan

Source: Author with Google Maps background

was to be relocated to the New Jersey shore (Newark) and with it a whole busi-ness economy. Backed with federal subsidies under urban renewal programmes, *revanchism* (Smith, 1979; 1996) was underway and the economic powers of the city were not going to lose the battle.

By 1968, a whole chunk of the waterfront had been declared "blighted" and designated as an urban renewal district (later named Southeast Brooklyn Bridge Urban Renewal Area) (Housing and Development Administration, 1968). Prior to that period and anticipating the change in land use to come with the completion of the World Trade Center, the DLMA encouraged the elaboration of the Lower Manhattan plan (1966). This plan focused on the potential of the waterfront: a great natural and unique place that the general public should take advantage of (Willis *et al.*, 2002: 57). It called for the water and shore to become a "shared public good". The claim over the area was to be realized through a massive redevelopment of luxury housing and plazas along the shore. The housing devel-opment did not occur however as a feasibility study showed that only a small fringe of workers in the area could earn sufficient to afford the place and that a percentage of the proposed housing should be affordable (Reed, 1969). However, the plan had put the waterfront at the forefront, revealing its edges as a great asset for future developments. What was needed then was a change in social profiles.

The Seaport area and its buildings would have been erased (as was most of lower Manhattan) at the call of progress and development, had intellectual figures like Peter Stanford, his wife, Norma and editorialist at the New York Times, Ada Louis Huxtable, not been involved in the fight for its preservation.

> When will New York learn its lesson? What will happen to the one last remaining neighborhood of these handsome Greek Revival buildings of the 1830's on Front Street from Coenties Slip to John Streets; the even earlier buildings on Fulton Street? Hemmed in by the new Stock Exchange on the south and the Brooklyn Bridge project to the north, the area will now be ripe for speculative purchase and destructive "upgrading". (Huxtable, 1964)

In 1967, with the help of friends concerned with maritime and New York history Peter Stanford put together South Street Seaport Museum, a non-profit corporation meant to preserve the area around Shermerhorn row (landmarked in 1968), a row of six joint houses dating from 1811. The plan they developed was meant to give the area the status of an open-air maritime museum (South Street Reporter, 1967). The idea was that it could provide "the telling of a valuable story and the addition of a valuable amenity to the city life" (South Street Seaport Museum, 1968). As capital was moving out from the buildings (not the land) thus pushing for their destruction, many believed in their safeguarding as witnesses to and common good of the urban history of the city: "New Yorkers will enter into the heritage of dreams and hard work that built their city" (South Street Seaport Museum, 1968). They also thought that the site had a role to play on the wider urban scale: it bore a strong symbolic function accumulating layers of history and being part of a wider cultural and social system.

By 1968 the City Council had approved the urban renewal plan aimed at restoring and rehabilitating the area south-east of the Brooklyn Bridge, including the Seaport, through a mix of preservation, pedestrian amenities, and new commercial and residential development. Preservationists had become more vocal and contestations over real estate forces erasing old urban fabrics were growing (Jacobs, 1961). At the same time, the US Mayors conference endorsed historical preservation as a tool for economic development. City authorities under Mayor Lindsay agreed to leave the area to South Street Seaport Museum by making it the unassisted sponsor for the Brooklyn Bridge Southeast Urban Renewal District (Lindgren 2014).

The Special District also established design controls aimed at enhancing the pedestrian experience and preserving clear views of ships docked at the Seaport and of the Brooklyn Bridge (Metzger, 2010).

Mayor Lindsay with David Rockefeller later announced the Manhattan Landing Plan (1972), which sought to create a 24-hour community in Lower Manhattan including South Street Seaport as a cornerstone for the development (Metzger, 2010). This plan envisioned the creation of high-density residential buildings along the waterfront including 6 million square feet of office space,

9,500 units of market-rate housing, a 1,000-car garage, and 400-room hotel. The project expanded Lower Manhattan by 88 acres of new platforms over the river (City Planning of New York, 2002?: 15).

Rousification and touristification

The 1970s crisis put a drastic halt to development plans and public expenditure (at all levels). It strongly damaged any development projects and parks' management, the only sector in the city that had until then mainly be financed by and managed with public money (Banarjee, 2001). This in turn brought a shift in the perception of public space. As their maintenance diminished, their degradation made them less appealing to the "public", and they soon became places of "marginalization" – leaving people to question their function and utility to the public. It led to private intervention (not-for-profit and businesses' organizations) to "sanitize" them (Banerjee, 2001).

By 1971 it was clear that South Street Seaport Museum could not bear the development of the district alone – at least not the type of development expected by City authorities, private interests and the upper-middle class of Manhattan. The neighbourhood had its own images and reputation, starting with the fish market and its surrounding crowd.[1] Moreover, the Museum was in financial disarray. Seaport Holdings owed millions to the banks, among them a 2.9 million loan it had made to the Museum (Lindgren, 2014). The South Street Seaport's financial situation became an opportunity for the banks and the Lower Manhattan development office which were eager to move forward with the Landing Plan. The banks purchased the air rights in a favourable deal and the City, backed with bonds it issued, purchased Shermerhorn Row and the blocks held by the Museum (Lindgren, 2014). The latter was to remain the main sponsor for the area but had to pay back 13 per cent to the City of leases and retail sells (Lindgren, 2014). To fit into the new development schemes, Peter Stanford was asked to come up with a new development project for the Seaport district. The expected orientation called for a profit-oriented and a market-based project with the removal of the fish market, regarded as having an "adverse impact" on the future development (Office of Lower Manhattan Development, 21 July 1975). With its specific ecological environment, the area was to play the "magical magnet for tourism and culture". It just needed some upscaling and sanitization.

Stanford came up with a new plan (1973), proposing a financially feasible balance of uses that would generate revenue to support the restoration of the area while allowing Museum activities (South Street Seaport Museum, 1973). The authors of the plan wanted to distinguish the envisioned development from a very conservationist approach and a modern development where the ships and the historical fabric would only serve as scenery purposes. They had something else in mind. The idea was to preserve the area as a "working district, with real activities and a genuine life of its own, while also being of interest to tourists and visitors" as it was "never elegant or particularly charming" (South Street Seaport Museum, 1973: 5).

Around that period, James Rouse had put waterfront redevelopments at the forefront of the public authorities' attention, notably with Faneuil Hall in Boston (1976) and Harbor Place in Baltimore (1980) (Figure 9.2). These projects were presented by the media as the new model to trigger urban Renaissance (*The Boston Globe*, August 25, 1980; *New York Times*, September 12, 1985; *The Sun*, June 24, 1990; *American Legion*, 1986). Capitalizing on their popularity and the then success of Baltimore's Inner Harbor, Rouse under his firm "Enterprise Development Company" contributed in the 1990s to the worldwide diffusion of the redevelopment model (Interview with M. Millspaugh). Festival marketplaces became the symbol of urban strategy pursued by city governments (Sigler, 2009) in order to attract back young suburbanites and young urban professionals ('yuppies') (Hall, 2002) to the city.

The Seaport area was to be his new playground, as he was approached by the new Chairman of the board of the Seaport Real Estate Committee, backed by Downtown Lower Manhattan Agency and City authorities (Lindgren, 2014). A master plan for the Seaport was designed by Benjamin Thompson. The proposed festival marketplace was composed of two new buildings and a pedestrian centre (Fulton Street) connecting Wall Street area to the East River. The first building which opened in 1983 was directly facing the Shermerhorn Row. It was conceived as a reminiscence of the old Fulton market that had been torn down earlier. Inside the visitor would find local merchants and artisan stands and specialized boutiques as well as a delicacy food court. The second building was erected on Pier 17 and completed in 1985. The architecture, the design and the scale of these two new steel and glass-structured buildings were

Figure 9.2 A view of Harbor Place, with the WTC and the National Aquarium
Source: Author, October 2015

to fit in the historical landscape of the Seaport. The ambitious emphasis on commercial development (approximately 250,000 square feet) far exceeded the earlier visions for the Seaport Museum, but the financial feasibility study predicted a strong market potential. The expanding workforce of lower Manhattan, the increasing number of tourists to New York and residents of the area would generate demand (Metzger, 2010: 33).

If South Street Seaport had been represented in the media as having largely preserved and "recycled" the old historic structures and fabrics associated with the port, contributing to the historical depth of the site and its "originality", numerous scholars have given critical accounts of the project (DeFillippis, 1997; Metzger, 2010). The intervention of the private sector imposed its own agenda on design, framing the whole district. As Andrew Hurley (2006) and Christine Boyer (1992) pointed out, references to social conflict and the industrial past had deliberately been ignored. Boyer (1992: 184) notes that the museum's historical imaginary transcended the physical building of the museum (Figure 9.3). It had been carefully composed and put into images, working as an historical enclave and as an architectural historical promenade for the tourists and the Wall Street white-collar workers. Referring to the place as a theme park, Michael Sorkin qualified this type of intervention as the end of public space (1992). Peter Hall depicted it as a "city as a stage" but without the real urban life (2002: 386). Looking at old photographs, hearing and reading various accounts on the social and cultural life depicting the place

Figure 9.3 View of the Seaport with the Museum Gallery from Water Street

Source: Maria Gravari-Barbas, 2004

before its transformation, one could only be struck by the radical changes in the urban setting. The upgrading and rehabilitation of the buildings, the design of the urban space, the choice of the outdoor furniture, the use of colours, the delimitation of public space and the signs, all gave a sense of homogenization, safeness and normalization, which in turn contributed to the smoothening and securing of the place and what certain scholars qualified as "sanitization" (DeFillippis, 1997; Hall, 2002).

The intervention by the developer altered the conception of public space. For instance, Rouse's company directly intervened on the public realm, modelling it according to urban design stances and marketing purposes (Figure 9.4). The Museum had lost its leadership as the main sponsor and was asked to adapt its logo to the festival market. At the time, the only remnant of the "transgressive" period was Sloppy Louise's – a restaurant where "locals" would go. The sanitized space was completed with the move of the Fulton Fish market in 2005 (but planned as early as 1975) to the Bronx. "The place lost its flavour although the smell remained for much longer" (Bridget, Interview, 2015).

It is however this careful attention to the urban design, architecture, functions and activities that made the commercial success of this project. In his account Goss (1996) argues that the success of festival marketplaces is linked to this very specific attention brought to the sanitization of the place and the constant interplay between past and present, creating illusions of an idealized and "cleaned" past, where people of all classes could be mixed (which was not necessarily the case)

Figure 9.4 South Street Seaport Marketplace's signage and new Fulton market building

Source: Maria Gravari-Barbas, 2004

without tensions in open spaces and where shoppers purchased goods from individual merchants. This illusion was carefully looked at by the choice of functions and activities mainly geared at entertainment and the selection, at the beginning of the project, of tenants and small retailers. Maria Gravari-Barbas (1998) underlines the importance of the attention given to the building's architecture, maintaining small scale and volume, respecting the morphological aspect of its environment and referring to the maritime history. The experiential aspect is the other dimension that added to the smoothening of the place. This was brought by enjoyment of historical revival with a promenade in a well-preserved heritage site, the presence of historical ships by the piers, the feeling of eating fresh food or enjoying an event by the water. These "urban experiences" would create the identity and the uniqueness of the place. Festival marketplaces appeared as an "uncomplicated destination" where nostalgia sustained the myth of a special place (Goss, 1996: 169) (Figure 9.5) – a place conceived as a stage where each element played a defined and specific role.

If contestation was palpable around the proposed development (most notably among preservationists) (Barbanel, 1985) no major comments arose around the leadership taken over public domain by private (for-profit) interests, under a 99-year lease, namely a developer whose first projects were shopping malls, contributing to the rise of suburbia. More than ever, with the withdrawal from public investments, were developers and private interests perceived by the general public and authorities as major contributors to improved urban quality. Public space

Figure 9.5 Concerts in front of the Marketplace building on Pier 17

Source: Maria Gravari-Barbas, 2004

was to be reclaimed and become "clean", "safe" and "diverting" (Gravari-Barbas, 1998). In turn, this meant that a whole fringe of the population, uses and commercial activities would lose their grip on the shaping of the place.

Turpitude and the new project

Coming to the Seaport in 2014, one would expect to come across Rouse's Festival market. However, walking down by the water, what will be found instead is Pier 17 under construction and rubble witnessing the destruction of Rouse's building.

At the beginning of 2000 the Seaport festival market started to run slow. The precepts of preservation had been primarily approached by architectural terms (Interviews, E. H. Planning Center MAS, 2015; H. B., Schack Institute of Real Estate, 2014) and no real consideration for the immaterial dimensions (its people, quirkiness, uses, etc.) had been looked at. The relationships and the dynamics the place had established with its users but also its adjacent environment had been underestimated. As many scholars have stressed, festival marketplaces were closed environments (Gravari-Barbas, 1998) with little trickle-down effects (Levine, 1987; Harvey, 2001) on other neighbourhoods. These places could easily be compared to entertainment and leisure as "tourist bubbles" (Judd, 1999) or a "touristscape" (Metro-Roland, 2012).

The "home-made" retailing principle that prevailed in Rouse's projects could not hold anymore and, contrary to his ethos, the place slowly turned into a shopping mall "like anywhere else" (Interviews, 2015).[2] The liberalization of the international trade (under the Doha cycle – WTA) in 2001 had put pressure on the price of goods made of American raw materials and manufactured in the country. The prior success of the marketplace had also increased land value and businesses were slowly no longer able to afford the rising rents (Interview, D.F., Groundswell, 2015). As early as 1984 small retailers had started to give way to business chains which by the 2000s were J.Crew, Ann Taylor, Aerosoles, etc., shifting the atmosphere of the place from an experimental to a commercial one. The project had less and less to do with history and the walls it was using. Comments over the changes induced with Rouse's intervention insisted on the vacuity of the place: the "branding of the shops", "bad food", "tacky architecture" and the "loss of social profiles" (Interviews, 2015).[3] By then, the festival market had mainly become a tourist destination as the shops, goods and cultural activities it offered would only cater for visitors' needs and had less to do with the daily life and the needs of residents.

If the branding and the association with a global network of festival marketplaces might well have been appealing for City authorities by bringing national and international attention, at the local level the project and the "standardization" of the public realm it created was felt as a loss of place, belonging and identity (Interviews, 2015).[4] These changes and renovations in the urban landscape raised issues about the means of conservation, the appropriation of history and the blurring lines between authenticity, reality and the imaginary. The "dream-house for contemporary capitalism" (Goss, 1996: 169) was a high price to pay for the lower social profiles and the residents that used the place in their daily life.

In 2001, 9/11 took away the Wall Street public, the tourists and the shopping mall. So did the 2008 debt crisis. The Rouse Company had been sold in 1996 to General Growth Properties, another shopping mall developer. The Howard Hughes Corporation had then become a subsidiary of General Growth. So when the latter declared bankruptcy in 2009, the largest in the country in terms of retail real estate, Howards Hughes became the new developer for the Seaport with a new lease agreement.

It took a few years after Super Storm Sandy (2012), which badly hit the Seaport and its economy, for Howard Hughes to come up with a new proposal to revamp the place. Since then, the developer had been working on the area with the fashionable New York based SHoP Architects firm. The redevelopment plan built on the idea of a 42-storey apartment building and added a massive shopping venue in the place of the former Rouse festival marketplace. Its development vision states:

> Pier 17 will feature a contemporary, sustainable design, conceptualized by SHoP Architects. With over 400,000 feet of retail space it will honor its historic roots as a bustling marketplace and influential port of trade centuries ago. The areas revitalization will also include lush open spaces with unmatched views of the Brooklyn Bridge, a rooftop venue with outdoor bars, restaurants and a retail environment complete with premier fashion brands, restaurants and a world-class market. (http://www.southstreetseaport.com/content/south-street-seaport/en/ABOUT.html)

However, implementing new buildings and proposing new functions does not necessarily make the place, its atmosphere and the "unparalleled" experience envisioned. The developer is thus using different tools in order to capture the attention of the public, users, residents and potential visitors he targets. Using design, place-making and urban quality, he accompanies his interventions with the use of images, storytelling and temporary activation of space.

Upgrading the skyline, history and the residents

Playing with time and history

The developer started with renaming the place in which he is involved and worked on its past and its future by the invention of a new tag accompanying the site branding. "*Seaport District New York's oldest new neighborhood*" erases all references to Rouse's festival market place or to the Seaport's Museum which, as we saw, played a major role in the historical preservation of the district (Figure 9.6).

This new branding takes possession of the district through banners, posters on the buildings in the hands of the developer as well as rectangular columns that punctuate the public space. This process asserts the presence of the developer

Figure 9.6 Seaport new branding

Source: Author, February 2015

over the public realm.[5] The images and the columns inform the visitor on the new branding. It tells what can be found or what will come. The grip on history also takes place through the control of what the site will become. For instance, some columns present the future project with elements and buildings that have not yet been validated through the democratic planning process (Figure 9.7). A newcomer to the site walking along Fulton Street would thus acknowledge in his imaginary this projected urban landscape as common ground. This was for instance the case with the 42-storey tower, which raised numerous criticisms and opposition from residents, users, and local representatives (Davidson, 2014).[6] One of the main

Figure 9.7 Column presenting perspectives of the envisioned project
Source: Author, February 2015

reasons was that it would considerably alter the view to the Brooklyn Bridge and the Seaport skyline. The view was considered a public good to be preserved but also a strong feature of the Seaport's identity (http://saveourseaport.org/; www. friendsofsouthstreetseaport.com/; Interviews with representatives of Water front Alliance, 2015 and Historic District Council, 2015).

This tower designed by the New York based SHoP architecture firm did indeed strongly alter the visual identity of the site by creating an important rupture in building scale. Analysed under architectural codes, the tower could be looked on as a flagship mixing fashion and architecture and becoming a visual reference of the site. However, as it strongly contrasted with the rest of the Seaport, some of the volunteers and members of the Seaport Museum, the activists of Save our Seaport, the Manhattan borough and the district representatives as well as part of the Community Board 1 could not refer to it as the symbolic identity of the Seaport. For them, this tower did not belong to the collective identity, as for others (developers, public authorities, retailers, etc.) it belonged to the register of images, referents that a global city should have and that speaks to a certain category of individuals, among whom the financial executives, transnational elites and the millennials.[7] Since then, the project to build a tower on this site was dropped but not the idea to build it elsewhere.

The history, its appropriation and control are part of the integral strategy being put together by the developer as *New York's oldest new neighbourhood* needs to have *its* own history. On the historical frieze erected along Pier 17 Hughes depicts the new urban development as a new chapter of the history of South Street Seaport. In this frieze the project happens to be the end result of an unfolding great epic narrating different stages the Seaport district went through. In this account the developer plays on the importance of the place as a symbol in the historical development of New York City. The justification for the redevelopment is done by mobilizing the idea of a lost *grandeur* that the proposed physical upgrading of the place should restore. It plays with the place's history and its narratives allowing it to give "flesh" and depth to its architectural interventions and decisions. The use of history legitimizes the project through the use of myth narration (Di Méo, 1996). Indeed the frieze uses photographs, precise dates, detailed facts and referred sentences that, put together, give a sense of story with a clear beginning and an end. Everything is set in order to remind the public of what this place will become, leaving out the issue of what it will not be or is not anymore. Reading the frieze, one learns very little about the place outside acknowledged common ground. If Rouse's festival marketplace is being celebrated as marking the "startling if sometimes contested urban renaissance in American urban life" nothing shows or explains its destruction as part of the new project. There is no word of the ups and downs of this tourist attraction.

The communication device tends to give a normative vision of the project, insisting on a set of values around the notions of progress, quality, amenities, urbanities, etc. By doing so the developer tends to essentialize his project as a natural and logical step, but he also overshadows the presence of others and the emotional and memorial relationships they might have with this territory.

This frieze associated with the other communication apparatus participates in the storytelling of the project, to a history in progress, which in turn gives a sense to the newcomers (residents, retailers, businesses) so that they participate and take part in the project. Laurent Matthey (2015) documented the change in paradigm in urban planning practices. According to his analysis, these practices have shifted from following rules and plans (as tools for regulation and urban development) to the use of values and story in order to build urban projects. This change, he stresses, should be linked to the management language and the entrepreneurial city. As Patrick le Galès defines it, the entrepreneurial city is characterized by:

> the competition and market's discourse, as well as the image and identity, the political prioritization of economical development and investments attraction, the flux of high social classes, the transformation of local government towards public-private partnerships which gives an important role to private actors in the definition of the general interest, issues to be dealt with in the city, the management modalities, the conception and implementation of projects. (2003: 287)

This story-telling process can thus be looked at as a marketing strategy promoting the project and the territory it takes place in. It becomes a powerful machine that narrates stories and creates imaginaries (Salamon, 2007) geared towards a selected public that will afford the amenities, services and luxurious accommodation planned on the waterfront.

Encapsulating everyday life

Festival market's architecture and urban landscape's celebration had failed to create a sense of identity and emotional links to the place. The current discursive process and interventions used by Howard Hughes carefully establish this reference by engaging with immaterial dimensions mobilizing emotions and representations, but also with the material by activating public space and vacant buildings with temporary interventions. These actions are usually referred to as temporary cultural events during which the Seaport would transform itself in a cultural stage with art, architecture and design exhibits, workshops and conferences, as well as in situ performances.

In their research on pop-up landscapes on Philadelphia's waterfront, Schaller and Guinand (2017) show the potentialities inherent in this type of temporary intervention. For instance, interim interventions change the representations one can have of a place, erasing the negative values it used to be associated with. It also alerts to the possibilities and the potentialities the space can offer in terms of functions, uses or events. This is particularly salient for the public but also for potential investors. As the type of interventions might arrest peoples' attention, they also play on notions of space and time. Pop-ups thus break through the routine time that people experience in the space of the city (Schaller and Guinand, 2017: 2). The built environment loses its sense of fixity and becomes novel and mutable (Campo 2002; Colomb 2012; Harris 2015). These actions also represent a good way to test out ideas and products.

Authors have shown that this means of intervention is more and more being used by public authorities and private actors as a way to strategically intervene

Figure 9.8 South Street Seaport becomes Seaport Culture District for a season

Source: Author, December 2015

and orient urban redevelopment (Tonkiss, 2013; Colomb, 2012). These interim interventions induce reinvestment. They have been recognized as an effective place-making strategy (Mould 2014; Oswalt, Overmeyer and Misselwitz 2013; Pagano 2013) which contributes to the establishment of a new relationship with city-users whose self-identities are in turn shaped by these new environments (Quaglieri and Scarnato, 2017). Consequently, these design interventions are increasingly being deployed to restructure the public realm. At South Street Seaport they have become part of the broader developer's scheme, a place-making trend in urban design aimed specifically at revalorizing urban space for aesthetic, cultural and experiential consumption (Zukin 1995).

As for the pop-up landscapes in Philadelphia, Howard Hughes announces his upcoming events through social media (Instagram, Twitter, Facebook, etc.) and word of mouth. The success of the recipe is to become the best (well-known) kept secret of things to do or visit in town. This media attention once again caters to a certain class category. As announced by the director of the Seaport Culture District at a conference during the Design Week at the Seaport district:

> By providing spaces the main idea was for each of these organizations to bring their own audience here, to give them the ability to have new space that would allow them to win in their own audience, bring people to part of the city that they have not been going to before. This district would get the value of people hopefully starting to say "Oh, you'll never believe where I was, there were these really cool things at the Seaport". (Sanders, 2016)

Although the developer aims at distinguishing its project from the "tourist trap" image with which Rouse's festival marketplace was associated, his precise commitment to bring the neighbourhood into the twenty-first century is strongly rooted in new demands for experiences imbedded in fashion, food and art – "creating an unparalleled New York experience and the most vibrant lifestyle destination in Manhattan". (http://www.southstreetseaport.com/LIFE-CULTURE/seaport-of-tomorrow.html).

For instance, the whole development project is presented as an experiment *per se*. By taking part in the various sponsored events and interventions, by reading the different narrations brought through the social media (and amplified by individuals' accounts) or the posters and columns, one gets the feeling of participating (and being invited to do so) in the project. As the developer states, he wants to "ensure the Seaport remains a top destination for Lower Manhattan residents and the millions of visitors who come to the city each year" (http://www.southstreetseaport.com/LIFE-CULTURE/seaport-of-tomorrow.html). However, these participants (visitors and residents) have been carefully selected. This is very explicit when looking at the people represented on the project's images used by the developer (Figure 9.9) or the new functions he proposes for the district:

> The redeveloped Seaport is building on its reputation as the next culinary hub. Acclaimed chef and restaurateur Jean-Georges Vongerichten is opening

a 40,000-square-foot seafood market inside the historic Tin Building as well as a seafood restaurant on Pier 17. Both Momofuku Group, led by David Chang, and By Chloe, a pioneering vegan restaurant, are launching their first food concepts in the neighborhood. These outposts, along with luxury cinema operator iPic Theaters, independent New York book store McNally Jackson, fashion brand Scotch and Soda and the Pier 17's rooftop, will ensure the Seaport remains a top destination for Lower Manhattan residents and the millions of visitors who come to the city each year. (http://www.southstreetseaport.com/ LIFE-CULTURE/seaport-of-tomorrow.html accessed 10.02.2017).

Moreover, the different narrations juxtaposed with the built environment speak directly to the users' ethos and emotions. The beautiful pictures narrating the daily-life of producers on the *Fulton Stall Market*'s building gives, for the person who buys these products, a sense of satisfaction by directly contributing to the sustainable local food movement of the Hudson valley (Figure 9.10). Once again, these types of services target a category of people with purchasing power who are keen on buying these types of products.

This communication strategy combines experiences with emotions. This marketing tool creates a sense of community and belonging within a given

Figure 9.9 Dining, entertainment, fashion, culture . . .

Source: Author, December 2015

Figure 9.10   Pictures of Hudson Valley farmers against the future building for the Fulton
         Stall Market

Source: Author, December 2015

geographical site with the strength, for the developer, to frame and shape it. As David Harvey (1997: 2) assessed in a critical essay: "the foreseen social class is more attracted to an 'image of community' than a community *per se*." For instance, by mobilizing values in accordance with sustainability, creativity, world references and technology, the developer gives a sense of progressive and positive development geared toward the future and strikes a chord of a specific public. The built environment then "contributes to transform and reproduce major ideological and structural conditions that mediate the everyday lives of individuals and communities" (Dickinson and Aiello, 2016: 1295). But this infatuation marginalized what exists and existed before. This dichotomy between an old and ordinary if not retrograde past and something noble, highend and dynamic was encountered at a Community Board 1 meeting which was opposing the testimonies of the new businesses and new residents against the actual Seaport community (CB1, 2 March 2015). As Gotham has shown in his account of tourism gentrification in Le Vieux Carré, New Orléans, gentrification is a multidimensional process (2005: 1114) which for some can be socio-cultural practice associated with progress and for others can represent a symbolic and material loss. Recently, the association *Save our Seaport - For sail not for sale* reacted vehemently against an advertisement Howard Hughes had put in the *New York Times* as the developer did not mention the maritime history in his vision of the "Seaport of tomorrow". *Save our Seaport* was then enjoining people to write to the *New York Times* to pay attention to "what this developer leaves out of its publicity" (http://saveourseaport.org/2016/06/09/write-a-letter-to-the-new-york-times/).

Indeed, the community rallied around the Museum and the ships still remain. The different relationships (material and immaterial) it had maintained with the buildings gave rise to a sense of collective ownership and belonging, which still plays out today. For instance, there was strong mobilization to protect the Museum, its ships and artefacts from the Sandy flooding. Today, volunteers are still very much engaged in the restoration of the ships and the museum and a small coalition of people under *Save our Seaport* and *Friends of South Street Seaport* (FOSS) is trying to counterbalance the developer's plan. This strong anchoring to the place by people who do not necessarily live in the area, brings out the important symbolic function the Seaport still plays today and which is being slowly replaced by the developer.

Conclusion

The privatization of Southbridge towers, units that were built under the Mitchell-Lama programme (1955) in order to spur the construction of middle-income housing (Chawalko, 2014) located North of the Seaport district, the purchase by the Howard Hughes Corporation of construction-rights to build a 1,000 foot tower at 80 South Street (Davidson, 2014), his recent acquisition of the Best Western Seaport Inn for 38,300,000 dollars (1 February 2016) or the selling, during summer 2016, of a development site to China Oceanwide Holdings (Kreuzer, 2016) show that Lower Manhattan is undergoing an important growth development. A report from State comptroller Thomas Di Napoli confirms that between 2000 and 2014 the residential population in the sector more than doubled (22,700 to 49,000 inhabitants) and more than one third of the families had household incomes superior to 200,000 dollars or more compared with 7 per cent citywide (Office of State deputy comptroller for the City of New York, 2016). The resurgence of the 9/11 area with the memorial monument and museum, the One World Trade Center, has boosted tourism adding a growing number of hotels. The area's economy and cultural environment is clearly changing and diversifying while social profiles are shifting. The wealth of this district outpaces the rest of the city (Office of State deputy comptroller for the City of New York, 2016). The Seaport is noticeably following this super-gentrification trend as it is expecting an important influx of new residents while at the same time is subject to real estate speculation. It also capitalizes on Lower Manhattan as a new destination product and venue. But Howard Hughes does not want the Seaport to become a tourist attraction. He aims to cater for the "super-rich financiers" (Lees, 2003), the visiting fashionistas, models or actors, *en vogue* artists and stylish young traders with an *art de vivre*. He unmistakably plays on the atmosphere of the site by adding amenities and services and by upgrading its urban landscape as a means to attract this new elite. As visitors are increasingly looking for new and different destinations and products, the developer packages this piece of land by presenting it as an ultimate product to experiment with and proposes how to engage with it.

Looking at Rouse's festival market and the new retail destination planned by the developer, one might consider the hypothesis that this new project is in fact not so distant from Rouse's one. Howard Hughes' Seaport district is solely adjusting to the shift in the post-Fordist and tourism economy increasingly geared at experiencing and consuming the immaterial.

However, different stories have shaped the Seaport: the natives' land, the slave trade, the mafia, Super-storm Sandy, etc. But very little traces remain. As the redevelopment moves forward a better understanding of the previous histories materializing through the landscape should be taken into account, since the built environment plays a communicative role (Dickinson and Aiello, 2016). It implies considering the values and meanings of the place for the people living and using it as such and for New Yorkers, not uniquely focusing on narrating a story of desire for desirable people. Bringing continuities rather than ruptures could be a better way to sustain the social and cultural wealth this place still holds. It would open the possibilities for its future development, else it may run the risk, like Rouse's project, of becoming an empty shell without much to experiment with.

Acknowledgement

This work was supported by the Swiss National Foundation for Scientific Research, early post-doc mobility under Grant P2LAP1_155177.

Notes

1 During the 1960s–1970s South Street Seaport witnessed the installation of various artists. It was also a place for prostitution and the mafia was very well known for controlling part of the Fulton fish market's economy.
2 This remark came many times in the discussion with the different interviewees.
3 This set of qualifiers came many times in the discussion with the different interviewees.
4 This impression and feeling came many times in the discussion with the different interviewees.
5 This marking of the place was also done by young recruits who scoured the place in order to announce up-coming events making visitors welcome to the Seaport. This in turn adds to the personification of the Corporation.
6 The Seaport had been put on the 2015 MAS (Municipal Art Society) watch list (see www.mas.org). That same year, it was also added to the most endangered historic place list of the National Trust for Historic Preservation (https://savingplaces.org/places/south-street-seaport#.V_5zfihQzZs).
7 By 2030 New York City is expected to increase by 1 million inhabitants (Schaller and Novy, 2010). Even though the purchasing power of New York's millenials has shrunk since 2000, 12 per cent of them still make a six-figure income (Ehrenfreund, 2016).

References

American Legion Magazine (February, 1986). 'Rebuilding the nation's inner cities. Interview with James Rouse', *The American Legion Magazine*.

Banarjee, T. (2001) 'The Future of Public Space', *APA Journal*. Winter, vol. 67, no. 1, pp. 9–24.

Barbarnel, J. (1985) 'Pavilion Rising at Seaport Draws Opposition', *New York Times*, 11 March 1985.

Bloom, N.D. (2004) *Merchant of Illusion: James Rouse, America's Salesman of the Businessman's Utopia*, Columbus: Ohio State University Press.

Boyer, M. C. (1992) 'Cities for Sale: Merchandising History at South Street Seaport', in Sorkin M. (ed.), *Variations on a Theme Park*, New York: Noonday Press, pp.181–204.

Bourdeau, P. (2012) 'Le tourisme réinventé par ses périphéries ?', in Bourlon F., Osorio, M., Mao, P., Gale, T. (eds), *Explorando las nuevas fronteras del turismo. Perspectivas de la investigacion en turismo*, Coyhaique, Aysen: Nire Negro, 31–48.

Butler, T. and Lees, L. (2006) 'Super-Gentrification in Barnsbury, London: Globalization and Gentrifying Global Elites at the Neighbourhood Level', Department of Geography, King's College, London, available at www.kcl.ac.uk/sspp/departments/geography/people/academic/butler/SupergentrificationinBarnsbury.pdf. [accessed 29 Jan 2017]

Campo, D. (2002) 'Brooklyn's Vernacular Waterfront', *Journal of Urban Design*, vol. 7, no. 2, pp. 171–199.

Chawalko, C. (2014) 'Debating Privatization: Southbridge Towers Vote', *Urban Omnibus*, 17 September.

City Planning of New York (2002 [date unconfirmed]) *Transforming the East River Waterfront*, New York: City of New York.

Colomb, C. (2012) 'Pushing the Urban Frontier: Temporary Uses of Space, City Marketing, and the Creative City Discourse in 2000s Berlin', *Journal of Urban Affairs*, vol. 34, no. 2, pp. 131–152.

Davidson, J. (2014) 'A Proposed Megatower on the Pier Would Wreck the South Street Seaport', *New York Magazine*, 23 December 2014.

Demarest, M. (1981) 'Living: He Digs Downtown', *Time*, 24 August 1981, available at http://content.time.com/time/magazine/article/0,9171,949385,00.html [accessed 29 Jan 2017]

DeFillippis, J. (1997) 'From a Public Re-creation to a Private Recreation: The Transformation of Public Space in South Street Seaport', *Journal of Urban Affairs*, vol. 19, no. 4, pp. 405–417.

Di Méo, G. (1996) *Les territoires du quotidian*, Paris: L'Harmattan.

Dickinson, G. and Aiello, G. (2016) 'Being Through There Matters: Materiality, Bodies, and Movement in Urban Communication Research', *International Journal of Communication*, no. 10, pp. 1294–1308.

Edensor, T. (2007) 'Mundane Mobilities, Performances and Spaces of Tourism', *Social & Cultural Geography*, vol.8, no.2, pp. 199–215.

Ehrenfreund, M. (2016) 'The Truth About Millenials in New York City', *Washington Post*, 27 April 2016 available at www.washingtonpost.com/news/wonk/wp/2016/04/27/the-truth-about-millennials-in-new-york-city/ [accessed 29 Jan 2017]

Feifer, M. (1985) *Going Places*, London: Macmillan.

Girard, A. (2013) 'Faut-il raccorder une théorie générale de la postmodernité à une théorie à moyenne portée du posttourisme?', in Bourdeau, P., Hugues François F. and Perrin B. (eds), *Fin (?) et confins du tourisme. Interroger le statut et les pratiques de la récréation contemporain*, Paris: L'Harmattan, p. 43–52.

Goss, J. (1996) 'Disquiet on the Waterfront: Reflections on Nostalgia and Utopia in the Urban Archetypes of Festival Marketplaces', *Urban Geography*, vol. 17, no. 3, pp. 221–247.

Gotham, K. (2005) 'Tourism Gentrification: The Case of New Orleans's Vieux Carré (French Quarter)', *Urban Studies*, vol. 42, no.7, pp. 1099–1121.

Gravari-Barbas, M. (1998) 'La "festival market place" ou le tourisme sur le front d'eau. Un modèle urbain américain à exporter', *Norois*, vol. 178, no. 1, pp. 261–278.

Gravari-Barbas, M. and Fagnoni, E. (2013) *Tourisme et Métropolisation. Comment le Tourisme redessine Paris*, Gravari-Barbas M. and Fagnoni, E. (eds), Paris : Belin, coll. Mappemonde.

Gunts, E. (1990) 'Popular Pavilions: Why Harborplace is such a smash', *The Sun*, 24 June 1990.

Hall, P. (2002). *Cities of Tomorrow*, Oxford: Blackwell.

Harris, E. (2015) 'Navigating Pop-up Geographies: Urban Space-Times of Flexibility, Interstitiality and Immersion: Navigating Pop-up Geographies', *Geography Compass*, vol. 9, no. 11.

Harvey, D. (1989) 'From Managerialism to Entrepreneurialism: The Transformation in Urban Governance in Late Capitalism', *Geografiska Annaler*, Series B, Human Geography, The Roots of Geographical Change: 1973 to the Present, vol. 71, no. 1, pp. 3–17.

Harvey, D. (1997) 'The New Urbanism and the Communitarian Trap', *Harvard Design Magazine*, Winter/Spring, no. 1, pp. 1–3.

Harvey, D. (2001) 'A View from Federal Hill', in Harvey D. (ed.), *Space of Capital: Towards a Critical Geography*, New York: Routledge, pp. 128–157.

Housing and Development Administration. (1968) *Brooklyn Bridge Southeast Urban Renewal Plan*, Lindsay's files, Archives of the City of New York, 28 October 1968.

Hospitalitynet (2016) 'RobertDouglas Advises on the $38,300,000 Sale of the Best Western Seaport Inn, New York, NY', hospitalitynet, 1 February 2016. Available at: www.hospitalitynet.org/news/4074078.html. [accessed 01 Feb 2017]

Hurley, A. (2006) 'Narrating the Urban Waterfront: The Role of Public History in Community Revitalization', *The Public Historian*, vol. 28, no. 4, pp. 19–50.

Huxtable, A. L. (1964) 'South of Brooklyn Bridge', *New York Times*, 10 February 1964.

Hydra, D. (2012) 'Conceptualizing the New Urban Renewal: Comparing the Past to the Present', *Urban Affairs Review*, vol. 48, no. 4, pp. 498–527.

Jacobs, J. (1961) *The Death and Life of Great American Cities*, New York: Random House.

Judd, D. (1999) 'Constructing the Tourist Bubble', in Judd R.J., Fainstain S. (eds), The Tourist City, New Haven, CT: Yale University Press, pp. 35–53.

Knafou, R. (2000) 'Les mobilités touristiques et de loisirs et le système global de mobilité', in Bonnet M. and Desjeux, D. (eds), *Les territoires de la mobilité*, Paris: PUF, pp. 85–94.

Kreuzer, T. (2016) 'City Planning Gives Howard Hughes Corp. Air Rights Go-ahead For 80 South Sreet So Sale to Chinese Can Proceed', *Downtown Post*, 8 February 2016.

Lees, L. (2003) 'Super-gentrification: The Case of Brooklyn Heights, New York City', *Urban Studies*, vol. 40, no. 12, pp. 2487–2509.

Le Galès, P. (2003) *Le retour des villes européennes : sociétés urbaines, mondialisation, gouvernement et gouvernance*, Paris: Presses de Sciences Po.

Le Galès, P. (2016) 'Neoliberalism and Urban Change: Stretching a Good Idea Too Far?', *Territory, Politics, Governance*, vol. 4, no. 2, pp. 154–172.

Levine, M. (1987) 'Downtown Redevelopment as Urban Growth Strategy: A Critical Appraisal of the Baltimore Renaissance', *Journal of Urban Affairs*, vol. 9, no. 2, pp. 103–123.

Lindgren, J. (2014) *Preserving South Street Seaport: The Dream and Reality of a New York Urban Renewal District.* New York: NYU Press.

Maitland, R. (2010) 'Everyday Life as a Creative Experience in Cities', *International Journal of Culture, Tourism and Hospitality Research*, vol. 4, no. 3, pp. 176–185.

Matthey, L. (2015) 'L'urbanisme qui vient', *Cybergeo: European Journal of Geography* [on line] Débats, Les valeurs de la ville. Available at: https://cybergeo.revues.org/26562 [accessed 01 Feb 2017]

Metro-Roland, D.M.M. (2012) *Tourists, Signs and the City: The Semiotics of Culture in an Urban Landscape*, Farnham: Ashgate Publishing Ltd.

Metzger, T. J. (2010) 'The Failed Promise of a Festival Marketplace: South Street Seaport in Lower Manhattan', *Planning Perspectives*, vol. 16, no. 1, pp. 25–46.

Mould, O. (2014) 'Tactical Urbanism: The New Vernacular of the Creative City: Tactical Urbanism', *Geography Compass*, vol. 8, no. 8, pp. 529–39.

Nix, C. (1985) 'Pier 17 Opens at Seaport with Fanfare of Trumpets and Fireworks', *New York Times*, 21 September.

Novy, J. (2011) *Marketing Marginalized Neighborhoods: Tourism and Leisure in the 21st-Century Inner City*, PhD thesis, Columbia University.

Novy, J. and Huning, S. (2008) 'New Tourism (Areas) in the "New Berlin"', Maitland, R. and Newman P. (eds), *World Tourism Cities: Developing Tourism off the Beaten Track*, London: Routledge.

Office of Lower Manhattan Development. (1975) *Letter from Claude Shostal to Edgar Fabber*, Beam's files, Archives of the City of New York, 21 July 1975.

Office of State Deputy Comptroller for the City of New York. (2016) 'The transformation of Lower Manhattan's economy', report 4, September 2016, New York: Office of the State Comptroller.

Oswalt, P., Overmeyer, K. and Misselwitz, P. (eds) (2013) *Urban Catalyst: The Power of Temporary Use*, Berlin: DOM Publishers.

Pagano, C. (2013) 'DIY Urbanism: Property and Process in Grassroots City Building', *Marquette Law Review*, vol. 97, no. 2/5, pp. 335–389.

Quaglieri, A. and Scarnato, A. (2017) 'The Barrio Chino as last frontier: the penetration of everyday tourism in the dodgy heart of the Raval', in Gravari-Barbas, M. and Guinand, S. (ed.) *Tourism and Gentrification in Contemporary Metropolises. International Perspectives.* London: Routledge.

Reed, A. P. (1969) *Lindsay's Files*, Archives of the City of New York, 6 January 1969.

Rojek, C. (1997) 'Indexing, Dragging and the Social Construction of Tourist Sights', in Rojek, C. and Urry, J. (eds), *Touring Cultures*, London: Routledge.

Salamon, C. (2007) *Storytelling. La machine à raconter des histoires et à formater les esprits*, Paris : La Découverte.

Sanders, J. (2016) Panel Discussion, *Placemaking for a More Livable City: The Temporary Activation of Urban Space*, 5 May 2016, Seaport Culture District.

Schaller, S. and Guinand, S. (2017) 'Pop-up landscapes, a new trigger to push up land value?', *Urban Geography*. Available at http://dx.doi.org/10.1080/02723638.2016.12 7671976719

Schaller, S. and Novy, J. (2010) 'New York City's Waterfronts as Strategic Sites for Analysing Neoliberalism and its Contestations', in Desfor, G., Laidley, J., Stevens, Q. and Schubert, D. (eds), *Transforming Urban Waterfronts: Fixity and Flow*, New York: Routledge, pp. 166–190.

Sigler, T. (2009) 'Mixed-use Megaprojects in the Global Context: Lessons from Las Vegas', *Human Geography*, vol. 2, no. 2, pp. 77–82.

Smith, N. (1979) 'Toward a Theory of Gentrification. A Back to the City Movement by Capital, not People', *Journal of the American Planning Association*, vol. 45, no. 4, pp. 538–548.

Smith, N. (1996) *The New Urban Frontier: Gentrification and the Revanchist City*, London, New York: Routledge.

Sorkin, M. (ed.) (1992) *Variations on a Theme Park*, New York: Noonday Press.

South Street Seaport Museum (1968) *Progress Report*, June, New York: South Street Seaport Museum.

South Street Seaport Museum (1973) *South Street Seaport Development Plan*. New York: South Street Seaport Museum.

South Street Reporter (1967) *Newsletter of Friends of the South Street Maritime Museum*, vol. 1, no. 2.

Stock, M. and Lucas, L. (2012) 'La double révolution urbaine du tourisme', *Espaces et sociétés*, vol. 3, no. 151, pp. 15–30.

Tonkiss, F. (2013) 'Austerity, Urbanism and the Makeshift City', *City*, vol. 17, no. 3, pp. 312–324.

Urry, J. (1990) *The Tourist Gaze*, London: Sage Publications.

Viard, J. (2000) *Court traité sur les vacances, les voyages, et l'hospitalité des lieux*, Paris: Editions de l'Aube.

Viard, J. (2006) *Eloge de la Mobilité. Essai sur le capital temps libre et la valeur travail*, Paris: Editions de l'Aube.

Willis, C., Willen P. and Rossant, J. (eds) (2002) *The Lower Manhattan Plan: The 1966 Vision for Downtown New York*. New York: Princeton Architectural Press.

Yarrow, A. (1988) 'Seaport: Sprightly at 200, Mature at 5', *New York Times*, 19 August 1988, p. C10.

Yudis, A. J. (1980) 'Faneuil Hall concept now in Baltimore', *The Boston Globe*, 25 August 1980.

Zukin, S. (1995) *The Cultures of Cities*, Cambridge: Blackwell.

10 Playing for/with time

Tourism and heritage in Greece and Thailand

Michael Herzfeld

Disentangling temporalities

A critical approach to the phenomenon of tourism – itself a complex and multi-faceted aspect of modernity and globalisation – requires a nuanced approach to the understanding of the role of time. This is not only because so much of tourism is about the visiting and presentation of so-called 'heritage' or 'historic' sites, or because visitors and locals may differ in their understanding of what is historically significant or even of what happened in the past, but also because the representation of historicity must also engage with the tactics of local inhabitants as they struggle to control part of that representational process. Residents may sometimes entertain very different conceptions of the past than do tourists. Moreover, their present-day trajectory may, and sometimes does, defy conventional wisdom. Condemned as they so often are to a 'slow train' of economic development (since tourism can be as much a means of marginalising poor and exotic groups as it can be for advancing developmentalist agendas of an encompassing nation-state), they may sometimes actually benefit, more or less by accident, from having initially remained at a depressed economic level during a period of national economic expansion.

In considering the tourist exploitation of vernacular or domestic architecture, a major motor of gentrification in tourism-dominated economies, at least five distinct kinds of temporality are in play. We ordinarily take time as a background phenomenon except when its management becomes a practical problem, as when we are late for an appointment or are forced to wait for someone else. But different kinds of temporality articulate our lives with larger events in quite distinct ways. We can begin by contrasting the formal time of official historiography as this is translated into visible markers of its passage ('monumental time' with the time of socially experienced histories that diverge from each other because they are refracted through the respective experiences of different historical classes as well as through everyday activity ('social time') (Herzfeld 1991). A subset of the latter kind is the rhythm, or *tempo* (Bourdieu 1977: 6–7), of social performance, one version of which is the agile 'playing for time' that has repeatedly – if not always permanently – saved local communities from implacable bureaucrats.

Then there is what we might call 'displaced time' – time from which we have become detached in our own imaginations. It has become a commonplace in critical anthropological writings, for example, that older styles of ethnography place entire populations in a time removed from the present and whatever passes for modernity. This is the phenomenon that Johannes Fabian (1983) has dubbed 'allochronism'. In their unconscious adoption of this temporal distancing, however, anthropologists have not been alone; they followed a trope that was well established in the industrial world – in journals such as *National Geographic* (Lutz 1993), in travel literature, and in nationalistic folklore studies (see Danforth 1984; Herzfeld 1987). Tourist operators have exploited this perspective with particular enthusiasm since it clearly sells to a population avid for a sense of retreating to an older, kinder epoch.

We might view this kind of activity as an attempt, stylised though it usually is, to recover some sort of *longue durée* from the remnants of the past now framed as desirable destinations. It is an exercise in time travel – but travel to an epoch that has been sanitised beyond recognition. Tourists who seek the authentic in a Beijing *hutong* equipped with a Jacuzzi and electronic access are either deliberately or unconsciously impervious to the irony of what they are doing. Their numbers increasing every year, they represent rich economic pickings for previously impoverished populations, who eagerly await their arrival and hope that it will come in time for them to exploit the romantically decaying built environment before it decays into dust altogether.

There is a double irony here. Against the remorseless march of time – the time that forever denies Ozymandias immortality and afflicts the entire range of architectural heritage despite the energetic claims of monumentality to a lien on eternity (Herzfeld 2009, 2015) – another kind of temporal contest lends urgency to the tourist operators' efforts. It is not only nature that threatens the past with erasure. Human agency, especially in the form of governments unsympathetic to potentially rebellious or subversive claims to heritage, may wreak terminal damage before conservationists can apply the brakes. There are many reasons for such official intransigence, and these are manifested in a variety of hostile processes: ethnic cleansing, eviction, eminent domain, commercial development, and gentrification, to name only the main ones (each of which can also appear in combination with one or more of the others). Their common feature is a drive toward homogeneity, which officials often attempt to explain as driven by 'renewal', 'beautification', or 'efficiency'. But such motivations also share a darker side: the erasure of living cultural alternatives to the monolithic visions of nationalist states and consumer economies alike. This returns us to the concept of 'monumental time', but it now appears more openly as, ironically, a variety of 'social time'. Real people impose it on others; real people suffer its effects. This is Ozymandias' secret: nothing ever fully resists corrosion over time. Clearing away a whole community to make way for some grandiose monumental assertion of permanence thus sacrifices human life to an illusion.

This is nowhere more dramatically true than in the framing of domestic architecture as national heritage. It is all very well to claim that houses of a particular era belong to the nation, or even to the entire world; but, when such claims lead to the official sequestration of houses because they are deemed too important to remain in the hands of ordinary people, deeply troubling ethical and legal questions emerge. To be sure, some people, lacking in awareness of the historical significance and aesthetic value of their places of residence, may destroy what for others are valuable testimonials to the historical past. In such cases, they may also be depriving themselves of the opportunity to extract economic advantage from their homes.

How, then, to strike a balance that will not displease all concerned? What lessons can past experience teach us about useful approaches to such thorny and conflict-ridden questions? How far do our conclusions pertain to ongoing debates in critical heritage studies, and to what extent must they be regarded as inherently provisional and therefore mutable? Are our answers the arbitrary realisations of capricious predilections or do they suggest an approach capable of generating a practicable and unforced consensus among residents, activists and scholars?

Self-gentrification

Tourist interest in 'ways of life' may be a peculiarly Eurocentric preoccupation; it is a search for an 'authenticity' that does not conform to the officialising language of national history, religious orthodoxy and social respectability but seeks places that reveal how real life is lived. To the extent that the potentially voyeuristic curiosity animating such tourism can be satisfied, populations that have resisted official assaults on vernacular culture, or that have managed to maintain a state of picturesque poverty, will benefit economically in ways that often also enhance the economy of the encompassing – and hitherto hostile – bureaucratic apparatus.

As Theodossopoulos (2016a) has emphasised, moreover, only a condescending assumption that natives of any sort lack agency and intellectual independence would lead us to assume automatically that their adaptations to tourism are necessarily a self-demeaning form of exoticism rather than a canny strategy to exploit the willing gullibility of visitors. There are situations in which adopting the trappings of a culturally or historically significant past may also lead to an increase in self-respect. Theodossopoulos, for example, documents the ways in which tourist interest led some Panamanian Emberá to reconsider how they had internalised rejection of older ways of clothing their bodies, with the result that a modified version of traditional costume returned, not only for the benefit of the foreign visitors, but also as an expression of pride in indigenous identity and traditions.

One example of such revalidation, in which agency is exercised by both state and local actors, lies in what I call self-gentrification, and it is primarily this line of thought that I take up here. The term recognises the close linkage between spatiality and subjectivity. Changes in house form, or in adaptations of existing living arrangements, often reflect ideological shifts in the apperception of the relationship between the individual and society (see Zhang 2006). An autonomous self emerges, capable of staking out a claim to agency in the face of an ideology that has hitherto denied it the possibility of doing so independently of official control. In post-Mao China, for example, the emergence of a highly entrepreneurial middle class is directly associated with spatial privacy, often in the form of gentrification (Non 2016a, 2016b; Zhang 2006, 2010).

In such situations, the moral imperative of subjugating the individual needs and identity to society gives way to an equally strong emphasis on the moral exaltation of the self (see Gershon 2011; Muehlebach 2012), and the organisation of domestic space responds to that change. Where to begin with the prevailing ideology was highly individualistic, on the other hand, existing, socially organised forms of individualised or kinship-based space demonstrate a remarkable persistence in the face of official disapproval (e.g., Hirschon and Thakurdesai 1970). In such societies, moreover, the pursuit of profit in combination with a strong sense of family loyalty may produce situations in which a family crams itself into very tight quarters in order to use the rest of its domestic space for commercial advantage. Using one's home as a hotel, for example, is a fairly common phenomenon on the Greek island of Crete, one of the two locations on which I focus in this chapter.

Prosperity often contributes to increased emphasis on the self. I do not mean this in the crass sense that greater wealth and consumerism make for narcissism, although some may find evidence for such a correlation in today's obsession with the 'selfie stick'. Rather, what I want to suggest is that, with greater prosperity and the creature comforts that it brings, people begin to acquire the means of improving their knowledge of the world around them and of increasing their educational reach. This may in turn be reflected in a more self-conscious focus on caring for their dwelling places. Not all poor people are heedless in these matters, to be sure, even when their habitations are shabby. But they may come to see the ability to make better use of those habitations, to refurbish them, and to make them into an attractive place for others to visit as a form of self-improvement.

It is this phenomenon that I call 'self-gentrification'. Non (2016a) proposes the related term 'gentrification from within', to make the useful point that some forms of gentrification, far from destroying the lives of residents, offer them the means of surviving as a community and with improved circumstances, and often with a diversification of the population. That, however, depends on the complaisance of the authorities, who may be more interested in removing poor people from potentially wealthy and wealth-attracting neighbourhoods. For that reason, too, there remains the continuing danger that the new demographic elements could constitute a Trojan Horse, letting new wealth in and allowing it to destroy existing forms of sociality.

In some cases, however, local populations do benefit from the arrival of the newcomers. This may be a matter of how well prepared and adaptable the existing locals are. If they are able to learn how to claim a key role in the management of architecture regarded by the authorities as heritage, and find in this cultural capital a means of self-transformation, their chances of staying on while also improving their living conditions may be enhanced. Matching Non's focus on the transformation of the real estate, I want to examine transformations, consciously sought or unconsciously undergone, that the politics of gentrification induce in the ways in which residents understand themselves as persons, as residents, and as citizens.

Two case studies

I take as my two primary examples the experiences of a small provincial town in the island of Crete, in Greece, and the old historic centre of the Thai capital, Bangkok, in both of which I have conducted ethnographic fieldwork lasting for a total of more than a year. These two cases, while certainly comparable, have distinctive and in some respects widely divergent histories. The Cretan case, moreover, hardly deals with a metropolis; it concerns a small provincial town, albeit with issues of conflict over the ownership of the past that resemble those of Athens in structural principle if not in historical detail (see, e.g. Caftanzoglou 2000). Taken together, however, they reflect some important aspects of how temporality and selfhood are managed in places where tourism is important (or potentially important). They display especially revealing aspects of the process that I call self-gentrification.

Greece and Thailand are both major tourist countries. Both offer alluring combinations of spectacular natural sights, large and attractive beaches, and historic monuments and interesting vernacular architecture. Although Greece is a European country, it has always carried the appeal of the exotic (even, it seems, for anthropologists, who were able to conduct research there long before it became commonplace to do so in more northerly European settings[1]); indeed, its Ottoman heritage has been used to revile its people as fallen from cultural grace (see Herzfeld 1987). Thailand, too, offers exotic cultural activities, and without the burden of having to play the role of cultural ancestor to the West, although, like Greece, it has often been subservient to Western political interests. Both countries make extensive play with their respective histories, carefully edited in accordance with nationalist ideologies that ostensibly deny such dependence; these histories are, concomitantly, canted towards the slanted expectations of those (mostly Western) tourists who are attracted by antiquity or by a more generic interest in 'culture'. Both have complex and politically powerful historic conservation regimes – the Archaeological Service in Greece, the Fine Arts Department in Thailand – that supervise the production of monumental history and to a significant degree determine tourist access to, and perception of, its material forms. My two main case studies concern one community in each country; I have described both in considerable detail in ethnographic monographs (Herzfeld 1991, 2016a).[2]

Local people who want to profit from the tourist industries are compelled to work within or around the cultural ideologies promoted by these state institutions. This is not to argue that citizens in either country always follow official ideology to the letter. Frequently they engage in a conceptual *bricolage* that suits their immediate needs rather than the historiography of the state. But the discursive framework to which they must ostensibly fit all their discourse about the past is that of officialdom. In Greece this particularly means a claim on the quintessence of European identity and infrangible continuity with the ancient past; in Thailand, it follows the periodisation associated with the various 'reigns' (*rachakan*) of the current dynasty.

Greece and Thailand have both historically been examples of what I have called 'crypto-colonialism' (Herzfeld 2002). Thailand, lacking Greece's lien on the European past, has instead adopted a civilisational discourse originating in Britain and France (Thongchai 2000). While tourists visiting Greece seek the foundational monuments of European culture, those who visit Thailand instead seek experiences that allow them a suitably sanitised and somewhat voyeuristic encounter with otherness. Thai entrepreneurs do not emphasise the European elements in Thai culture even though the example set by the Thai monarchy, in particular, reproduces the cultural accoutrements of the European bour-geoisie (Peleggi 2002a). In that context, the production of romantically exotic ruins becomes, for the historically uninformed European visitor, a rich source of business (Peleggi 2002b). A trickle-down economic effect reinforces this paradox of an exotic archaeology held in place by European-derived conceptual structures – the paradox that characterises the condition of crypto-coloniality, beautifully illustrated by the use of vaguely Greek columns in a wide vari-ety of edifices from the old Grand Palace to the imposing modernity of the Rachaprasong Amarin mall.

Between the two cases to be considered here, therefore, there is a considerable degree of contextual divergence. The difference is further accentuated by the fact that the local people in the Greek case – the historic Old Town of Rethemnos – are mostly owners of their properties; the conservation authorities express no desire to expel them from their domiciles even were that to be legally possible (which it is not), but do try to persuade them to be better, historically more aware guard-ians of their piece of the patrimony. While they were initially resistant to these efforts, as we shall see, they have largely come to accept the state's intervention as an economic life-saver, especially now that the Greek state is embroiled in what promises to be a protracted struggle for recovery from the heavy burden of its national debt. Rethemnos is, relatively speaking, a successful and prosperous town by current standards in Greece – a major transformation of what had been a genteel but extremely poor backwater less than half a century ago.

In the Thai case, the residents of an embattled urban enclave in the dynastic old city of Bangkok called Pom Mahakan (see Herzfeld 2016a) are now, in a strictly legal sense, squatters, those who had title to their houses having partially accepted the meagre compensation that was the first move by the authorities to get them to move out. The city bureaucrats have in fact been trying for a quarter of a century to replace the community with an empty lawn, a so-called 'public park' (*suan sattharana*); the residents' resistance to those efforts, successful up to the time of this writing but latterly undermined by growing exhaustion and by the authorities' intransigence, has largely taken the form of a concerted representa-tion of their cultural heritage as demonstrably truer to Siamese tradition than the legalistic intransigence of the city bureaucrats. They occupy a space deeply satu-rated with the symbolism of official Thai historiography, with its periodisation in terms of the reigns of the kings of the present (Chakri) dynasty, and dotted with some of its most venerable Buddhist institutions. The bureaucrats' efforts are also guided by a Western-derived model of beautification that has lately also led them to modify some of Bangkok's most appealing tourist attractions, including at least three prominent markets. One of these, a famous flower market, was located at Pak Khlong Talad – ironically, the port of entry for the earliest imports of Western goods.[3]

While I have drawn a partial parallel between the two countries in terms of their somewhat similar dependence on the Western colonial powers and their successors, especially during the time of the Cold War, they appear to differ culturally in the extent and modality with which they confront repression. Greece, which claims to have been the birthplace of democracy, has resisted the crypto-colonial administrative doctrine through which that legacy was used contrastively to represent the country as a disgrace to its antecedents, and seems latterly to have shed some of its effects. Thailand, however, has not yet achieved such transcendence. The cultural dimensions of this difference might appear to lie partially in contrasted ways of facing repression. Greeks have a long tradition of aggressive interaction, particularly between males, and this pattern translates easily into open resistance to military rule and to the colonial forces and local quislings who served to enable the country's beholden condition. Thai concepts of male aggression – which certainly resembles those of Greece in certain interpersonal characteristics (see Pattana 2005) – are muffled in ordinary social life by the attributes of compromise, concern not to inconvenience others (*kraengjai*), and a remarkable capacity for avoiding precisely the kinds of noisy confrontation that Greek masculine codes demand.

But explanations in terms of gender conventions are misleading. We should not forget that women as well as men are actively engaged in resisting authority in both countries, even though women in both countries are generally not expected to display the same kinds of aggressive attitude as men. Nor should it be forgotten that at certain times Thais, stereotypically famed for their smiling gentleness,[4] have nevertheless demonstrated for their political rights no less raucously than Greeks. Indeed, the student demonstrations of 1973 – which led to the fall of a military regime – inspired their Greek coevals to rise up against their own military junta.[5] Thais have also shown a remarkable capacity for organising protest on the ground (see, e.g. Missingham 2003; Sopranzetti 2012a). So we should not set too much store by stereotypical portrayals of difference, especially as, in the present discussion, it is the Thais of Pom Mahakan who are stoutly resisting official policy while the Greeks of Rethemnos have gained most through cooperation and negotiation. Those stereotypes have their uses in the political games under consideration; they are not explanations of either tactics or outcomes.

The real difference, I suggest, lies instead in the dominant political mode at a given time rather than in some putative cultural difference conceived in essentialist (Geertz 1973: 234–254) terms. In Greece, the military junta of 1967–74 energetically pursued tourist development, both as a means of satisfying a restive population and as a way of seeking legitimacy abroad, and often at the expense of the physical and social environment. Their dictatorial approach, combined with an almost farcical commitment to a vision of reviving the ancient glories (which, it was soon clear, they did not really understand), meant that residents defied conservation regulations at their peril; the charge of despoiling the national patrimony was not one to be lightly risked. When the junta fell from power and democratic institutions rapidly took hold, the idea that 'historic houses' were of value was not immediately welcomed or understood, and, especially after the ascent to power of the moderate socialist Andreas Papandreou, local conservatives could even claim that the state was now in the hands of 'communists' who did not understand the meaning of private property!

Four decades and more later, the situation in these two sites exhibits a stark contrast. The residents of Pom Mahakan are today poorer and more under-employed than Rethemniotes were half a century ago. Unlike the Rethemniotes, they are still (at the time of writing) fighting what may be the last, desperate battle for their community's survival, their adaptation of the rhetoric and historiography of the nationalist model of 'culture' (*watthanatham*) their only real defence against a powerful, if internally divided, establishment. To make that tactic work, they have played for time – holding off the violence of eviction through legal delaying tactics even when they are sure the courts will side with the authorities; organising community and mounting a watch on the old city wall; and occasionally reducing bureaucrats to impotence through their skilled management of interaction in the present and the interpretation of the past.

The residents of Rethemnos are in very different condition. They promote the historical uniqueness of their town in terms that do not contradict the neo-Hellenism, now also notably softened and diluted, of the state. They seem to have weathered the ongoing crisis of the euro more successfully than their compatriots in the capital, Athens. They complain, to be sure, that conditions have deteriorated – a pharmacist, for example, told me that basic medicines were in short supply as recently as 2013 – but they also recognise that their erstwhile poverty, which made it impossible to demolish their old houses before the authorities noticed the tourist potential those houses represented, saved what is now a relatively reliable and lucrative resource. It is as though they had unwittingly played for time against their own lack of appreciation and desire to modernise at all costs, and are now able to reap the economic benefits of inhabiting a historically important site that attracts many tourists, while becoming increasingly knowledgeable about, and invested in, its scholarly significance as well.

I now turn to the two cases in more detail, to demonstrate how this contrast emerged and what lessons we can draw from it. In Rethemnos artisanship offers a telling example of how a profession can benefit from tourism even as it locks its practitioners into a humble class identity (Herzfeld 2004[6]). The supply of apprentices has now largely dried up, in part because of greater prosperity and educational opportunities, and the importation of cheap tourist goods has moved the economic emphasis into the retail arena. Even the brief visits that I have made in recent years, however, have convinced me that the population and its social relations, while changing under conditions of relative prosperity, do not appear to be suffering the kind of uprooting that we see in the Bangkok community. They have already benefited from tourism; the people of Pom Mahakan desire a similar salvation, but with much less cause for optimism. Although we may perceive remarkably similar dynamics in the two communities' respective experiences of refashioning subjectivities, the differences will suggest the importance of understanding local cultural values as well as overarching political dynamics if we are to make sense of the very different conditions in which these two communities find themselves today.

Rethemnos and the recovery of historical consciousness

I conducted research in the Old Town of Rethemnos in the 1970s and 1980s. At the time of my first attempts at research there, in 1970, I was mostly interested in the local practice of ornamenting the plaster-covered houses of the poorest residents with an inventive set of designs usually scratched into the surface with a makeshift aluminium comb (Herzfeld 1971). This way of ornamenting the houses of a poverty-stricken town's poorest residents seemed to most observers to be a mark of deep penury, even by the standards of a Greece still smarting economically from the savage deprivations of World War II and the Civil War that followed on its heels. That seemed to be confirmed by the comment of one resident, who described the wavy lines on his house as portraying his own blood running in the walls. Today, these designs must vie with graffiti, both overlaying clearly Venetian cornice shapes and thereby earning the ire of the establishment scholar-bureaucrats of the Archaeological Service.

At the time of my original research, Rethemnos was a neglected seaside town, isolated from the two major towns to the east (Iraklio) and west (Chania), respectively, by a main road that curved inland and away from it; by its lack of a manufacturing base; and by its self-perception as a genteel place that had fallen on hard times and had suffered culturally (according to the prevailing nationalist ideology) from having been a predominantly Muslim town. Described in lyrical terms but with a clear recognition of its depressed economy by two major Greek novelists (Nakou 1955[7]; Prevelakis 1972), it hosted an exceptionally large flock of Greek Christian migrants who had been relocated there in 1924 following the exchange of populations between Greece and Turkey mandated by the Treaty of Lausanne. It was, in other words, a town displaced both by its own internal fall from grace as a once glorious 'city of letters' when under Venetian rule during the Renaissance on the one hand, and by associations with the Turkishness both of those who had departed and, allegedly, of the refugees who had come to take their place on the other. In the nationalistic and isolationist climate of post-Lausanne Greece, these were hurtful images. They were matched by the dilapidation and desperation that seemed to characterise both its residents' self-view and the feelings the little seaside town evinced in others.

Yet today, amid a crisis that has battered the Greek national economy and has seriously depressed the living conditions of most citizens, the Old Town of Rethemnos is, at least in appearance, one of the country's few urban spaces that do not look abandoned, where shops have not been boarded up, and where people seem to have some prospects of keeping their lives more or less on track. I do not want to over-state the contrast, but it is palpable, pervasive, and dramatic nonetheless. This once poverty-stricken if proud town, a dusty and eroding complex of some of the most glorious examples of domestic-vernacular and monumental architecture of the era of Venetian colonial expansion in the Mediterranean, has been able to exploit its architectural glories in large measure because, during most of the period of my research, it lacked the economic resources to demolish and replace its older buildings.

An internal comparison within Crete is also instructive. Iraklio and Chania are both Cretan cities endowed with some architecture from the same period. Leaving aside for the moment the extremely convenient fact that Rethemnos also boasts an unusually attractive sandy beach right in the centre of the north side of the Old Town, neither of those cities presents such a well-preserved architectural complex. In Iraklio, one major church of the Venetian period was destroyed by order of the military junta during the 1970s, while the junta's relentless nationalism also led to the neglect of what were misleadingly described as 'Turkish' – that is, Ottoman – remains. At that point, the current investment in the preservation of Ottoman monuments that we see in Rethemnos as elsewhere would have been well-nigh unimaginable.

The junta was bent on the economic development of the country, but went about it in exactly the wrong way – by disfiguring the Cretan coastline with hideous, modernist hotel complexes, built without the restraint of zoning laws or environmental protection legislation, and polluting the seaside. Some tasteful and relatively small hotels were built during this time. But the overall approach was to treat tourism as an immediately exploitable and inexhaustible resource within an economy that lacked, and still lacks, major industrial infrastructure. The junta was prepared to kill the goose that laid the golden egg in its heedless hunt for easy finance, even if that meant destroying old houses and monuments or surrounding them with high-rise monstrosities. There were small exceptions here and there – the Venetian harbour in Chania remains a gem – but they were few and far between.

Such policies were not unpopular with local people. Even after the collapse of the junta, a few of the more culturally myopic residents of Rethemnos's Old Town, mostly wealthy individuals who saw the small Venetian and Ottoman homes as unsaleable and unfit for industrial development, destroyed these properties, using ingenious ruses to escape official condemnation and even to entangle officials in unwitting complicity (see Herzfeld 1991: 251). They played with time in a way not, apparently, anticipated by the conservation regime (for bureaucrats have their own devices for buying time and distorting its flow): they deliberately accelerated the process of decay. Their strategy was not directed at evicting tenants – apparently they had none – but simply at freeing up their real estate for what they saw as profitable development. The term 'development' (*anaptiksi*) had acquired considerable currency during the junta years; although its association with the junta's policies brought it into disrepute – as has happened to a more limited extent in Thailand – in the post-junta years, modernity was seen as desirable by rich and poor alike. The difference was that the poor did not have the means to acquire it.

The rich, by contrast, bought time at what appeared to be a good price – but they quickly found themselves on the wrong side of history. Historic housing became fashionable and thus also valuable, and attracted the attention of the more educated tourists the authorities were keen to attract. Poorer citizens such as the landlords of the house where my wife and I stayed during the 1986–87 academic year – economically and socially they are representative of the vast majority of the Old Town's roughly 6,000 residents at that time – could not afford to destroy their houses; had they done so, they would have made themselves homeless. Thus, although they did not consciously play for time, or with it, they discovered that, unlike their fellow citizens who were more directly concerned with immediate profit, they had inadvertently profited from a temporality of which they could hardly have been aware – and gradually came to reap a handsome economic, social, and cultural profit into the bargain.

This was possible because Rethemnos was a rare exception to the pattern of destruction in the name of development. A pro-junta mayor of Rethemnos, Dimitrios Archontakis, an erudite schoolmaster with locally acknowledged expertise in the ancient classics, turned out to have been remarkably prescient, at a time when the military dictatorship was already in disarray, in imposing what at the time was a highly unpopular conservation straitjacket. Indeed, he was so successful that, once the economic benefits of his policy had become apparent, he was able to get re-elected (which he needed to do in order to col-lect his mayoral pension) despite his junta-related past, a fatal blemish for most political ambitions, and despite a local population that tended strongly toward left-wing politics and that in 2007 elected a PASOK (socialist) mayor at a time when that party's popularity had begun to collapse nationally.[8] The people of Rethemnos judge their mayoral candidates on a range of criteria, party mem-bership and past associations being only part of the picture. In 1978, only four years after the fall of the military regime, Archontakis won the mayoralty with 51.9 per cent of the vote.[9]

This was very much a result of the economic boost that his conservation-ist policies seemed to generate. He also appealed to a gradually growing class constituency; the rising *embourgeoisement* of the population was apparent in a number of indicators, including, interestingly for our purposes, a sudden increase in overseas tourism among those who had hitherto been members of a genteel but poor shopkeeper and artisan population. These people, forced by poverty to remain in the Old Town rather than building and living in new quar-ters in the new areas already beginning to burgeon in its immediate hinterland, were discovering a new identity. This is the process that I propose to call self-gentrification; it is as much about self-making as it is about the refurbishment of space and architecture.

Our landlords, specialty bakers who were running the town's sole workshop for the production of *filo* and *kataifi* pastry, exemplify this process. When we first came to know them, in 1985, they were emerging from a respectable state of poverty fairly typical of the artisan class and were beginning to earn just enough money to contemplate renovating their home and workplace, a beautiful Venetian mansion with a portal dated 1609 and with a relatively well-documented history. At that time, their renovations, again representative of similar changes occurring throughout the Old Town, were mostly about comfort (and perhaps a small measure of bourgeois aesthetics). They, like their neighbours, showed no interest in the historical significance of their house and its spectacular façade ornamentation, although they were intrigued by our open admiration. Warm and hospitable hosts, they were popular in their section of the Old Town, and were locally well-regarded as hard workers and as sociable neighbours and tradespeople. Their financial instincts certainly proved sound. Within a few years, they were able to dower their daughter with an apartment in the new area (the daughter and her husband were later able to upgrade to a magnificent villa in the adjacent countryside), and began to travel abroad. Again, this was a far from unusual pattern.

But the social climb on which our landlords and others like them had embarked was about more than money; it was also about cultural capital (Bourdieu 1984). Overseas travel – formerly a luxury that few provincial Greeks had been able to enjoy – served to mark their ascent in a discreetly visible way. One of their destinations, coincidentally, was Thailand, and their observations of that country's food practices offered a particular resonance inasmuch as they were also involved in a food-related industry. They had been producing the basic pastry but had hitherto disdained selling the sweetmeats themselves; indeed, they would have had little time to do so since their main business lay in supplying the major hotels with their products. About the time of their social ascent, however, they also began to make the sweetmeats for sale, sometimes persuading tourists to try their wares and watch them being made by hand (and getting their establishment featured in at least one glossy magazine abroad and another at home in Greece). One by one, the hotels that had been their major customers meanwhile turned to much cheaper factory-made products, no doubt calculating that most foreigners would not have the discernment to taste the difference. Selling to tourists, which began as a sideshow, acquired genuine economic value as these hard-working bakers approached a retirement that still seems to recede with every year of the unremitting labour needed to sustain their newly enhanced lifestyle.

Increasingly accustomed to the idea that foreigners would experience the hand-crafting of pastry as fascinating and exotic, they now began to see that once irritating curiosity as economically productive; visitors would feel obliged to sample and, often, to buy the sweetmeats. The bakers' enhanced understanding of this exoticism paralleled a growing awareness of the significance of the architecture of their house. Having wondered why anyone would want to preserve their old doorway a scant fifteen years earlier, they now managed to get the wooden part replaced by a replica approved by the Archaeological Service and arranged authentic-looking internal wooden beams at strategic spots – and boasted knowledgeably of having a Venetian house.

Such pride in historical background was a significant transformation in terms of the knowledge and ambitions of two unassuming, highly intelligent but poorly educated people who had earlier simply been trying to make a living. They had learned to understand their role as one of heritage guardianship, a selfhood more closely calibrated to a state that had itself become steadily less exclusively fixated on the classical past and more interested in the post-classical evolution of national and local culture.

Their children benefited from and built on this enhanced and increasingly cosmopolitan sophistication; the son became a university-trained civil engineer, while the daughter ran a day-care centre. But the parents' visit to Thailand was perhaps the most startling rupture with their past. Formerly the passive recipients of tourist attention, they now not only sought tourist-based income but used it to emulate European tourists in turn. Their rising class status was especially evident in their discomfort with the idea of eating Thai street food – a fastidious bourgeois response.[10]

Today, Rethemnos has a lively tourist scene. Greek tourism in general seeks to benefit from the country's poverty; it is a cheaper destination than, say, Italy, in most respects at least. But overall this strategy has not saved significant parts of the population from near-destitution. Although Rethemniots certainly have experienced a substantial downturn in comparison to the early 1990s, they, by contrast, are still living in an economic situation that promises some sort of viable, liveable future. Now they could afford to visit exotic places where fate has been a great deal less kind and directly emulate those Scandinavian, Dutch and German visitors who still gaze on their activities as equally exotic and mired in picturesque poverty.

Difference and a view of Bangkok

What did our landlords see in Bangkok? Mostly, predictably, they visited the prescribed monuments such as the Grand Palace. But individual tourists often stray from those places into more vernacular spaces. Such is the case with people visiting the Temple of the Golden Mount, who often, beforehand or afterward, because they are lost or because they are curious, wander into the small nearby community of Pom Mahakan where I have been conducting research since early 2003 (Herzfeld 2006; see also Ariya n.d.; Bristol 2007, 2009). Such a humble setting would have been invisible to our Rethemniot landlords, however, not least because the main profession of the Pom Mahakan residents is the production and sale of street food – the very activity that provokes their disgust as well as the disdain of government officials, whose values are predicated on the imitation of bourgeois taste promoted by the monarchy at least since the reign of Rama V (see Peleggi 2002a), and who have repeatedly declared holy war on street vendors and popular markets. It may be that a preference for 'European' lifestyles has played as much of a role as economic constriction in inducing our former landlords to prefer European destinations such as Paris and Prague for their more recent peregrinations. The Thai authorities would presumably see this preference for the visible order and cultural style of European capitals as a vindication of their beautification campaign; most tourists from Western Europe and North America would staunchly disagree.

Had our former landlords visited Pom Mahakan, however, they might have been struck by a sense of familiarity. There, too, culture and history are commodities that play a central role in self-gentrification. The Pom Mahakan community bought time against the threat of comprehensive eviction by constituting itself as a site of representative Thai culture. At the same time, its leadership, perhaps unconsciously imitating the monarchy, has built for itself a cultural profile that closely resembles the bourgeois culture from which most of the NGO activists hail (see also Delcore 2003). Not only has the community resorted to fairly predictable legal blocking devices such as protracted courtroom appeals, but it has presented itself, in ways that range from regarding its spirit shrines as shrines to the ancestors of all Thais to a highly visible deployment of its internal cultural diversity and artisanal production, as the quintessence of Thailand. Its unifying strategy has been to make any attempt to destroy the community look like sacrilege and treason in a single toxic package.

Here, in short, we have, not gentrification (there are no wealthy would-be residents ready to pick at the architectural carrion), but something not unlike the *self*-gentrification we have already seen in Rethemnos. Self-gentrification is a concept that resonates with the residents' own rhetoric of self-development (*kan patthana tua eng*) – a term with which they seemed to be comfortable even as they ostensibly shared my distrust of development *tout court*. The rhetoric of self-gentrification draws a variety of background histories into a narrative that at one and the same time adheres to the official historiography of smooth dynastic progression and also suggests a principled resistance to bureaucratic authority held by the residents to have betrayed the basic principles of the older Siamese and royalist model of Thainess.

In this strategy, the most powerful trump card is often tourism. Thailand, like Greece a crypto-colonial state heavily dependent on tourism to buttress its crisis-assaulted economy (and also, in this regard, resembling Greece during its own earlier period of military dictatorship), can ill afford to undermine its own tourist industry – although the ill-advised beautification campaign of the Bangkok authorities threatens to do just that. The residents of Pom Mahakan have crafted a policy that would turn their community into a 'living museum' (significantly, they often use the English term[11]). Because they recognise that they are fighting an already well established if misguided beautification campaign (see Askew 1996 for a useful early discussion), they counter with a similarly developmentalist rhetoric. The subtle shift from 'development' to 'self-development', however, masks the nurturing of an independent subjectivity that increasingly sees resistance to the formal bureaucracy as lying at the very core of Thainess, even though it also seeks social redemption in the achievement of consumerist ambitions (see Sopranzetti 2012b). In respect of their resistance to authority, the residents of Pom Mahakan perhaps resemble the mountain dwellers of the swashbuckling villagers of the Cretan mountain villages than they do the genteel inhabitants of the Old Town of Rethemnos; but their consumerist ambitions more closely resemble the recent transformation of the latter into true bourgeois.

Throughout the more than two decades of their struggle, the residents of Pom Mahakan have nevertheless identified themselves, accurately enough in economic as well as social terms, as poor people. Their unemployment rate is high, and most of those who do have regular work are engaged in professions, such as street food-hawking, that are spectacularly precarious. But that poverty has released their energies into community-building in ways that could have produced a vernacular paradise – a money-spinning tourist enterprise designed to contrast with, and yet also to reinforce, the surrounding panorama of palaces and temples.

Poverty and power: heritage, hope, and the lessons of comparison

We can thus see that in both cases, that of the Old Town of Rethemnos and the very different situation of Pom Mahakan, a period of poverty while the rest of the encompassing nation-state was racing ahead economically may inadvertently provide an excellent long-term investment in a tourist future. In Rethemnos, delaying tactics or effects (depending on whether the delays were intentional) allowed time for a nationally resonant heritage discourse to trigger a change of policy; in Pom Mahakan the outcome appears to be less hopeful, but the tactic of buying time at least gave the residents a chance to press their cause.

Only a cynic would actually propose the maintenance of poverty as a solution to the destruction of urban communities and their heritage. But authorities make a grave mistake when, as in the case of Pom Mahakan, they make the very different assumption that poverty reduces communities' capacity to unite against official intransigence. A past marked by extreme poverty, having reduced the capacity of local populations to destroy what eventually becomes their most valuable resource, may school local populations to a remarkable degree of resolution; the experiences of middle-class communities in Thailand and Greece suggests that these communities have much greater difficulty in achieving the kind of unity that makes concerted political activity possible. Evidence is accumulating that poverty and the perceived oppression by the European Union are leading formerly middle-class Greeks toward a new, defiant solidarity at local and national levels alike (Knight 2015: 121–131; Rakopoulos 2016; Theodossopoulos 2016b), and in some cases are also leading both middle-class and less affluent Greeks toward solidarity with immigrants who might otherwise find themselves the targets of extremist violence (Papataxiarchis 2016; Rozakou 2016). Where local official-dom seems relatively benign and non-confrontational and people are partially protected from the impact of a weakening national economy, as in Rethemnos, the pursuit of economic comfort and cultural redemption may pose few challenges to official moral authority. When, however, as in Pom Mahakan, officialdom threatens violent action, those echoes easily become parodic and challenge the right of the bureaucrats to speak for the national polity.

In both processes, tourism has been a catalyst. It offers models of a way of being that Greeks in particular, nurtured on a diet of official claims to a quintessentially European identity, can channel into an aggressive competition over cultural capital. Suddenly the agonistic relations of yore (see Peristiany 1965) become the solidarities of the present (Herzfeld 2016b: 201–202; Rakopoulos 2016: 143). The state, satisfied that at least the material patrimony will not be damaged, encourages this development. But in Thailand, where the emulation of the tourists' consumerist habits has triggered alarm and fear among the rapidly swelling ranks of the bourgeoisie, those who openly parody and criticise the official renditions of national being may stand a better chance of being heard. In both cases, the transformations wrought by involvement with tourism generate new understandings of selfhood and subjectivity. Those understandings undermine the hierarchies on which officialdom has built its authority. They entail a reconsideration of the nation's place in the world and threaten the power of the crypto-colonial elites that would prefer to fossilise those subjectivities in perpetuity.

The local cases thus cannot be viewed independently of their larger political contexts. While tourists usually prefer to ignore or even excuse the political injustice that has made their presence possible, scholars cannot afford to do so. Gentrification can realise the economic and cultural goals of the poor, especially when it occurs as a result of internal dynamics. But that kind of self-salvation seems to depend on factors that keep residents firmly in place – the *hukou* (home residence) system in the Shanghai described by Non (2016b), ownership of the relevant real estate by even poor residents in capitalist Greece. The 'self-development' that their long, painful resistance to authority has engendered among the people of Pom Mahakan may actually produce a form of gentrification in the long run – but only if the residents are allowed to remain in place, building on the extraordinary resilience and solidarity that long opposition to authority has fostered.

If the people of Pom Mahakan are removed,[12] the authorities will have destroyed a form of self-governance that would have attracted an educated and interested tourist audience, and will also have actively discouraged a process capable of sustaining the emergence and reinforcement of social justice and a pluralistic understanding of history. The repressive regularity of formal periodisation will have erased the temporal play that makes argument, and therefore justice, both attainable and real, and would displace the residence from the economic benefits of access to the tourist trade. This scenario offers a stark contrast with the Rethemniot experience as it has already evolved. The residents of the Old Town of Rethemnos are today reaping the cultural, economic, and social benefits of the political and conceptual earthquake following the fall of the junta in 1974, a political shift that significantly reduced both the cultural marginalisation of the Greek working classes and the stultifying neo-classicism that had hitherto undergirded the state's sclerotic control of the past and enhanced the agency of local residents in the shaping of tourism in their town.[13]

Acknowledgement

This chapter has been substantially revised and enlarged since its conference version appeared in Chinese translation (by Yang Jiaojiao) in Xiang Cheng-yu and Qiu Yun-zi (eds), *Heritage Tourism and Cultural China: Tourism Summit Forum 2014 Volume* (Harbin: Hei Long Jiang People's Press [2016], pp. 13–19. Some of these materials also appeared in an earlier form in Michael Herzfeld, 'Heritage and the right to the city: When securing the past creates insecurity in the present', *Heritage and Society* 8: 3–23. I am indebted to Hei Long Jiang People's Press and Taylor & Francis, respectively, for permission to re-use these materials in the present chapter. I also wish to thank volume editors Maria Gravari-Barbas and Sandra Guinand for their sage observations and for their calm patience.

Notes

1 Among the best-known pioneers in this area are J.K. Campbell (1964) and Ernestine Friedl (1962).
2 For a somewhat different but equally sympathetic perspective on the Thai case, see Thanaphon (2007). Excellent architectural studies have been published by Bristol (2007, 2009) and Chatri (2003; see also Chatri 2012). Studies of the Greek case have mostly been formal analyses of the Venetian and Ottoman architectural heritage (e.g. Dimakopoulos, 2001).
3 See, e.g. Kaewta (2016); Sasiwan (2016). The latter also discusses the Pom Mahakan case.
4 Consider, in this context, the tourist slogan, 'Land of Smiles', which completely elides the ironic uses to which Thai smiles are often put.
5 Remarkably little has been written about this connection. One exception is Kallivretakis (2004); see also Herzfeld (2002: 920).
6 In this study, I offer a detailed analysis, based in part on the close examination of videotaped artisan–apprentice interactions and of life histories and other ethnographic data, of the ways in which the training of apprentices inculcates class attitudes that are disapproved of by wealthier and more formally educated citizens.
7 An excellent account of this work and its role in Nakou's life is provided by Tannen (1983).
8 This mayor, Giorgos Marinakis, was re-elected in 2014 with 73.15 per cent of the vote, a major achievement that shows just how remarkable the achievement of his conservative predecessor was in regaining power against such strongly pro-socialist and anti-junta odds in the local context.
9 See www.eklogika.gr/page/elections/yp_rethymnis and www.rethymno.gr/munici pality/history-municipality/history-of-major.html (accessed 9 May 2014).
10 Street food is plentifully available in Rethemnos as elsewhere in Greece. In the past, however, it was considered less acceptable to eat such food on the street than tourism has perhaps made it to do so today.
11 The use of English terms to 'officialise' policy-related terms and concepts is a widely shared device across the Thai political spectrum. Words like 'participation' displace the Thai *kan mi suan ruam* when the idea is to emphasise adherence to globally authoritative discourses. The habit is widespread not only in academic and NGO circles but, by contagion, in local communities seeking to emphasise their legitimacy and respectability.
12 They were again threatened with immediate eviction in April–May 2016. At the time of going to print, the eviction order had not yet been fully executed.
13 On the linke between agency and the uses of authenticity in a tourist context, see Theodossopoulos 2013.

References

The author has followed the Thai convention for naming Thai authors in the references and citations by using each author's first name as the principal reference element.

Ariya Aruninta. n.d. 'Controversies in Public Land Management Decision-makings: Case Study of Land Utilization in Bangkok, Thailand', unpublished paper, Available at http://www.land.arch.chula.ac.th/data/file_20090921165036.pdf (accessed 13 Feb 2017).

Askew, M. (1996) 'The Rise of Moradok and the Decline of the Yarn: Heritage and Cultural Construction in Urban Thailand', *Sojourn*, no. 11, pp.183– 210.

Bourdieu, P. (1977) *Outline of a Theory of Practice*, Translated by Richard Nice, Cambridge: Cambridge University Press.

——. (1984) *Distinction: Critique of the Judgement of Taste*, Translated by Richard Nice, Cambridge: Harvard University Press.

Bristol, G. (2007) *Strategies for Survival: Security of Tenure in Bangkok*, Case study prepared for Enhancing Urban Safety and Security: Global Report on Human Settlements 2007. Available from www.unhabitat.org/grhs/2007 [accessed 29 Jan 2017]

——. (2009) 'Rendered Invisible: Urban Planning, Cultural Heritage, and Human Rights', in Logan, W., Mairead, N., Craith, M., and Langfield, M. (eds), *Cultural Diversity, Heritage and Human Rights*), London: Routledge, pp.117–134.

Caftanzoglou, R. (2000) 'The Sacred Rock and the Profane Settlement: Place, Memory and Identity under the Acropolis', *Journal of Oral History*, no. 28, pp. 43–51.

Campbell, J.K. (1964) *Honour, Family, and Patronage: A Study of Institutions and Moral Values in a Greek Mountain Community*, Oxford: Clarendon Press.

Chatri Prakitnonthakan (2003) 'Pom Mahakan: Anurak roe thamlai prawatisaht?', *Silapawatthanatham*, February 2003, pp. 124–135.

——. (2012) 'Rattanakosin Charter: The Thai Cultural Charter for Conservation', *Journal of the Siam Society*, no. 100, pp. 123–148.

Danforth, L. M. (1984) 'The Ideological Context of the Search for Continuities in Greek Culture', *Journal of Modern Greek Studies*, no. 2, pp. 53–87.

Delcore, H. (2003) 'Nongovernmental Organizations and the Work of Memory in Northern Thailand', *American Ethnologist*, no. 30, pp. 61–84.

Dimakopoulos, I. (2001) *Ta spitia tou Rethimnou: simvoli sti meleti tis anayenisiakis arkhitektonikis tis Kritis tou 16ou ke tou 17ou eona*, 2nd edition, Athens: Tamio Arkheoloyikon Poron ke Apallotrioseon.

Fabian, J. (1983) *Time and the Other: How Anthropology Makes its Object*, New York: Columbia University Press.

Friedl, E. (1962) *Vasilika: A Village in Modern Greece*, New York: Holt, Rinehart & Winston.

Geertz, C. (1973) *The Interpretation of Cultures*, New York: Basic Books.

Gershon, I. (2011) 'Neoliberal Agency', *Current Anthropology*, no. 52, pp. 537–555.

Herzfeld, M. (1971) 'Cost and Culture: Observations on Incised Cement Decorations in Crete', *Kritika Khronika*, no. 23, pp. 189–198.

——. (1987) *Anthropology Through the Looking-Glass: Critical Ethnography in the Margins of Europe*, Cambridge: Cambridge University Press.

——. (1991) *A Place in History: Social and Monumental Time in a Cretan Town*, *Princeton:* Princeton University Press.

——. (2002) 'The Absent Presence: Discourses of Crypto-colonialism', *South Atlantic Quarterly*, no. 101, pp. 899–926.

——. (2004) *The Body Impolitic: Artisans and Artifice in the Global Hierarchy of Value.* Chicago: University of Chicago Press.

——. (2006) 'Spatial Cleansing: Monumental Vacuity and the Idea of the West', *Journal of Material Culture*, no. 11, pp. 127–149.

——. (2009) *Evicted from Eternity: The Restructuring of Modern Rome, Chicago*: University of Chicago Press.

——. (2015) 'Heritage and Corruption: The Two Faces of the Nation-State', *International Journal of Heritage Studies*, no. 21, pp. 531–544.

——. (2016a) *Siege of the Spirits: Community and Polity in Bangkok, Chicago*: University of Chicago Press.

——. (2016b) 'Critical Reactions: The Ethnographic Genealogy of Response', *Social Anthropology* no. 24, pp. 200–204.

Hirschon, R. and Thakurdesai (1970) 'Society, Culture and Social Organization: An Athens Community', *Ekistics*, no. 30, pp.187–96.

Kaewta Ketbungkan (2016) 'Vendors Paralyze Traffic with Floral Blockade to Protest Eviction', *Khao Sod English*, 2 July. www.khaosodenglish.com/news/bangkok/2016/07/02/vendors-paralyze-traffic-floral-blockade-protest-eviction/ [accessed 21 August 2016].

Kallivretakis, L. (2004) 'Apopse tha yini Tailandhi': I erminia enos 'eksotikou' sinthimatos tis ekseyersis tou Politekhniou', *Takhidhromos*, no. 246, pp. 46–51.

Knight, D.M. (2015) *History, Time, and Economic Crisis in Central Greece*. Basingstoke: Palgrave.

Lutz, C. (1993) *Reading National Geographic*, Chicago: University of Chicago Press.

Missingham, B.D. (2003) *The Assembly of the Poor in Thailand: From Local Struggles to National Protest Movement*, Chiang Mai: Silkworm Books.

Muehlebach, A. (2012) *The Moral Neoliberal: Welfare and Citizenship in Italy*, Chicago: University of Chicago Press.

Nakou, L. (1955) *I kiria Do-re-mi*, Athens: Difros.

Non Arkaraprasertkul (2016a) 'Gentrification from Within: Urban Social Change as Anthropological Process', *Asian Anthropology*, no. 15, pp. 1–20.

——. (2016b) *Locating Shanghai: Globalization, Heritage Industry, and the Political Economy of Urban Space in a Chinese Metropolis*, PhD dissertation, Department of Anthropology, Harvard University.

Papataxiarchis, E. (2016) 'Being "There": At the Front Line of the "European Refugee Crisis"', *Anthropology Today*, vol. 32, no. 2, pp. 5–9 and vol. 32, no. 3, pp. 3–7.

Pattana Kitiarsa (2005) '"Lives of Hunting Dogs": "Muay Thai" and the politics of Thai masculinities', *South-East Asia Research*, no. 13, p. 57–90.

Peleggi, M. (2002a) *Lords of Things: The Fashioning of the Siamese Monarchy's Modern Image*, Honolulu: University of Hawaii Press.

——. (2002b) *The Politics of Ruins and the Business of Nostalgia*, Bangkok: White Lotus.

Peristiany, J.G., ed. (1965) *Honour and Shame: The Values of Mediterranean Society*, London: Weidenfeld & Nicolson.

Prevelakis, P. (1972) *The Tale of a Town*. Translated by Kenneth Johnstone, London: Doric Publications.

Rakopoulos, T. (2016) 'Solidarity: The Egalitarian Tensions of a Bridge-Concept', *Social Anthropology*, no. 24, pp. 142–151.

Rozakou, K. (2016) 'Socialities of Solidarity: Revisiting the Gift Taboo in Times of Crises', *Social Anthropology* no. 24, pp. 185–199.

Sasiwan Mokkhasen (2016) 'Vanishing Bangkok: What is the Capital Being Remade Into, and For Whom?", *Khao Sod English*, 5 July. www.khaosodenglish.com/news/bangkok/2016/07/05/reorganization/ [accessed 21 August 2016].

Sopranzetti, C. (2012a) *Red Journeys: Inside the Thai Red-Shirt Movement*, Chiangmai: Silkworm Press.

———. (2012b) 'Burning Red Desires: Isan Migrants and the Politics of Desire in Contemporary Thailand', *South East Asia Research*, no. 20, pp. 361–79.

Tannen, Deborah. 1983. *Lilika Nakos*. Boston: Twayne.

Thanapon Watthanakun (2007) *Kahnmoeang roeang phoen thi: pholawat thang sangkhom khawng chumchon (Koroni soeksah: Chumchon Pom Mahahkahn* [Politics of place: Social dynamics of a community (case study: the Pom Mahakan community)], Bangkok: 14 October, Scholarly Institutional Foundation.

Theodossopoulos, D. (2013) 'Emberá Indigenous Tourism and the Trap of Authenticity: Beyond Inauthenticity and Invention', *Anthropological Quarterly*, no. 86, pp. 397–426.

———. (2016a) *Exoticisation Undressed: Ethnographic Nostalgia and Authenticity in Emberá Clothes*, Manchester: Manchester University Press.

———. (2016b) 'Philanthropy or Solidarity? Ethical Dilemmas about Humanitarianism in Crisis Afflicted Greece', *Social Anthropology*, no. 24, pp. 167–184.

Thongchai Winichakul (2000) 'The Quest for "*Siwilai*": A Geographical Discourse of Civilizational Thinking in the Late-Nineteenth-Century and Early-Twentieth-Century Siam', *Journal of Asian Studies*, no. 59, pp. 528–49.

Zhang, L. (2006) 'Contesting Spatial Modernity in Late-Socialist China', *Current Anthropology*, no. 47, pp. 461–84.

———. (2010) *In Search of Paradise: Middle-Class Living in a Chinese Metropolis*, Ithaca, NY: Cornell University Press.

Part IV

Forms and expressions of tourism gentrification

A critical analysis

11 Tourism and urban changes

Lessons from Lisbon

*Teresa Barata-Salgueiro, Luis Mendes
and Pedro Guimarães*

Introduction

Over the past few years the central areas of many cities have witnessed remarkable changes. This dynamic has allowed them to become both lively and vibrant as a result of the hotel industry, retail, leisure facilities and tourism in general. Berlin is the most impressive case of this in Europe (Bader and Bialluch, 2008; Füller and Michel, 2014).

Some years ago, Martinotti (1993) proposed a classification of cities using 'stages' based on the importance of non-residents or city-users. The dynamics of Portugal's larger cities are increasingly associated with the number of non-residents, users or investors. It has also been acknowledged that tourism has played a larger role in the shaping of urban space on a scale that was hitherto unknown (see Ashworth and Dietvorst, 1995; Neuts *et al.*, 2012).

Lisbon has been experiencing significant changes in its economy, its image and its social environment as a result of increasing numbers of tourists and visitors, as well as foreign students who visit or live in the city. Policies of urban regeneration and improving the culture offered by the city have favoured and boosted this attractiveness. Using this city as a case study, we aim to show how tourism has contributed towards the rehabilitation of some of Lisbon's central districts, as well as highlighting how it has changed the lives of some who live there by offering a range of opportunities aimed at improving the physical conditions of the area and by making it possible to start new businesses. We also point out some of the new problems caused by tourism which require public attention.

We argue that the growth of tourism is part of the strategy followed by neo-liberal governments to strengthen the urban economy and to improve its competitive position in regard to the spatial division of consumption as Harvey (1989) has indicated, as well as favouring capitalist accumulation (Britton, 1991). Moreover, the effects of tourism closely resemble those caused by gentrification. It is therefore challenging to study the touristification of cities as a vehicle of gentrification.

In order to find empirical evidence we chose two kinds of central neighbourhoods in Lisbon where the process of touristification has been marked differently.

The first is the metropolis' traditional retail and services centre. The second is the historical working-class residential neighbourhoods located in the vicinity of this central business district (CBD), which is now home to restaurants, cafés and nightspots, as well as accommodation. In the case of the former, the changes are exemplified by regeneration based on tourism and retail. In the second, the process is more diverse, although business opportunities geared towards tourism are numerous. In both cases, changes in urban policies favour the gentrification process that existed even before the recent boom in tourism.

The review of national and international literature and analysis of planning documents have allowed the theoretical framework that contextualises the case studies to be built. Data collection involved the acquisition of official statistics and the analysis of tourist accommodation on offer in the neighbourhoods, which was chosen through searches on the Airbnb and Booking.com sites and through interviews with municipal technicians and real estate agents. Regular surveys conducted as parts of projects which we were previously engaged in and local observations complete the information.

Policy and urban management

As the end of the twentieth century approached, the state – shaped by the neoliberal reference framework – reinforced the competitive rationales of a spatial nature (Peck and Tickell, 2002). The new urban policy was geared far more towards the market and therefore affected by paradigms based on promoting consumption, inter-city competitiveness, and the proactive role of private stakeholders in the city's planning process and production (Hall and Hubbard, 1996; Brenner *et al.*, 2013; Rossi and Vanolo, 2015). Urban policies began to include explicit economic goals favouring different strategies and paths of development. One – very clear in Lisbon from the beginning of the 1990s – pinned its fortunes on tourism and consumer culture.

The shift in the city's management is clearly visible both in planning and the policy for the old areas, along with measures to attract investors to real estate. Strategic planning became increasingly important, emphasising city marketing to a higher degree, promoting large-scale flagship projects and organising events aimed at attracting visitors. At the same time, significant changes to the rehabilitation policy for the old areas of the city took place. The first programme aimed at funding the rehabilitation of housing in Portugal dates back to 1976. Nevertheless, the first integrated urban rehabilitation programme did not appear until 1985. Rehabilitation at that time focused on improving housing conditions, investing in the public space, seeking to sustain the number of settled residents as well as attracting new ones. Following this, periods occurred in which rehabilitation focused mainly on physical appearances and on renewing buildings, while other periods had broader goals which intertwined with strategic planning and depended upon the participation of a large number of actors, including local residents.

The *Strategic Charter* for the city (2010–2024) set out the main goals for its development (CML, 2009). Following this, the city council approved a document of orientation in 2011: the *2011–2024 Rehabilitation Strategy for Lisbon* (CML, 2011). This document recognises that the policies that had been employed up until then had failed because the municipality simply did not have either the financial or technical means to replace actors in the private sphere. As an alternative, this *Strategy* proposed to attract private investment for the purpose of rehabilitation and tackled the question from a market-oriented point of view. A new paradigm for action was envisaged along two main axes. The first axis would place the emphasis on pre-existing heritage and adopt the '3R-philosophy': *Reuse* empty buildings, *Rehabilitate* buildings in a state of disrepair to be rented out, and *Regenerate* priority neighbourhoods, thus upgrading the 'consolidated city', i.e. areas built some time ago that are already established in physical and social terms. The second axis was concerned with the roles played by various actors: the government, the municipality and private investors, where responsibilities were kept separate and investment in rehabilitation was made attractive.

The city government adopted a diversified strategy towards the built environment but it is worth underlining the important role given to the private actors in the so-called rehabilitation process. As a matter of fact, this process is closer to the concept of regeneration as defined by Roberts (2000), which nowadays has been widely disseminated. So, in this text we shall be using the terms 'regeneration' and 'rehabilitation' synonymously. It is still important to stress that urban regeneration has become an essential political tool whereby land that has been devalued may be reintroduced onto the market, thus ensuring enhanced value and allowing it to overcome the rent gap studied by Smith (1979, 1987). Urban regeneration aimed towards tourism is a good example of this kind of process which is very well illustrated in the case of Lisbon. Similarly, Abe (2012) thinking about Japanese and European cities speaks about a period of 'regeneration urbanism' during the 1990s and the first decade of the twenty-first century in which urban redevelopment encouraged tourism.

In Portugal, the laws regulating the housing market have acted as a brake on advancing gentrification by hindering the eviction of tenants from their homes (Mendes, 2014). Only after the tenancy law had been amended in 2012 by the conservative government as a result of the challenging conditions of the international bailout of the country did evictions become easier, affecting both families and commercial premises. This explains the recent acceleration of gentrification.

The touristification of the central city

The relationship between tourism and urban areas is by no means recent (Mullins, 1991; Ashworth and Page, 2011). However, nowadays urban tourism is in full swing due to world-wide urbanisation and internationalisation (Galdini, 2007), changes in mobility (Williams and Hall, 2002), and the new economic impetus owing to investment in urban regeneration. This has been accompanied by important investments made to preserve built heritage and to promote cosmopolitan environments, particularly in the historical districts (Wilson and Tallon, 2012; Leite, 2013; Delgadillo, 2015).

In this text we assume that touristification is a process based on the change that is caused by an increase in the volume of tourists visiting a particular place. The impacts make themselves felt in at least three domains: accommodation, retail facilities and public space. There was a broadening of the variety of accommodation available, with the arrival of hostels and apartments for short-term rentals in buildings that have been fully or partially rehabilitated for this purpose. In terms of retail and services, Lisbon's examples point to changes in the functional composition of the neighbourhoods and the emergence of new specialisations. Together with the growth of the tourist industry, there has also been an increase in the pre-existing array of restaurants and other places to eat and drink, as well as a whole panoply of entertainment spots and places devoted to culture.

Public space has also undergone changes. The trend has been to extend leisure areas and pedestrian thoroughfares as well as the quality of spaces. The consumer rationale means that squares and other pedestrian areas are now filled with outdoor cafes and restaurants, replacing free public space with a compulsory consumer place as Cócola Gant (2016) describes in relation to Barcelona. In Lisbon, reducing free public space has not been as drastic as in Barcelona because many of the pavement cafés were made at the expense of motor vehicles, whether in the form of roads or parking spaces. Besides, the municipality has recently provided many open spaces and cultural facilities in the *Baixa* district and along the riverfront in areas that had been previously closed off to the public. The local policy towards the valorisation of public spaces is aimed at the residents in general, but the facilities provided such as greenery and urban furniture also serve tourists and other visitors.

The above-mentioned impacts of tourism follow changes in the built heritage with either new buildings or rehabilitating pre-existing constructions, as happened in the historical areas of Lisbon. Touristification reinforces the trend to upgrade these areas from social as well as economic aspects. In fact, the regeneration of the inner city favours the movement of affluent, usually young, middle-class residents into run-down areas, sometimes triggering gentrification (Smith, 1996; Slater, 2013).

Following the foreign examples and combining statistical data, fieldwork and interviews, Mendes (2008, 2014), Rodrigues (2010) and Pavel (2015) claim that until 2001 the gentrification process in Lisbon was fragmented and modest, thus it was not shown in the statistical data. However, Rodrigues identifies important trends in the social recomposition of Lisbon between 1980 and 2001, which points to gentrification processes being already under way. More recently, Barata-Salgueiro (2016) and Mendes (2016) – working with the data collected by the 2011 census – have shown much clearer signs of the process under way in some central districts of the city.

For decades, Lisbon's gentrification was characterised by the stage one "pioneering gentrification", according to Clay's model (1979). New residents, often with better access to financial resources and higher cultural and social capital, move into traditionally working-class neighbourhoods. Typically, Lisbon's gentrification begins with families, singles, artists or professionals seeking small spaces available in rundown neighbourhoods that offer environments suitable for alternative lifestyles. This first wave corresponds almost to an "embryonic process of gentrification", according to Mendes (2008), which develops slowly and intermittently.

Recently, the situation has been changing very quickly. Gentrification is expanding with great speed and also quite aggressively due to intense touristification. Lisbon is now reaching stage three of gentrification, after having experienced stage two during the past ten years, as Mendes (2014) explains in-depth. Rents increased dramatically and class struggle between gentrifiers, tourists and older residents became most pronounced. Media attention developed as physical changes become more evident and private capital flowed in, as there now exists a growing consensus that the area is a "safe investment" (Mendes, 2016).

There has been some discussion lately about whether touristification is a kind of gentrification since both processes often share common traits with one another. In Lisbon's inner city districts the two processes ('pure gentrification' and 'touristification') appear together so it is difficult to separate them. While current academic research on urban gentrification has not extensively considered the role of tourism, an interest in tourism and neighbourhood change has emerged and various studies have begun to analyse the two processes in parallel. The concept of tourism gentrification has therefore come to light in recent literature (Wilson and Tallon, 2012; Liang and Bao, 2015).

Based on Gotham (2005) but applied to Lisbon specificities, Mendes (2016) defines tourism gentrification as the transformation of popular and working-class neighbourhoods of the inner city into places of consumption and tourism sites, so that the recreation, leisure or accommodation functions gradually replace the traditional residential and retail functions, thus emptying the neighbourhoods of their original population. Escalating rents prevent poor families and immigrants from settling there.

In some districts that have been shaped by the new urban cultures and aesthetic concerns, as in Lisbon's *Bairro Alto*, both gentrifiers' practices and the habits of new tourists looking for the authenticity of the city experience *as* residents become similar. They are both the root cause of complaints and conflicts that emerge when residents' daily living spaces are shared by the activities that tourists engage in (Hiernaux and González, 2014). These conflicts are constantly reported in newspapers in which residents often complain about the noise and the dirt during the evenings and at night (Pavel, 2015; Nogueira, 2015). More recently, the complaints have been almost the same as those that Füller and Michel (2014) refer to regarding Kreuzberg, Berlin: rising rents and the reduction in the number of flats available to rent.

In Lisbon's central and historical quarters, regeneration is not being undertaken with permanent residents in mind, but rather, it is aimed at temporary residents – namely tourists – because real-estate investment sees tourism as an opportunity for more lucrative business. We do not have information which would enable us to compare Lisbon with the situation found by Zhao *et al.* (2009) in Nanjin, where the real-estate developers played the leading role in the progression of urban tourism gentrification. But the real-estate brochures, newspapers and some interviews conducted with municipal technicians point to the presence of large developers shouldering small investments in Lisbon. The large developers first buy the whole building to renew as a hotel or as upscale apartments. The small investors run one or two flats bought or rented as short-term rentals or as hostels.

We argue that the shift in capital flows to the real estate market of tourist accommodation in much of the central districts, combined with the political trend to increase urban regeneration, reinforces the importance of activities aimed towards tourist consumption and encourages direct and indirect displacement and gentrification. Marcuse's (1985) analysis of displacement, which supports the concept of indirect displacement, is particularly important to our argument in terms of understanding the effects of excessive touristification in the eventual production of gentrification.

In the CBD, significant changes have also been made to retail businesses, as Mitchell and Kirkup (2003) have shown. The same is documented in recent literature on Lisbon and Porto (see Guimarães, 2016; Barata-Salgueiro, 2015; Cachinho, 2015; Fernandes, 2013).

Touristification and gentrification associated with urban regeneration may contribute to similar changes in the urban landscape and its use. It is difficult to distinguish the effects of tourism from the general process of urban change and commodification due to the similar demand for the urban experience amongst the new urban tourists and the potential gentrifiers from the middle classes, as Füller and Michel (2014) also point out. The characteristics of touristification are similar to those described for gentrification in terms of capital investment, social and functional upgrading, replacing the population, a decrease in the number of permanent apartments, the increase in real-estate prices, and changes in the appearance and day-to-day living of the area.

Tourism-led gentrification: City centre retail and the colonisation of historical neighbourhoods

According to the Census data (INE), Lisbon lost 260,000 people (32 per cent of its population) between 1981 – the highpoint of Lisbon's population – and 2011. But during the last inter-census period it recorded a small increase (of 1.2 per cent) in the number of resident families, from 241,000 to 243,900. Statistical evidence indicates a change in the demographic trends of Lisbon as a result of the gentrification of the inner city, even if it has not been developed to the same extent when compared with other cities.

Between 2001 and 2011 the demographic loss was much lower and several central wards experienced an increase in the number of residents. What is more significant is that this rise largely involved young adults (between 25 to 39 years of age) with college degrees (Figure 11.1). In this figure we have highlighted the city districts which lost more than 45 per cent of their residents between 1981 and 2011; they practically correspond to the historic city. Within these

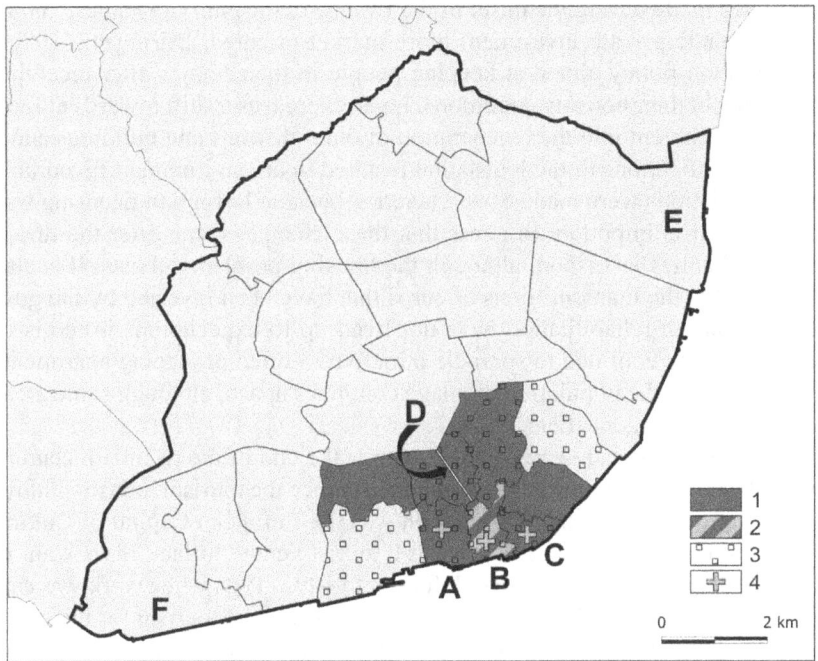

Figure 11.1 Demographic dynamics in central Lisbon

Source: INE, Census data. Cartography by Joaquim Seixas

Notes: Legend: 1 – Districts which lost 45 per cent or more of their residents between 1981 and 2011; 2 – Baixa Chiado is the only district which saw its population increase between 2001 and 2011; 3 – A rise larger than 25 per cent of the proportion of young adults (of 25 to 39 years of age) in the residential population (2001–2011); 4 – The largest increases in the proportion of adults with college degrees (more than 100 per cent) (2001–2011).

Districts studied: A- *Bairro Alto*; B - *Baixa-Chiado*; C- *Alfama*. Other places referred to by the text: D- Av. da Liberdade; E- Parque das Nações area where the 1998 International Exhibition took place; F- Belém, a hotspot for tourism

we then represented the rise in the proportion of young adults and adults with college degrees in the first decade of the twenty-first century. The advance in higher education is very important in the city (an increase of 55.7 per cent). However, the increase surpasses the 100 per cent mark in the CBD and the two historical neighbourhoods – *Bairro Alto* and *Alfama* – where gentrification and touristification have been quite dynamic.

Nevertheless, despite the increase in *Baixa* district, the historical neighbour-hoods nearby continue to be drained of their inhabitants, although at a lower rate. Situated to the east of the city centre, the *Alfama* district lost 22 per cent of its residents between 2001 and 2011. To the west, *Bairro Alto* lost almost one-fifth during the same period, despite the fact that one of its wards recorded gains between 1991 and 2001. This has been explained by the rehabilitation process which has been ongoing since the second half of the 1980s, and the changes in urban policies made during the onset of the twenty-first century (Mendes, 2008), which have made private investment more market-oriented. During the 1980s, the rehabilitation policy aimed at keeping people in their homes after receiving improvements in their housing conditions. Later, there was a shift towards attracting private investment into the regeneration of older districts and buildings and a number of modifications to the legislation resulted in an environment favourable to this aim: evictions were made easier, taxation became lower and licensing was made quicker. It is important to stress that these changes came after the disappearance of shanties in Lisbon, although the housing problem persists. It is also recognisable that the many millions of euros that have been invested by the government into urban rehabilitation have not lived up to expectations in terms of outcomes. In being confined to sporadic initiatives – often producing apartments that are too small – the population drainage continues apace, although some areas have attracted young gentrifiers.

Another important change registered towards the end of the twentieth century links with the city's economic base and the importance the tourism industry enjoys in its planning documents. In 1994, Lisbon was the European Capital of Culture and this event helped to change the city as a tourist venue. In fact, if we want to speak about tourism we need to go back further to 1992–1994 when work was carried out to develop the city's masterplan. The importance of tourism for the city's economy had already been underlined in these documents which pointed to the development of specialised tourist niches such as meetings and events and established a coordinating body to draw up strategic guidelines for tourism. In order to do so, the Lisbon Tourism Association (*Associação de Turismo de Lisboa, ATL*) was set up in 1997; it was a private agency running on public funds and its purpose was to develop tourism, promote Lisbon as a tourist destination and give information and support to tourists. Building large-scale facilities also appeared in the plans, like an Exhibition Park, a Centre of Congresses, and a sports pavilion for competitions and cultural events. The facilities were built within the regeneration framework of a former industrial zone where the 1998 International Exhibition was held (see Figure 11.1 for relative location).

Afterwards, strategic plans geared towards tourism followed, also including other works, and a concerted effort was made to promote the city. Since 2007, plans for increasing tourism at a national, regional and local scale have mushroomed and tourism has now been integrated into most of the spatial policy instruments. The plans always refer to the need to promote Lisbon as a brand and the advantages of specialising in segments that ensure that demand is sustained throughout the year. To meet this requirement they point to the diversification of the attractions the city has to offer, exploring the complementary characteristics of the city and the region.

Lisbon's traditional city centre (*Baixa* and *Chiado*) used to concentrate the country's and the city's most important administrative, financial and commercial functions until the mid-twentieth century. From then on, owing to the explosive growth in services and problems with traffic congestion, the city-region underwent important changes and a CBD expansion developed towards the North. The growth in the built area, the restructuring of tertiary centres and the appearance of superstores and shopping centres in the final 20 years of the twentieth century, have all led to the transformation of a monocentric metropolis into a polycentric structure with significant social-spatial fragmentation (Barata-Salgueiro, 1997). At the same time, there was progressive disinvestment in the centre that saw many buildings being abandoned and much of the area's retail aging and frozen in a time warp.

The presence of empty dwellings in the city rose from 9.2 per cent in 1991 to 15.5 per cent in 2011. In the neighbourhoods studied this percentage varies from 26.8 per cent in *Bairro Alto* to 35.0 per cent in *Baixa-Chiado* in that same year.

Over the past few years, we have witnessed increased public and private investment in the rehabilitation of buildings, retail establishments and public spaces. During this time, the *Baixa-Chiado* recovered its importance thanks to a process in which tourism and retail played a relevant role working as an anchor for the area's vitality. The different municipal planning tools were crucial in this operation. In 2010, the detailed plan for the *Baixa* district was passed and acted as a catalyst for the rehabilitation of privately owned buildings. It involved bringing together public works in terms of infrastructure, cultural facilities (two museums), improved roadways and enhanced public spaces. It is worth stressing that the plan linked with important public works programmes for the riverfront. Taken within this framework, the whole central area became particularly attractive to real estate investment for tourist and retail purposes. Recently, laws have been passed that speed up licensing and works on the one hand, and on the other, have seen the remodelling of taxation on real-estate investment funds and a special tax regime for non-regular residents was created. Local government has also been active in city marketing. For instance, a good example is the Lisbon Shopping Destination platform that unites the different retail areas in the city under one umbrella.

Tourism and urban regeneration has progressed considerably in Lisbon. According to Lisbon Tourism Office, the number of passengers arriving in the city has registered a 65 per cent growth in the last ten years. In 2014, Lisbon airport received 18 million passengers, and pleasure cruisers docked with about half a million visitors. In the same year, the hotel industry recorded 9 million overnight stays, i.e. 20 per cent more than in 2013 (INE, 2016).

Tourism in the city has been developed consistently and to a remarkable degree during the last twenty years, evidenced by the numbers of tourists and the hotels on offer. As can be seen in Figure 11.2, this growth was sharper after 2008/2009, keeping pace with the increase in hotel offerings. In the last seven years, 73 more hotels were built, of which 39 are four- or five-star rated. Most of these are situated in the new CBD. Only 10 per cent of the hotels are located in the historical centre.

At the same time, building rehabilitation devoted to tourist accommodation in the city's historical centre was witnessed. In keeping abreast with developments in low-cost travel to Lisbon airport and the rise of young tourists, we have seen an increase in the availability of accommodation that is not situated in the traditional hotel sector. In 2013, 54 hostels were available, 63 per cent of which opened

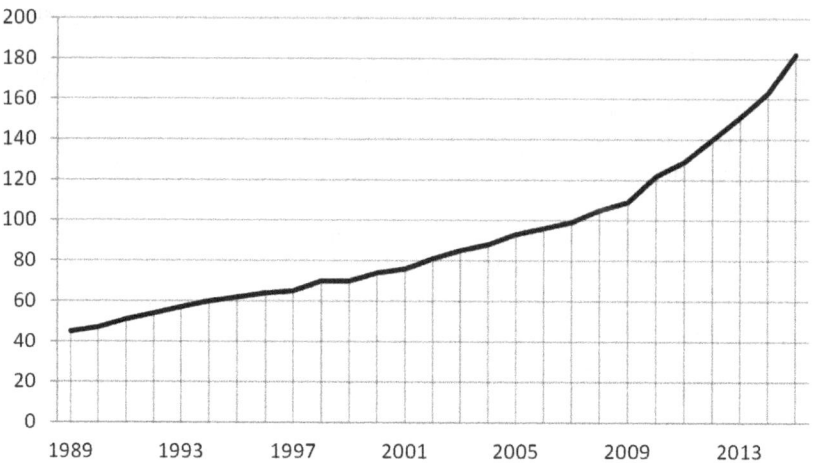

Figure 11.2 Number of hotels in Lisbon, between 1989 and 2015

Source: Associação de Turismo de Lisboa (2015). Design by the authors

their doors after 2009. Contrary to the hotels, the hostels are mostly located in the old centre and its environs (72 per cent), mainly in the quarter lying to the west (Guimarães, 2015).

Another kind of accommodation which significantly expanded are short-rental apartments and rooms. There is no complete data on the number and location of apartments and rooms to rent in the city, although from November 2014 – when the municipal register became obligatory – until the end of April 2016, 4,350 units were duly registered in Lisbon. This number includes 3,974 apartments, 333 guest houses and 43 houses. An analysis of Airbnb shows that the most significant availability is of T0 and T2 apartments that range from studios up to 2 bedroomed apartments (1 to 6 beds). In this category, the above-mentioned website advertised 1,004 apartments at the beginning of August 2015. The sample collected for our database covers 421 flats which equates to 41.9 per cent of the total on offer. There were 306 advertisements for apartments able to accommodate between 8 and 10 people, and sometimes more than 10. From these we collected information from 147 (48 per cent). We have decided to use only these two size categories because there were many repetitions in the intermediate ones.

The map showing the location of the apartments revealed a dense concentration in the historical quarters (Figure 11.3), with slight differences when the size of the apartment is taken into account. *Alfama* offers mostly T0 to T2 studios, while on the other side of the city centre, in *Bairro Alto*, the offer is more diverse as this quarter has a more varied building stock. The concentration of the accommodation business in *Bairro Alto* is best seen if we consider the proportion of beds per capita. Basing findings on a field study, Pavel (2015) has calculated that this neighbourhood offers the capacity to receive one temporary resident for every two

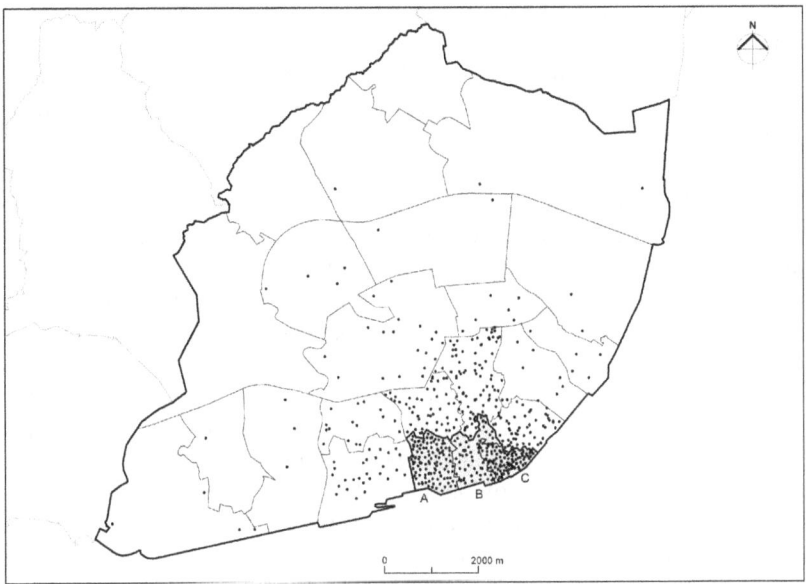

Figure 11.3 Short-rental apartments in Lisbon

Sources: Sample by the authors taken from the Airbnb website, August 2015. Cartography by Leando Gabriel

Notes: Legend: Each dot represents 1 apartment; A- *Bairro Alto*; B- *Baixa-Chiado*; C- *Alfama*

permanent residents. A survey conducted by the authors on February 2016 on the website Booking.com shows that for *Alfama* this relationship is 2.2. This means that for one temporary resident there are 2.2 permanent residents. This calculation does not include the massive supply of private apartments found, for example, on the Airbnb website. These results resemble the ones Cócola Gant (2016) has found in the Gótic district of Barcelona, where the density of hotel beds per capita reaches 1.6, i.e. the number of hotel beds is almost equal to the number of inhabitants living in the neighbourhood. These figures lead us to question the carrying capacity of the historic districts and the sustainability of their social fabric in the face of such strong pressure posed by touristification.

It is also worth mentioning that contrary to what is often said, students and tourists get along relatively well with the local residents in the historical neighbourhoods of Lisbon (Bastos, 2015). The same cannot be said about gentrifiers from the upper classes and condominium residents that have little to do with the neighbourhood's life or its public spaces, as research on socio-spatial fragmentation has shown (Barata-Salgueiro, 1997, Malheiros *et al.*, 2013). Apart from housing, they limit themselves to consuming 'reserved' spaces such as clubs, restaurants and more sophisticated shops (Rose, 2004; Davidson, 2010; Bridge *et al.*, 2012), often not located in the vicinity.

The problem seems to lie in the balance, or rather, a lack of balance. When the number of tourists reaches a certain threshold, they end up bothering each other and frustrating their expectations. They have come in search of experiences and to enjoy the authenticity of the place yet what they find are apartments that have been standardised by a global IKEA style, restaurants full of tourists like themselves and chain stores or other kinds of stores that are a feature of every city.

Our findings back up the data published by LaSalle (JLL), which used the Booking.com website as its source and calculated that 85 per cent of the tourist apartments were situated in the historical centre, including the expanded CBD. According to JLL (2015: 6), this trend should continue to grow because "foreign and local investors and developers continue to buy buildings in Lisbon's historical centre so as to rehabilitate them for this specific purpose", namely investors holding "Golden Visas". This visa is an easier way of obtaining residential permits in Portugal provided that the third-country nationals make investments higher than a specified amount (it was first set at one million euros and afterwards was dropped to 500,000 euros), or create a certain number of jobs, though to a great extent investments are made in the form of buying real estate.

At the same time, the retail structure was adapted to the changes in the buildings' conditions and their occupancy. In the historical quarters, this meant a sharp drop in the number of local corner-shops and an upgrading of bars and restaurants, as the displacement caused by tourist facilities may be both residential and commercial (Porter and Shaw, 2009). The changes in the retail structure of the neighbourhoods affects the lives of residents on a daily basis, forcing their indirect displacement – a process that Davidson (2008) calls the "neighbourhood resource displacement". Nowadays, the clientele is more cosmopolitan and more numerous. The number of low-quality souvenir shops has also mushroomed. Likewise, another change linked this time with the arts appeared (art galleries, designer goods, or high-quality designer crafts), mainly in *Bairro Alto*, which got close to becoming a 'creative place' (Costa, 2007; André and Gabriel, 2016).

After a period in which there was a steep decline – extending from the 1970s to the 1990s – the city centre of Lisbon was greeted with new life and new specialisations that catered to a variety of clientele and where tourists mingled with other visitors. Therefore, the data recently collected by our project REPLACIS (Barata-Salgueiro and Cachinho, 2011; Barata-Salgueiro, 2015) in regard to the centre's main retail axes and its comparison with earlier data, has allowed us to identify deep-seated, high-speed changes in the retail structure. The first evident change is related to the increase of luxury and prestige offices along the rehabilitated *Avenida da Liberdade,* the avenue that projected the CBD northwards. Here many hotels, costly apartments, prestige offices and luxury retail stores are concentrated; the latter are visited mostly by foreigners (from Angola, Brazil, Russia and China), as JLL (2014) suggests. Until 2002, luxury goods were marketed in multi-brand shops, but since 2014, 63 per cent of the luxury brands had shops of their own situated along the *Avenida da Liberdade*. Today, this percentage is even higher due to new shops having opened. In 2013 and 2014 alone, 22 new shops opened up.

The *Baixa-Chiado* district has been classified by the municipality as a heritage site and has been the subject of intensive rehabilitation during the last few years, and has also experienced significant changes in retail. Between 2007 and 2015, the number of retail units increased but almost half the shops that existed in 2007 could no longer be found. However, attesting to the feasibility of the district, the shops were replaced by newer ones, in greater numbers. The process is partially connected to having bolstered the importance of the area, thus making it more attractive to private investment. At the same time, the amendments to the law simplifying the eviction of long-term tenants led to the upgrading of retail. One of the outcomes of this process is that a large number of chain stores exist in *Baixa-Chiado*, particularly the international chains that are located preferentially in the *Chiado* which now has the highest retail rents in the city (Table 11.1).

The retail structure shown in Figure 11.4 clearly reveals that this is the metropolitan core which has a marked presence of stores selling personal goods, most of which offer clothing and footwear. Nevertheless, there are differences between what is on offer in *Baixa*, with its medium quality, mass-tourism focused brands such as Stradivarius and H&M, and the more prestigious and costly brands such as Hugo Boss and Nike found in *Chiado*. However, when taking into account the entire central area, between 2007 and 2015 there was a decrease in the number of stores selling personal goods, mostly due to an increase in the following three categories. The first, called "Other" in Figure 11.4, includes tourism-focused stores (selling souvenirs and crafts). Despite the fact that they are not devoted to selling one specific kind of merchandise, their specialisation comes from the target-customer which they attract. Although most of these shops belong to the category of ethnic retail and sell low-quality goods, the rise in tourism has also encouraged the opening of new shops that sell traditional products such as cork-based goods.

The second category that experienced important growth includes hotels and restaurants which are closely linked to the rising importance of tourism. Predominantly in the *Baixa* area, hotels and other kinds of accommodation, countless pavement cafés and restaurants that serve fast-food or standard cuisine which is disguised as being traditionally Portuguese – and caters mainly to tourists – have flourished. Finally, the third category that has dramatically increased refers to services offering "body and health care"; they include pharmacies and the

Table 11.1 Retail change between 2007 and 2015 in downtown Lisbon

	Baixa	Chiado	Total
Total number of units (2007)	143	119	262
Total number of units (2015)	142	145	287
Units closed since 2007	68	55	123
New units in 2015 compared to 2007	70	77	147
National chain stores (2015)	19	12	31
International chain stores (2015)	16	42	58

Source: Data compiled by the authors

Note: The shopping centre is not included

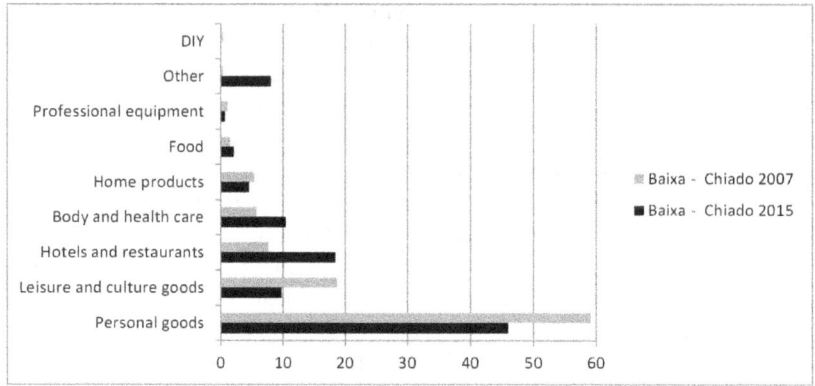

Figure 11.4 Baixa-Chiado: its functional retail structure and the changes witnessed
     between 2007 and 2015

Source: Data compiled by the authors. Design by the authors

Note: The shopping centre is not included.

increasingly important sector of herbal products and wellbeing products. This
increase relates to new concerns for health within new lifestyles, not only gentrifiers, but the population in general. It is not linked to tourism. Its importance in
the central core shows how the phenomena are intertwined.

Discussion and conclusion

The case studies revealed deep changes in the recent dynamics of the inner city.
Underpinning the dynamics has been both the city's strategic planning – which
has turned tourism into one of the pillars of the economy – and several urban
policy measures favourable to urban regeneration. Both phenomena have been
subject to fluctuation as they adjusted to political cycles. Thus, while embracing neoliberal management philosophies, they also addressed social concerns and
undertook collaborative forms of planning.

The examples taken from Lisbon's historical neighbourhoods and its retail
core show the importance of how tourism has fuelled urban change. They allowed
us to identify most of the factors described by Davidson and Lees (2005) when
they discussed gentrification. There was significant capital investment in buildings that had undergone rehabilitation as well as in new businesses opening up,
predominantly aimed at tourism. Contrary to the first wave of recovery work that
has produced apartments to sell or rent to long-term residents, during the last few
years real estate rehabilitation has been effectively for tourist accommodation.
Indeed, investment in tourist accommodation is held responsible for important
rehabilitation or renovation work being done but the rising rents prevent the
lower and middle classes from entering the market. Thus, the capacity to house
permanent residents in these central districts is threatened.

Tourism, or in more general terms, the consumer and leisure economy in which tourism is included, has provided an opportunity for various enterprises to emerge, bringing new activities to these neighbourhoods, creating jobs for local residents, business opportunities for different categories of investors and attracting the attention of outsiders. The issue of local employment received praise from the local interviewees (Benis, 2011) and in the mass media (Bastos, 2015). For this reason, tourism is viewed as a solution for poor neighbourhoods, although most of the jobs created are precarious and poorly paid (Gibson, 2009).

Apart from rehabilitating the buildings, the growth of the hotel industry and changes in retail being relevant aspects of landscape change, it is worth stressing that touristification has also helped to bring life and buoyancy to the neighbourhoods. If a few years ago the *Baixa* area was literally deserted at night, today the situation is completely different because the hotels and restaurants have helped to turn it into a dynamic spot. Also helping the neighbourhood's renaissance has been the work done to upgrade the public space undertaken by the municipality. It is undeniable that the centre of Lisbon has gone back to being a vibrant area in which liveliness has bubbled over into the historical zones which, nevertheless, has given rise to residents' complaints about the noise and disturbances at night.

More than seeing an upgrade in terms of income, the most glaring change caused by both touristification and gentrification in Lisbon has been cultural, caused by the arrival of people who have different lifestyles. These include students, artists or 'new tourists'. There is social change due to the coexistence of permanent and temporary residents, and simple users which have quite different features in the different places we have studied. In the historical districts we find traditional working-class residents, some immigrants, gentrifiers, students and tourists. The social environment in these historical quarters has become more diverse and multicultural, and varied lifestyles are shared which has broken the homogeneous situation that was prevalent before.

Examples of renewing old local traditions may be found where they have exercised a dynamic effect on the resident community, although they have also been put at the disposal of tourism. The tourist movement provides a clientele for some of those activities and stores, although it has also paved the way for an upsurge in the number of globalised souvenir shops. This fact also poses a threat to the neighbourhood's identity and the improbable likelihood of the traditional residents continuing to live there. The changing character of the neighbourhoods occurs because their residents have been displaced, because their networks, social fabric and community spirit built up over time and giving them their identity, has been destroyed. Downtown retailscape happens because the area's shops have disappeared – some of which were very old, almost landmarks, which made the place unique; they have now been replaced by hotel lobbies, large chain stores and souvenir shops that are common in all of the tourist spots. This process has threatened the place's identity and has caused it to lose its differentiating factors, leading it to resemble what is known as a clone town (Warnaby, 2009).

In regard to facilities, there is no doubt that it is appropriate to talk about an upgrade. In the main residential neighbourhoods, tourism and the arrival of creative young people and entrepreneurs have been responsible for the appearance of many shops and some services connected with the arts, culture, retail and leisure that occupy a higher rank than those that existed previously. In the commercial core, this upgrade is only too obvious with luxury establishments, an increase in chain stores frequently operating using international capital and in the new formats or concept stores. But co-existing with these upmarket places are the "inferior" quality outlets targeting tourists, which are often run by people from third countries, mainly Bangladeshi nationals.

Finally, and in regard to displacement, a tendency has been noted whereby the population and economic activities have taken over because the previous occupants have left, expelled by the new rent levels and the urban changes undertaken or the alterations inflicted on the neighbourhood's social and lifestyle trends. The number of direct evictions of residents was modest until fairly recently in the cases we studied. This was the case for two reasons: the large number of apartments lying empty and the time it takes to evict elderly residents from their homes in conformity with the law. Nevertheless, displacement in terms of economic activity has been very high, mainly in the city centre area, as a result of amendments made to the laws on tenancy and rehabilitation.

Changes in the neighbourhoods' atmosphere was responsible for the departure of some inhabitants, particularly the pioneering gentrifiers who were displeased with a certain tourist thematisation, and the former residents who did not wish to return once they had settled elsewhere. Lacking the conditions required for their daily lives – such as convenience stores and local networks of social relationships – people left because they found that the place had become uncomfortable and felt they had become estranged from it.

Rising real-estate prices after rehabilitation work was carried out prevented some people from continuing to live there or others – who had had access to it before – from moving in. Indeed, in the cases we studied, indirect displacement was much more meaningful. It was caused by the spiralling cost of real-estate which had made these new places inaccessible to the children of the older residents, and even to the young middle-class, especially because the depression made employment extremely precarious. Likewise, many of the independent retailers or the traditional workshops no longer had the capacity to compete with the large chain stores in rent bidding.

We have seen that touristification has produced significant changes in the urban space. It led to the rehabilitation of buildings in order to increase the availability of accommodation and the services required to cope with the demands of increasing numbers of visitors. The rising competition between cities and tourist venues called for public intervention so as to set up conditions for the regeneration of the cultural realm and the public space in particular. In this sense, we may say that what happened involved a process of tourism-led regeneration owing to the fact that tourism acted as the catalyst to a larger set of interventions that produced the inner city urban space. These changes have had similar impacts to those of gentrification (a process that is also expanding) so that, in line with other studies undertaken on this topic, we might conclude that a process of tourism gentrification also occurred at the same time, causing the direct displacement of many shops in the city centre and in the historical quarters, as well as the indirect displacement of residents and businesses.

Nevertheless, we doubt whether the concept of tourism gentrification is more enlightening and is more informative than the concept of tourism-led gentrification for two reasons. Firstly, tourism and gentrification grow in the same areas. Secondly, in cities like Lisbon there is such a huge availability of tourist accommodation that, in a short while, it will become excessive. This will lead to a reconversion of many of these apartments into long-term rentals. In having been rehabilitated and well located the target will be the middle and upper classes. Given this scenario, accommodation for tourists may well be simply a stage in the gentrification process, although it will be a decisive factor in speeding up gentrification on a widespread scale owing to the real-estate products available and the stakeholders involved.

To summarise, it is worthwhile stressing two points that have come up in the literature and what we found in Lisbon. Tourism-led regeneration is better understood within the global circulation of capital in search of investment opportunities, making wagers on real-estate rehabilitation and on business activities connected with tourism. From the hotel chains, travel agencies to retail and other kinds of business linked with leisure, they all invest in the city which has been transformed into a commodity at the service of tourism.

Regardless of the flexible nature of the concept, this type of change is helping to entrench the differences in accessing urban space, segmenting residential spaces, leisure spaces and retail spaces, denying the right to the *polis* to an ever-increasing number of former users and leading to potential conflicts between: underprivileged inhabitants versus the wealthy; permanent residents versus travellers and tourists; specialised shops versus global chain stores; local characteristics versus the global machine that standardises everything and places it at the service of the market.

References

Abe, D. (2012) 'Tourism, Gentrification and Neighborhood Management in Regenerated-Cities: Towards a "Post-regeneration Urbanism"', [Online], in Proceedings of the *6th Conference of the International Forum on Urbanism*, 25–27 January 2012, Barcelona. Escola Técnica Superior d'Arquitectura de Barcelona, pp. 1–6, [Accessed 15 February 2016], Available from: http://upcommons.upc.edu/bitstream/handle/2099/12560/C_213_3.pdf.

André, I. and Gabriel, L. (2016) 'A mercadorização das artes', in André, I., Gabriel, L. and Estevens, A. (eds), *Atlas das Utopias Reais: Criatividade, Cultura e Artes*, Lisbon: Le Monde Diplomatique, pp. 36–41.

Ashworth, G. and Dietvorst, A. (eds) (1995) *Tourism and Spatial Transformations Implications for Policy and Planning*, Wallingford: CAB International.

Ashworth, G. and Page, S. (2011) 'Urban Tourism Research: Recent Progress and Current Paradoxes', *Tourism Management*, vol. 32, no. 1, pp. 1–15.

Associação de Turismo de Lisboa. (2015) *Evolução de unidades hoteleiras e capacidade na cidade de Lisboa*. Lisboa: Observatório do Turismo de Lisboa.

Bader, I. and Bialluch, M. (2008) 'Gentrification and the Creative Class in Berlin-Kreuzberg', in Porter, L. and Shaw, K. (eds), *Whose Urban Renaissance? An International Comparison of Urban Regeneration Strategies*, London & New York: Routledge, pp. 93–102.

Barata-Salgueiro, T. (1997) 'Lisboa metrópole policêntrica e fragmentada', *Finisterra. Revista Portuguesa de Geografia*, vol. 32, no. 63, pp. 179–190.

Barata-Salgueiro, T. (2015) 'City Centre Temporalities Induced by Retail', in D'Alessandro, L. (ed.), *City, Retail and Consumption*, Naples: Università degli studi di Napoli 'L'Orientale', pp. 25–34.

Barata-Salgueiro, T. (2016) 'Impacts of Gentrification and Urban Tourism on Retailing', in Olivera, P., Gasca, J. (ed.), *City, Retail and Consumption. Proceedings of the 5th International Seminar*, Mexico City.

Barata-Salgueiro, T. and Cachinho, H. (ed.) (2011) *Retail Planning for the Resilient City: Consumption and Urban Regeneration*, Lisboa: Centro de Estudos Geográficos.

Bastos, J. (2015) 'A ver passar os turistas', *Expresso*. [Online]. 1 August 2015. [Accessed 12 January]. Available from: http://expresso.sapo.pt/sociedade/2015-08-01-A-ver-passar-turistas.

Benis, K. (2011) *Vielas de Alfama. Entre Revitalização e Gentrificação. Impactos da 'gentrificação' sobre a apropriação do espaço público*, [Online], Dissertação de Mestrado em Estudos Urbanos em regiões Mediterrâneas, Lisboa: Universidade Técnica de Lisboa, [Accessed 22 January 2016], Available from: https://www.repository.utl.pt/bitstream/10400.5/3659/1/VIELAS-DE-ALFAMA-20110916.1.pdf.

Brenner N., Peck, J. and Theodore, N. (2013) 'Neoliberal Urbanism: Cities and the Rule of Markets', in Bridge, G. and Watson, S. (eds), *The New Blackwell Companion to the City*, Oxford: Wiley Blackwell, pp. 15–25.

Bridge, G., Butler T. and Lees, L. (eds) (2012) *Mixed Communities: Gentrification by Stealth?* London: Polity Press.

Britton, S. (1991) 'Tourism, Capital, and Place: Towards a Critical Geography of Tourism', *Environment and Planning D: Society and Space*, vol. 9, no.4, pp. 451–478.

Cachinho, H. (2015) 'Avenida da Liberdade: From the Bourgeoisie Promenade to the Showcase of International Capital', in D'Alessandro, L. (ed.), *City Retail and Consumption*, Naples: Università Degli Studi di Napoli 'L'Orientale', pp. 35–47.

Câmara Municipal de Lisboa. (2009) *Carta Estratégica de Lisboa 2010–2024. Um compromisso para o futuro da Cidade*, [Online], Lisboa: Câmara Municipal de Lisboa, [Accessed 10 May 2016], Available from: www.cm-lisboa.pt/municipio/camara-municipal/carta-estrategica.

Câmara Municipal de Lisboa. (2011) *Estratégia de Reabilitação urbana de Lisboa -2011/2024*, [Online], Lisboa: Câmara Municipal de Lisboa. [Accessed 10 May 2016], Available from: /www.cm-lisboa.pt/fileadmin/VIVER/Urbanismo/urbanismo/Reabilitacao/estrat.pdf.

Clay, P. (1979) *Neighborhood Renewal: Middle-Class Resettlement and Incumbent Upgrading in American Neighborhoods*, Massachusetts: Lexington Books.

Cócola Gant, A. (2016) *Apartamentos turísticos, hoteles y desplazamiento de población*, [Online], [Accessed 3 January 2016], Available from: http://agustincocolagant.net/apartamentos-turisticos-hoteles-y-desplazamiento-de-poblacion/.

Costa, P. (2007) *A cultura em Lisboa: competitividade e desenvolvimento territorial*, Lisboa: Imprensa de Ciências Sociais.

Davidson, M. (2008) 'Spoiled Mixture: Where Does State-led "Positive" Gentrification End?', *Urban Studies*, vol. 45, no. 12, pp. 2385–2405.

Davidson, M. (2010) 'Love Thy Neighbour? Social Mixing in London's Gentrification Frontiers', *Environment and Planning A*, vol. 42, no. 3, pp. 524–544.

Davidson, M. and Lees, L. (2005) '"New-build Gentrification" and London's Riverside Renaissance'. *Environment and Planning A*, vol. 37, no. 7, pp. 1165–1190.

Delgadillo, V. (2015) 'Patrimonio urbano, turismo y gentrificación', in Delgadillo, V., Díaz, I., Salinas, L. (eds), *Perspectivas del Estudio de la Gentrificacíon en México y América Latina*, Coyoacán: Instituto de Geografía, UNAM, pp. 113–132.

Fernandes, J. (2013) 'Muitas vidas tem o centro e vários centros tem a vida de uma cidade', in Fernandes, J. and Sposito, M.E. (eds), *A nova vida do velho centro nas cidades portuguesas e brasileiras*, Porto: CEGOT, pp. 31–43.

Füller, H. and Michel, B. (2014) '"Stop Being a Tourist". New Dynamics of Urban Tourism in Berlin-Kreuzberg', *International Journal of Urban and Regional Research*, vol. 38, no. 4, pp. 1304–1318.

Galdini, R. (2007) 'Tourism and the City: Opportunity for Regeneration', *Tourismos: An International Multidisciplinary Journal of Tourism*, vol. 2, no. 2, pp. 95–111.

Gibson, C. (2009) 'Geographies of Tourism: Critical Research on Capitalism and Local Livelihoods', *Progress in Human Geography*, vol. 33, no. 4, pp. 527–534.

Gotham, K. (2005) 'Tourism Gentrification: The Case of New Orleans' Vieux Carré (French Quarter)', *Urban Studies*, vol. 42, no. 7, pp. 1099–1121.

Guimarães, P. (2015) *O planeamento comercial em Portugal - os projectos especiais de urbanismo comercial*, PhD Dissertation in Geography, Lisbon: Universidade de Lisboa.

Guimarães, P. (2016) 'An Evaluation of Urban Regeneration: The Effectiveness of a Retail-led Project in Lisbon', *Urban Research and Practice*. Available at: http://dx.doi.org/10.1080/17535069.2016.1224375 [accessed 01 Feb 2017]

Hall, T. and Hubbard, P. (1996) 'The Entrepreneurial City: New Urban Politics, New Urban Geographies?', *Progress in Human Geography*, vol. 20, no. 2, pp. 153–174.

Harvey, D. (1989) 'From Managerialism to Entrepreneurialism: The Transformation in Urban Governance in Late Capitalism', *Geografiska Annaler. Series B, Human Geography*, vol. 71, no. 1, pp. 3–17.

Hiernaux, D. and González, C. (2014) 'Turismo y gentrificación: pistas teóricas sobre una articulación', *Revista de Geografía Norte Grande*, no. 58, pp. 55–70.

INE. (1981, 1991, 2001, 2011) [Online], [Accessed 16 June 2016], Available from: http://censos.ine.pt/xportal/xmain?xpgid=censos2011_apresentacao&xpid=CENSOS.

INE. (2016) *Dados estatísticos do turismo*, [Online], [Accessed 16 June 2016], Available from: www.ine.pt/xportal/xmain?xpid=INE&xpgid=ine_base_dados.

John Lang LaSalle. (2014) *Lisbon Street Shoping. A afirmação do comércio de rua em Lisboa*, Lisbon: John Lang LaSalle.

John Lang LaSalle. (2015), *Apartamentos Turísticos em Lisboa. Um Mercado em Crescimento*, Lisbon: John Lang LaSalle.

Leite, R. (2013) 'Consuming Heritage. Counter-uses of the city and gentrification', *Vibrant. Virtual Brazilian Anthropology*, vol. 10, no. 1, pp. 165–189.

Liang, Z. and Bao, J. (2015) 'Tourism Gentrification in Shenzhen, China: Causes and Socio-spatial Consequences', *Tourism Geographies*, vol. 17, no. 3, pp. 461–481.

Malheiros, J., Carvalho, R. and Mendes, L. (2013) 'Gentrification, Residential Ethnicization and the Social Production of Fragmented Space in Two Multi-ethnic Neighbourhoods of Lisbon and Bilbao', *Finisterra. Revista Portuguesa de Geografia*, vol. 48, no. 96, pp. 109–135.

Marcuse, P. (1985) 'Gentrification, Abandonment, and Displacement: Connections, Causes, and Policy Responses in New York City', *Journal of Urban and Contemporary Law*, no. 28, pp. 195–240.

Martinotti, G. (1993) *Metropoli. La nuova morfologia sociale della città*, Rome: Il Mulino.

Mendes, L. (2008) 'A Nobilitação Urbana no Bairro Alto: Análise de um Processo de Recomposição Sócio-espacial', Master's Thesis in Geography, Lisbon: Universidade de Lisboa.

Mendes, L. (2014) 'Gentrificação e políticas de reabilitação urbana em Portugal. Uma análise à luz da tese do rent gap de Neil Smith', *Cadernos Metrópole*, vol. 16, no. 32, pp. 487–511.

Mendes, L. (2016) 'What Can Be Done to Resist or Mitigate Tourism Gentrification in Lisbon? Some Policy Findings & Recommendations', in Glaudemans, M. and Marko, I. (eds), *City Making & Tourism Gentrification*, Tilburg: Stadslab, pp. 34–41.

Mitchell, A. and Kirkup, M. (2003) 'Retail Development and Urban Regeneration: A Case Study of Castle Vale', *International Journal of Retail & Distribution Management*, vol. 31, no. 9, pp. 451–458.

Mullins, P. (1991) 'Tourism Urbanization', *International Journal of Urban and Regional Research*, vol. 15, no. 3, pp. 326–342.

Neuts, B., Nijkamp, P. and Leeuwen, E. (2012) 'Crowding Externalities from Tourist Use of Urban Space', *Tourism Economics*, vol. 18, no. 3, pp. 649–670.

Nogueira, R. (2015) 'Lisboetas sentem-se cada vez mais acossados pelos turistas', *O Público*, [Online], 1 June 2015, [Accessed 12 February 2016], Available from: www.publico.pt/local/noticia/lisboetas-sentemse-cada-vez-mais-acossados-pelos-turistas-1697332.

Pavel, F. (2015) *Transformação Urbana de uma Área Histórica: o Bairro Alto. Reabilitação, Identidade e Gentrification*, PhD Dissertation in Arquitecture, Lisbon: Universidade de Lisboa.

Peck, J. and Tickell, A. (2002) 'Neoliberalizing Space', in Brenner, N. and Theodore, N. (eds), *Spaces of Neoliberalism: Urban Restructuring in North America and Western Europe*, Oxford: Blackwell, pp. 33–57.

Porter, L. and Shaw, K. (eds) (2009) *Whose Urban Renaissance? An International Comparison of Urban Regeneration Strategies*, London & New York: Routledge.

Roberts, P. (2000) 'The Evolution, Definition and Purpose of Urban Regeneration', in Roberts, P. and Sykes, H. (eds), *Urban Regeneration – A Handbook*, London: Sage publications, pp. 9–36.

Rodrigues, W. (2010) *Cidade em Transição. Nobilitação Urbana, Estilos de Vida e Reurbanização em Lisboa*, Oeiras: Celta Editora.

Rose, D. (2004) 'Discourses and Experiences of Social Mix in Gentrifying Neighbourhoods: A Montreal Case Study', *Canadian Journal of Urban Research*, vol. 13, no. 2, pp. 278–316.

Rossi, U. and Vanolo, A. (2015) 'Urban Neoliberalism', in Wright, J. (ed.), *International Encyclopedia of the Social & Behavioral Sciences* (2nd edition), London: Elsevier, pp. 846–853.

Slater, T. (2013) 'Gentrification of the City', in Bridge, G. and Watson, S. (eds), *The New Blackwell Companion to the City*, Oxford: Blackwell, pp. 571–585.

Smith, N. (1979) 'Toward a Theory of Gentrification: A Back to the City Movement by Capital Not People', *Journal of the American Planning Association*, vol. 45, no. 4, pp. 538–548.

Smith, N. (1987) 'Gentrification and the Rent Gap', *Annals of the Association of American Geographers*, vol. 77, no. 3, pp. 462–465.

Smith, N. (1996) *The New Urban Frontier: Gentrification and the Revanchist City*, London: Routledge.

Warnaby, G. (2009) 'Look Up! Retailing, Historic Architecture and City Centre Distinctiveness', *Cities*, vol. 26, no. 5, pp. 287–292.

Williams, A. and Hall, C. (2002) 'Tourism, Migration, Circulation and Mobility. The Contingencies of Time and Place', in Hall, C. and Wiliams, A. (eds), *Tourism and Migration, New Relationships between Production and Consumption*, Netherlands: Springer, pp. 1–52.

Wilson, J. and Tallon, A. (2012) 'Geographies of Gentrification and Tourism', in Wilson, J. (ed.) *The Routledge Handbook of Tourism Geographies*, London: Routledge, pp. 103–112.

Zhao, Y., Kou, M., Lu, S. and Li, D. (2009) 'The Characteristics and Causes of Urban Tourism Gentrification: A Case of Study in Nanjing', *Economic Geography*, no. 8, pp. 1391–1396.

12 The rent gap re-examined

Tourism gentrification in the context of rapid urbanisation in China

Zeng-Xian Liang

Introduction

Since first developed by Smith (1979), the rent-gap theory has attracted considerable attention in the literature on gentrification and continues to be discussed and criticised today. Some researchers have further developed the rent-gap theory (Clark, 1987; Clark and Gullberg, 1991); others have proposed a new theory of a value gap (Ley, 1996), or have pointed out the limitations of the rent-gap theory or even rejected it (Bourassa, 1993). In the last decades, gentrification has expanded from Western countries to emerging countries such as China (He, 2010). As a consequence and manifestation of globalisation and neoliberal urbanism (Lees, Slater and Wyly, 2008), gentrification has become a global phenomenon (Davidson, 2007), accompanied by a global process of class (re-)formation (Butler, 2002). The geographical spread of gentrification is widely known as "generalized gentrification" (Smith, 2002). Today, a complex relationship exists between gentrification, globalisation and neoliberal urbanism in China (He, 2010). Strong state intervention has promoted gentrification in many rapidly urbanised metropolises in China. This involves a series of market reforms and policy initiatives, policy interventions and governmental investment, as well as the mobilisation of important financial resources (He, 2007). For example, most old housings in inner cities are owned by local housing authorities, while individual owners possess only use rights. This separation of housing-use rights, housing ownership and land ownership leads to fragmented property rights, which hinders gentrification. To tackle this process and promote gentrification, the state provides urban land for re-settlement, implements land assembly and relocates original residents (He, 2007). Thus, public authorities play an important role in urban gentrification which can be coined as 'state-sponsored gentrification'.

Since 2000, gentrification has been observed in rural areas as well as metropolises in China. New kinds of gentrification are emerging, such as tourism gentrification. The latter is used to describe the transformation of a middle-class neighbourhood into a relatively affluent and exclusive enclave marked by developing tourism and corporate entertainment. This phenomenon is framed as super gentrification (Gotham, 2005). However, tourism gentrification in rapidly urbanising China presents features distinct from those observed in Gotham's (2005) early study of New Orleans' Vieux Carré (Liang and Bao, 2015). According to our observation, tourism gentrification in China is neither a complete super gentrification nor an urban regeneration process as New Orleans' Vieux Carré. Tourism gentrification in China shows new characteristics that can be linked to new town development or historical preservation processes. For example, Shin (2010) showed that historical preservation accompanied by real-estate investment contributed to tourism gentrification in Nanluoguxiang (Beijing) leading to a new case of tourism gentrification in China. According to Liang and Bao (2015), this process is part of a broader transformation of lower-class to upper-class social spaces. However, some questions remain.

First, how should tourism gentrification be defined in emerging countries such as China, where its context and causes are so different? Gotham (2005) described tourism gentrification as a kind of super-gentrification, referring to the transformation of middle-class neighbourhoods into affluent and exclusive enclaves. Tourism gentrification in China shares features of super-gentrification and 'new-build' gentrification but it also renews historical areas (Liang and Bao, 2015) and may rise from brownfield areas.

The second issue of interest concerns the rent gap defined as the disparity between the potential ground rent level and the actual ground rent level. Once the rent gap is wide enough within a given community, gentrification may occur. According to the classical rent-gap theory gentrification involves reinvestment in inner cities after an initial phase of suburbanisation (Smith and Williams, 1986). Suburbanisation and gentrification are very much linked in the US. The dramatic suburbanisation of the urban landscape provides an alternative geographical locus for capital accumulation which encourages reinvestment in inner cities (Smith, 2005: 37). In the US, suburbanisation decreases investment in inner cities both directly and indirectly. This leads to a decrease in actual ground rent. Meanwhile, suburbanisation contributes to urban expansion, which has historically increased potential ground rent in inner cities. Therefore, suburbanisation is the key form of gentrification occurring in the US. However, in modern China, tourism gentrification is not preceded by counter-urbanisation or suburbanisation. This is an important issue for further investigations in research on tourism gentrification in China and may provide insights for gentrification in other emerging countries.

This chapter investigates the specific nature of tourism gentrification in China. It aims at exploring the different features of tourism gentrification in the context of China's rapid urbanisation.

Re-conceptualising tourism gentrification

The evolution of gentrification

Many new forms of gentrification have arisen in today's era of generalised gentrification. The significance of these new developments demands a country-specific re-examination of gentrification in the modern era. In the postwar period, terms such as "brownstoning" (Bunting, 1987; Engels, 1994), "homesteading", "white painting", "white-walling" and "red-brick chic" (Lees *et al.*, 2008: 6) were used to describe the emerging phenomenon of gentrification. These terms concerned gentlemanly activities and behaviour rather than referring to an influx of gentlemen (Liang and Bao, 2015). Glass (1964) incorporated socio-spatial transformation to produce a now widely accepted classical definition of gentrification. Glass described the displacement of urban working-class communities in inner-city London by predominantly middle class communities. After World War II, members of the European and North American middle class community started to resettle in inner cities which were culturally rich and diverse and offered land and housing with higher investment value. However, in the age of generalised gentrification, this back to the (inner) city movement has spread in different forms from Western countries to emerging countries. Gentrification has occurred in markedly different ways over time in different cities and neighbourhoods (Smith, 2002).

The different forms emanating from Glass' definition of gentrification can be characterised as follows. First, the concept of gentrification has been extended to encompass not only middle-class 'back to the city movements' after counter-urbanisation or suburbanisation, but also 'stay in the city movements' whereby middle-class urbanites relocated to or re-aggregated in inner cities.

Second, gentrification has occurred not only in inner cities but in waterfront areas (Hoyle, 1988), suburbs (Badcock, 2001; Hackworth and Smith, 2001), rural areas (Hines, 2007; Phillips, 2010; Hines, 2011) and historical areas (Shin, 2010). Third, gentrification has involved not only the renewal of old houses but also the upgrading of community infrastructure and facilities, along with the construction of new upper-class consumption spaces (Smith, 2005). Fourth, unlike traditional gentrification, which occurs in low-income neighbourhoods, 'new-build' gentrification takes place in urban brownfields. Fifth, gentrifiers are not always middle class; they may belong to the new middle class (Ley, 1996), the super-rich (Gotham, 2005) or the transnational global elite (Butler and Lees, 2006); alternatively, they may be students (Smith and Holt, 2007; Hubbard, 2008).

Some new forms of gentrification have been identified as extensions of gentrification, such as rural gentrification (Phillips, 2004; Phillips, 2010; Hines, 2011), new-build gentrification (Davidson and Lees, 2005; Davidson and Lees, 2010; He, 2010), super-gentrification (Lees, 2000, 2003; Butler and Lees, 2006), studentification (Smith and Holt, 2007; Hubbard, 2008) and tourism gentrification (Gotham, 2005; Zhao *et al.*, 2009). Glass' definition refers only to the most traditional and easily recognisable forms of gentrification (Lees, 2000).

The nature of tourism gentrification

What are the characteristics of tourism gentrification, and how is tourism gentrification related to other forms of gentrification? As all other forms of gentrification, the transformation of social space and its landscape is the core process of tourism gentrification. However, Gotham (2005) points out a unique set of institutional connections between tourism development, real-estate industry and local institutions. This unique set of institutional connections can be identified in the concomitant processes of globalisation and localisation. These connections present a challenge to traditional explanations of gentrification that emphasise the driving role of production-oriented and consumption-oriented factors. During tourism gentrification, tourism and tourism-related investment promote the development of local tourism and lead to an increase in property value. Tourism development and investment not only reorganise the local economy but lead to demographic changes (Mullins, 1991, 1994). These demographic changes can be divided into different phases, according to social-class structure. As gentrification is likely to be a continuous process, the main social status of a gentrified community will change from working class to middle class, rich and even super-rich. Tourism gentrification attracts primarily individuals from the middle-class, super-rich individuals as well as a fixed proportion of members of the low-income class, most of whom provide services for tourism enterprises (Liang and Bao, 2015). Within the tourism gentrification logic, the redevelopment of physical infrastructure, landscape and facilities are also promoted by tourism development. This process may entail commercial revitalisation, historical preservation, protection of neighbourhood integrity, construction of tourism and cultural facilities, renewal of old houses, improvements to living environments, street and square theming, architectural restoration or transformation of buildings into lofts, offices or exhibition spaces. Shifts in cultural references and lifestyles are another critical aspect of tourism gentrification. The reorganisation of local economies leads to an increase in consumption spaces, such as clubs, bars, shopping centres, hotels, theme parks and restaurants. Many of these emerging consumption spaces cater to the needs of upper-class immigrants as well as visitors. Gentrified areas may look like fantasy cities (Hannigan, 1998). The upper-class individuals are gentrifiers whose lifestyle choices tend to determine the cultural production of social space. Images and symbols that evoke romance, inspire nostalgia or advertise music, dancing or shopping have long attracted visitors in Chinese cities. Developers may use their spatial imagination to envision communities that cater to the cultural-consumption preferences of these new upper-class gentrifiers which they also use as a mean to build their own cultural class identity. They also influence the space development and its amenities as developers look at gaining profits by satisfying the needs or attracting the upper-class gentrifiers. In turn, low-class residents are indirectly forced to leave as the community's social spaces do not cater to them.

According to Gotham (2005), tourism gentrification is the transformation of a middle-class neighbourhood into a relatively affluent and exclusive enclave combined with tourism venues. This is equivalent to the super-gentrification that constituted the third wave of gentrification in North America (Hackworth, 2002; Smith, 2002). In China, however, tourism gentrification appears to occur not only in some inner-city areas (Zhao *et al.*, 2009; Shin, 2010) with low-income residents, but also in middle-class neighbourhoods and even in brownfield sites without any local residents. The process of tourism gentrification in historical areas also differs between China and North America. In China, tourism gentrification in historical areas has occurred through the simultaneous processes of landscape renovation and the recovery of historic buildings, consumption-space construction and real-estate development. In most urban brownfields, tourism gentrification is characterised by the development of new-build tourism projects, luxury residential areas and other facilities. Compared to the renewed characteristics of Gotham's Vieux-Carré new-build development is thus the salient characteristic of tourism gentrification in China (Liang and Bao, 2015) which is not always the case for super-gentrification. The definition of tourism gentrification should then be revised to accommodate the new conditions of countries such as China.

Gentrified areas are mostly "fragments" of a city. Most scholars may choose a neighbourhood or a community with a clear boundary to conduct research on gentrification. In China, the urban social space was controlled in Danwei System to achieve communism and modernisation in the period of planned economy from 1949 to 1993 (Ma and Bray, 2006). Community thus mostly refers to an area with relatively clear boundaries which is governed by one street office (the lowest government in China) and developed by the same authority. To illustrate the process and characteristics of tourism gentrification in China, two communities have been chosen: Shenzhen and the Beijing Overseas Chinese Towns (OCTs) communities. The study investigates communities rather than cities, because communities are highly heterogeneous and tourism development in China normally occurs within these communities. Therefore, tourism gentrification is best observed at a community level.

Gentrification is not an outcome of the transformation of social space, but the very process of transformation. Liang and Bao (2015) developed an analytical framework to look at tourism gentrification. The factors addressed in this framework are actors (those involved in gentrification), background (where gentrification begins), space and landscape (where gentrification ends), and domain of gentrification and displacement (how gentrification occurs). This framework enables detailed analysis of distinct new forms of gentrification and takes into account the dynamic of globalisation.

Tourism gentrification in Shenzhen OCT

Nowadays, China celebrates its thriving neo-liberal urbanism (He and Wu, 2009) and has engaged in globalisation (Liang and Bao, 2015). Globalisation and neo-liberal urbanism have fundamentally changed the context of gentrification in large cities across China (He, 2010), especially Shenzhen. Shenzhen is a large city in southern China close to Hong Kong. Since the 1980s, rapid urbanisation has transformed Shenzhen from a traditional village into an international metropolis (Wang, Wang and Wu, 2009). Shenzhen, with an average of 10.78 million inhabitants and 1600.20 billion yuan GDP in 2014 is the third largest metropolis in mainland China, following Beijing and Shanghai. Over the last 30 years or more (from 1979 to date), many communities in Shenzhen have undergone gentrification processes. The Shenzhen OCT (Overseas Chinese Town Holdings Company) community has experienced three phases of tourism gentrification, as shown in Table 12.1.

Table 12.1 Temporal analysis of tourism gentrification in the Shenzhen OCT community

Period	Tourism development/investment	Demographic	Physical infrastructure/landscape/facilities	Culture/lifestyle
Before 1981 Pre-gentrification	Traditional farm with agricultural industry.	Natural growth of population of 1986, most are peasants, a few are returned overseas Chinese.	Traditional rural landscape. A few living facilities.	Classic rural culture of Lingnan in China. Rural lifestyle in collective farm.
1981–1988 Pre-gentrification	Introduced labour-intensive factories. Built working-class residential areas and living facilities.	Increasing population of more than 10,000. Original peasants transformed into workers. Most immigrants are working class.	A planned working-class community. Urban landscape with basic living facilities.	Working class culture and organized activities. Urban lifestyle.
1989–1997 Primary tourism gentrification	Economic transition. Developed theme parks and tourist-related projects. Developed separate high-class residential areas. Increasing value of properties.	Increasing population. More than 15,000 middle-class immigrants. Small amount of displacement in working-class residential areas.	Tourism landscape and attractions. High-class residential areas with full living facilities. Space consumption. Gated community.	Chinese traditional culture production of space. Modern urban lifestyle. Self-consciousness about traditional cultural.
1998–2013 Mature tourism gentrification	Developed or renewed tourism-related facilities and theme parks. Transformed factories into creative and consumption space. Comprehensive development of high-class residential areas. Increasing value of properties.	Increasing population of more than 50,000. More than 5,000 super-rich and high-level middle-class immigrants. Larger amount of displacement in working-class and low-level middle-class residential areas. Second-home owners.	Tourism landscape and attractions. High-class residential areas with luxury living facilities. Space consumption, theming and upgrading.	Western culture fashion and nostalgia in cultural production of space. Modern and diversified urban lifestyles. Cosmopolitan.

Source: Liang and Bao (2015: 471)

Pre-gentrification phase

Before 1978 the land now occupied by Shenzhen OCT housed a traditional farming community, comprising five villages, 418 households and 1,986 people. Most of these residents were peasants and a few were returning Chinese from Southeast Asia. The residents depended heavily on agriculture and had limited living facilities. The community's landscape and lifestyle was similar to that of other rural communities in south China. The government's 1978 reform and opening-up policy attracted a large inflow of capital to China. In 1981, in order to develop the urban economy the farming community was transformed by the local government into an economic-development zone (see Figure 12.1). Its goal

Figure 12.1 Map of Shenzhen OCT

Source: Author

was to attract external investment for the creation of labour-intensive factories and to encourage local economic development. Some of these factories were funded by Hong Kong and overseas Chinese investors who not only provided new capital and technology but also supplied new urban facilities, cultural and lifestyle options. During this phase, part of the farming community was transformed into three open working-class residential areas for the factories' migrant workers. These residential areas were plain, characterised by a high density of small houses without elevators or particular decoration. The planned working-class community had basic living facilities and a public space for daily activities. In 1985, the OCT Group, which is authorised by the China State Council and engages mainly in tourism development, especially theme parks, was established. The OCT Group, as the primary land developer, was a governmental body but functioned as an enterprise administrator. It needed to attract external investment and manage social affairs to meet residents' demands. Between 1986 and 1988, as a means to attract overseas enterprises the OCT Group improved the area's living environment and created new investment opportunities. New working-class housing such as the Eastern Residential Area and new facilities such as the East Market, with convenience stores and some open spaces for recreation were constructed. Shenzhen OCT became a working-class community with a complete set of high-quality facilities but without any tourism dimension (see Figure 12.2). According to the community governor, this community at the point of construction comprised more than 10,000 people, most of whom were migrant workers from various regions of China. During this phase, traditional community life was transformed into a modern system of organised activities with a predominance of urban lifestyle.

Primary phase of tourism gentrification

Two significant changes led to the second phase of gentrification in Shenzhen OCT. The first was the Chinese central government's policy promoting external investment in specific industries, such as the tourism industry, which brought

Figure 12.2 Current view of Guanghua Street (a) and Guangqiao Street (b)

Source: Author's own collection, March 2011

fierce competition to the OCT Group from other development zones. The OCT Group had thus to consider forms of economic development other than light industry. The second significant change was the housing reform initiated in China in 1994, when China's State Council began to promote housing's commercialisation and marketing. The government encouraged domestic enterprises to invest in high-end housing for the new urban rich while continuing to mandate that State-owned enterprises provide welfare-oriented public housing for low-income families. At the same time, the holders of welfare-oriented public houses were able to buy their property for a very low price, turning social housing into private houses that could be sold. These elements provided a basis for gentrification.

From 1989 onwards Shenzhen OCT transferred its economic focus from light industry to tourism. Theme park development was taken more seriously. The reasons are as follows: first, in the late 1980s theme parks were considered as a symbol of modernisation and urbanisation in China, and most large cities had plans to build theme parks. Secondly, theme parks could meet the increasing leisure and entertainment demand of urban residents in rapidly developing cities, especially for those lacking tourism resources. Thirdly, for most urban residents as well as some rural migrant workers, visiting a theme park was considered a new experience of city life. Theme parks were also a means and opportunity for Chinese people to enjoy historical culture and worldwide landscape in their city without having to travel far or abroad.

In 1989, the OCT Group invested 100 million yuan in a theme-park project named Splendid China (Bao, 1994). The park successfully recovered its initial investment within a year. Tourism subsequently became an important part of the local economy. The OCT Group built two other theme parks: China Folk Culture Village (110 million yuan; opened in 1992) and Window of the World (560 million yuan; opened in 1994). Shenzhen OCT experienced a new wave of tourism-related developments with the opening of Huaxia Art Centre in 1991, the Seaview Hotel in 1992, the Tourism School in 1993, the Yanhanshan Hotel in 1993, Yanhanshan Country Park in 1993 and He Xiangning Art Museum in 1997. These tourist attractions transformed the OCT from a light-industry zone and working-class community into a famous tourist resort.

During that same period, the OCT was gradually transformed by the emergence of a cluster of high-class residential areas. These new residential areas, such as Oriental Garden (1987), Seaview Garden (1990), Guihua Garden (1993), Hubin Garden (1994), Zhonglv Plaza (1996), and Huiwen and Lihai Garden (1997) were equipped with banks, bookshops, fitness plazas, supermarkets, car parks, kindergartens, elementary schools, sports centres and cultural art centres (see Figure 12.3). All the houses in these new-build residential areas were rapidly sold. The number of newly built houses suggests that more than 15,000 middle-class people moved into the OCT area between 1989 and 1997(Liang and Bao, 2015). Most of the early buyers were investors, managers or senior professional technicians from Hong Kong and Macau along with a few new-rich individuals from Mainland China and high-level managers of the OCT Group. To cater to the newcomers' preference for traditional Chinese culture, the developers constructed

a  b

Figure 12.3 Current view of Huiwen and Lihai Garden (a) and Guihua Garden (b)
Source: Author's own collection, March 2011

classical Chinese gardens and decorated the houses in a traditional looking style. According to Bao's research (1996), the Shenzhen OCT received 6.47 million visitors in 1994. Among them, more than 80 per cent were domestic market, and the rest were visitors from abroad (mostly from Hong Kong). The Shenzhen OCT also became a famous tourist attraction for Asian overseas visitors during that period. Tourism development increased the value of commercial space and consequently housing prices, leading to the small-scale displacement of working-class residents. This phase of tourism gentrification in the OCT shares characteristics with new-build gentrification.

Mature phase of tourism gentrification

In the late 1990s following rapid urbanisation Shenzhen's economy was based primarily on the service industry and its society exhibited post-industrial features. Light industry moved completely out of the inner city as factories in this sector were unable to afford the increasing cost of renting land. The OCT then became one of Shenzhen's urban centre and the potential value of its land increased. A major industrial transformation occurred in the OCT. Much of its light industry was relocated, leaving behind factory buildings and equipment as well as numerous former employees who needed to make new living arrangements. Some sought new jobs in the OCT area but most followed the factory activities.

In 1996, the OCT Group decided to stimulate further economic transformation by expanding tourism and upscale residential areas in the community. In 1997, the OCT Group registered on the Shenzhen Stock Exchange and obtained considerable capital. As a result, the economic development of the OCT no longer relied on external investment; the OCT Group invested directly in tourism and real-estate industries. In 1998, the first phase of the construction of Shenzhen Happy Valley, in which the Group invested 800 million yuan, was completed. According to Botterill (1997: 46), when a city becomes a post-industrial society, theme parks are developed to cater to members of the middle class. Shenzhen Happy Valley is merely a new kind of park facility offering numerous rides, films on fashion and, most importantly, a theme park about happiness, fantasy and modernity.

Other tourism-related industries were rapidly developing at the same time. Some of the old tourism facilities and enterprises in Shenzhen OCT, such as the Shenzhen Bay Hotel (renamed the InterContinental Shenzhen after its renovation) and the Seaview Hotel, were upgraded or renovated. The old factory buildings and equipment yards were transformed into cultural and consumption spaces such as the OCT Temporary Art Centre and OCT-LOFT. These renovation projects provided a platform for communication between artists and attracted many visitors. The OCT Bay, in which the Group invested 2 billion yuan, opened to both visitors and local residents in 2011. It is an urban entertainment complex that combines a large-scale shopping centre, offices, luxury clubs, a cultural centre, creative hotels, IMAX cinemas and cultural shows, meeting the luxury consumption needs of the super-rich. The community has three theme parks: Shenzhen Happy Valley, Window of the World and Splendid China Folk Village (the latter combining the Splendid China and China Folk Culture Villages theme parks). The community was officially designated a state 5A-class tourist attraction in 2007.[1] Window of the World and Shenzhen Happy Valley received 3.60 million and 3.30 million visitors in 2014 and ranked respectively 14th and 17th in a list of popular tourist sites for the Asia-Pacific region (TEA and AECOM, 2015). Shenzhen OCT now receives nearly 8 million visitors each year.

The ongoing housing reform was another force driving tourism gentrification. Since 1998, the OCT Group invested in expensive luxury residential areas with high-quality facilities, such as Portofino (with its Water Bank unit, Xiangshanli unit, and Swan Castle unit), the new Jinxiu Garden and the expanded Oriental Garden. Both sale prices and rent in these new residential areas are higher than in any previously built areas. According to a report produced by various real-estate agencies in 2011, housing prices in the Water Bank unit of Portofino ranged from 105,000 yuan (about 17,000 US dollars) to 218,000 yuan (about 35,000 US dollars) per square meter (Liang and Bao, 2015). The *per capita* disposable income of Shenzhen residents in 2011 was only 36,505 yuan (less than 6,000 US dollars). Only members of the upper middle class or the super-rich could afford these new houses. As the cost of living in the OCT continues to increase, most working-class residents and some members of the lower middle class have been displaced. According to statistics provided by the OCT Group, the OCT housed more than 50,000 people in 2011, of whom more than 5,000 were super-rich residents of the Swan Castle or Water Bank units of Portofino (Liang and Bao, 2015). As these rich newcomers enjoy the combination of local and European culture and lifestyle, the residential areas were built imitating the Italian style (see Figure 12.4). Public spaces are designed with luxury decoration, and upscale living facilities are restricted to rich residents only. Moreover, products and services provided in these residential areas are expensive. The community gradually became a complete tourist resort with a fully facilitated residential community. This mature phase of tourism gentrification shares features with both new-build gentrification and super-gentrification. The process of

Figure 12.4 Current view of Water Bank unit of Portofino Garden (since 1998)

Source: Author's own collection, March 2011

tourism gentrification in the OCT community is ongoing and may enter a new phase in the future. Tourism and tourism-related developments have played a vital role in this process.

Tourism gentrification in Beijing OCT

Beijing is emerging as a global city (Gu *et al.*, 2015) defined as a post-industrial society (Sassen, 2001) in which the reconstruction of social space is dominated by the middle class. Since China's reform and opening-up policy of 1979, Beijing has experienced rapid economic development and urbanisation. The city's gross domestic product increased from 10.88 billion yuan in 1978 to 2133.08 billion yuan in 2014 (Beijing Municipal Statistics Bureau, 2015). Beijing has become modernised and urbanised within a very short time. It is a rapidly changing city in which gentrification happens suddenly and extends widely. In this context, tourism gentrification in the Beijing OCT community can be regarded as rapid new-build gentrification (see Table 12.2).

Pregentrification phase

Beijing OCT is a new-build community with an area of 1.5 square kilometres constructed on land occupied before 1949 by four traditional farming villages: Houfengcun, Nandashanzi, Qiugezhuang and Liuzuotun. The original landscape and lifestyle of these villages were similar to those of other rural communities in north China. Most of the local residents were peasants who depended on agriculture and animal husbandry. Before 1949, the villages' combined population did not exceed 200 inhabitants. In the mid-1950s, the area was designated by the local government as a heavy industry development zone. The implemented plan and policy attracted large factories, such as the Beijing Coke Chemical Factory, the Beijing Chemical and Dyestuff Factory (BCDF) and the Beijing Glass Factory (BGF). The Beijing Coke Chemical Factory was so large that it provided Beijing

Table 12.2 Temporal analysis of tourism gentrification in the Beijing OCT community

Period	Tourism development/ investment	Demographic	Physical infrastructure/ landscape/facilities	Culture/lifestyle
Before 2000 Pre-gentrification	Traditional working-class residential areas and living facilities affiliated to heavy industrial district.	Population of 1,504 in the location, but more than 26,455 around. Most are workers and peasants.	A planned working-class community. Urban landscape with basic living facilities.	Working-class culture and organized activities. Urban lifestyle.
2001–2005 Pre-gentrification	Economic transition and heavy industry emigration. Community light industry recession. Investment declines.	Declining population of less than 1,000 in the location and 15,000 around. Most emigrate following the heavy industry.	A recessionary working-class community. Some building demolition and some land becomes brownfield sites.	Working-class culture and organized activities. Urban lifestyle.
After 2006 Rapid tourism gentrification	Developed tourism-related facilities and theme parks. Developing brownfield sites into creative and consumption space. Comprehensive development of high-class residential areas. Increasing value of properties.	Population of 12,934 (2012) in the location and more than 30,000 in the surrounding area. Most are new middle-class or super-rich newcomers. The local working-class residents relocate within the community.	Tourism landscape and attractions. High-class residential areas with luxury living facilities. Space consumption, theming and upgrading.	Western culture, fashion and nostalgia in cultural production of space. Modern and diversified urban lifestyles. Cosmopolitan.

Source: Author's own analysis

with 80 per cent of its domestic and commercial gas. It had been the largest commercial gas and coke base in China since 1959. The Beijing OCT then received
more and more industrial immigrants and supporting facilities.

The development of heavy industry around Beijing OCT site attracted labour-
intensive factories to the area, in turn encouraging local economic development.
Working-class communities were built for the newcomers and the reconstruction of the area's urban facilities, culture and lifestyle displaced local social
space. Most local peasants gave up farming and turned to work in light industry,
which supported some heavy-industry factories like BCDF and BGF and also
met the newcomers' lifestyle needs. Local peasants contributed to construction,
transportation, household services and community services, forming a collective
community economy. These changes attracted migrant workers from surrounding provinces, with most of the workers taking up heavy manual work such as
house building and settling down within the zone. Some resettled in the aforementioned villages. The population of these four villages increased from less
than 200 inhabitants in 1949 to 1,507 in 1992.

Before 1999, during the late phase of industrialisation, the social space of the
villages exhibited an industrialisation trend (see Table 12.3); the average family
size decreased, and from 1992 to 1999, the total population of the above four villages remained stable, at approximately 1,500 inhabitants. However, the number
of households increased sharply from 431 to 678. During the same period, the
area of cultivated land decreased from 244.5 acres in 1992 to 174.6 acres in 1999.
This decline was especially noticeable in Nandashanzi. Before 2000, the Beijing
OCT site was a high-density residential area inhabited by working-class individuals with jobs in heavy industry, migrant workers from surrounding provinces
and local low-income residents. These working-class residential areas comprised
high-density small houses without elevators or elaborate architectural features.

Table 12.3 The evolution of the site of the OCT community in 1992 and 1999

Year	Village	Households (N)	Population (N)	Cultivated area (acre)
1992	Houfengcun	186	564	61.1
	Nandashanzi	48	268	77.4
	Qiugezhuang	111	330	58.2
	Liuzuotun	86	345	47.8
Total		**431**	**1,507**	**244.5**
1999	Houfengcun	231	519	58.4
	Nandashanzi	152	358	14.8
	Qiugezhuang	174	355	69.8
	Liuzuotun	121	272	31.6
Total		**678**	**1,504**	**174.6**

Sources: Editorial Committee for Chaoyang Chorography (1993). Beijing Chaoyang District Name
Chorography. Beijing: Beijing Publishing House, 528–529; Editorial Committee for Chaoyang
Chorography (2000). Beijing Chaoyang District Name Chorography. Beijing: Beijing Publishing
House, 125

The planned working-class community had basic living facilities and a public space for daily activities.

In 2000, Beijing began to intensify its bid to host the Olympic Games and oriented economic development accordingly. In the 2000 *Beijing Industrial Relocation Plan within Third and Four Ring Road Area* and the 2008 *Action Plan for Beijing Olympic Games*, the Beijing municipal government clearly stated that more than 200 heavy-industry factories, such as the Beijing Coke Chemical Factory and most of the factories surrounding the Beijing OCT community, would be relocated outside Beijing by 2008. In addition, numerous light-industry factories supporting the heavy industry were to be removed or transformed. The effects of this industrial shift on local social space were substantial. First, many local residents left to staff the relocated factories, creating a new social structure. The community's main social class had changed: its main members were no longer working class but specialised work forces from the third sector economy. Second, the shift in the principal sector of the economy broke the industrial chain. Most original enterprises were either transformed or geared towards new markets, their workers were laid off, transferred to other factories or required to learn new skills to meet the needs of the new enterprises. Most former factory employees shifted towards the service industry. Third, the industrial landscape was rebuilt to cater to new needs. Many old factories and other facilities were transformed; some were pulled down and became brownfield sites awaiting development. This industrial shift contributed to the rapid and abrupt transformation of social space from an industrial to a post-industrial environment. Fourth, the local culture varied with the changing social class structure in the post-industrial society. The culture of the working class was no longer popular. People tended to become more individualist and cater for their own needs in life. This drastic reconstruction of social space broke the economic, social, spatial, interpersonal and cultural links between the industrial and the post-industrial society.

Rapid tourism gentrification

In the area around the Beijing OCT community, the pillar industry had to be rebuilt and social space reconstructed. The local government sought to turn to the tourism industry because traditional industrial sectors were not encouraged by local government in this area. Tourism and tourism-related development played a vital role in this process. In 2001, the local government cooperated with the OCT Group (the same developer as in the Shenzhen OCT community) to develop a theme park, a series of tourist-related facilities and high-class residential areas in the Beijing OCT. Approximately 2 billion yuan (about 305 million US dollars) were invested in a theme park named Beijing Happy Valley, and even larger investments were made in other projects (OCT Group and Chaoyang District Government, 2002. The newly developed site was named Beijing OCT, and offered a theme park, a theatre (Beijing OCT Grand Theatre), a shopping plaza (Vecchio Square) and two high-class residential areas (Jinchan Garden and Jinchan Nanli).

In 2006, the first phase of Beijing Happy Valley was completed. This new kind of theme park proved to be a successful tourism investment, helping investors as well as visitors. In 2014, Beijing Happy Valley received more than 3.34 million visitors (TEA and AECOM, 2015) and became the most popular theme park in Beijing. As of 2015, the total investment in the Beijing OCT Theatre was 350 million yuan (approximately 54 million US dollars). The new Theatre can hold up to 1,600 spectators and is suitable for big shows, operas, symphonic orchestral performances and corporate events. The Beijing OCT Theatre has become one of the most popular urban tourist attractions in Beijing, and has stimulated night-time economic development within the community. Vecchio Square is a themed shopping plaza that integrates supermarkets, brand restaurants, chain hotels, children's entertainment, luxury clothing stores, spas, beauty shops and other high-class life facilities. Vecchio Square was designed as an upper-class consumption space for middle-class and super-rich immigrants and visitors, but not for working-class residents.

The above two high-class residential areas are new-build gentrified communities located on former brownfield lands on the east of Beijing Happy Valley. House prices in Jinchan Garden are higher than those in most local new-build housing communities. According to statistics provided by Ganji Net (http://bj.ganji.com, a leading second-hand housing exchange network in Beijing), the average house price in Jinchan Garden costs between 38,218 yuan (5,820 US dollars) and 43,681 yuan (6,650 US dollars): higher than the average house price in the Fatou business district (to which Beijing OCT belongs). In addition, most houses in Jinchan Garden are large (more than 100 square metres), with luxury decor and high property-management fees (see Table 12.4).

The selling features of Beijing OCT are its Western culture and lifestyle. Houses in this area are much more expensive than other local houses with similar characteristics. Only the rich or the super-rich can afford them. Immigrants choose to purchase houses here mainly because they enjoy the worldwide fashion without the interference of local culture. Consumption spaces have been upgraded with luxurious items and services, even community services are very expensive for the working classes. The community has gradually become a complete tourist resort with a fully facilitated residential community. Local low-income residents have been displaced by the rich as they could not afford the high and rising living costs. The new-build Beijing OCT has gradually become a mature tourism-gentrified community.

Re-examining the rent-gap theory in China

Rent gap produced during China's rapid urbanisation

Several Chinese scholars have insisted that a rent gap can be observed in China's history (He, 2007; He and Wu, 2009; Liang and Bao, 2015). The country's rapid urbanisation provides a special context for tourism gentrification in Beijing OCT. Since China's reform and opening up in 1979, the devolution of administrative

Table 12.4 The status of Beijing OCT community in 2015

Residential area	Property management company	Plot ratio	Greening rate	Property management fee	Population	Demographics
Jinchan Garden	Beijing OCT property management co., LTD	2.34	70%	2.60 yuan RMB Square meters per month	3,000–4,000	Migrant buyers are the majority, most from other provinces, such as Inner Mongolia, Shanxi, Shandong and Guangdong; Beijing buyers are not more than 20%; Middle-class or super rich with higher income; Most take it as s second house, fewer rent out.
Jinchan Nanli	Beijing Senhe property management co., LTD	2.1	30%	1.35 yuan RMB Square meters per month	8,000–10,000	Local residents, relocation households and immigrant tenants are the majority; Some take it as s primary house, some rent out; Middle class with multiple income.

Sources: Editorial Committee for Chaoyang Chorography (1993). Beijing Chaoyang District Name Chorography. Beijing: Beijing Publishing House, 528–529; Editorial Committee for Chaoyang Chorography (2000). Beijing Chaoyang District Name Chorography. Beijing: Beijing Publishing House, 125

power to city authorities and the fiscal reform of central and local governments encouraged city governments to engage in entrepreneurial behaviour. To obtain an advantage in the market-based competition between cities, the entrepreneurial city government is broadly responsible for urban planning, land demolition, inhabitant resettlement and infrastructure construction. Most importantly, the government combines many land property rights to provide developers with a brownfield site ready for further development. Therefore, gentrification in China usually occurs on a large scale and is based on efficient government–enterprise cooperation. Most development problems can be solved using this cooperation framework. For instance, most types of gentrification are associated with residential displacement, land-status change (normally from agricultural to commercial), land requisition, demolition and planning, community construction and infrastructure financial support. With these incentives from local government, all major issues can be solved legally, quickly and effectively. This differs from the progressive gentrification occurring in Western countries.

In addition to cooperation between local government and developers, there are two more factors leading to the production of a rent gap without prior suburbanisation. First, time constraints on reinvestment in one place are important in China, albeit overlooked by many Western scholars. According to the law in many countries, approval must be obtained from the local government or community-planning commission for property construction, renovation or modification. No investment in newly built houses is permitted within a certain legal time-frame. If a house has been newly built or repaired, no investment should be made in changes to its structure or function. This constitutes a time constraint on house reinvestment. In most Western contexts, re-investment in inner cities has occurred slowly. In the Vieux Carré, gentrification began in the 1960s and has lasted for more than 50 years (Gotham, 2005). As gentrification is slow, most reinvestment plans in Western countries conform to time constraints. In China, however, community reconstruction and renewal happens quickly. Some houses are repaired for a second time within 10 or even less than 5 years. In some cases, housing repair even contravenes time constraints. As a result, these constraints on reinvestment hinder further development. China's rapid urbanisation in the last three decades provides the most important context for gentrification. In fact, the phenomena of contravening time constraints among China cities are widespread due to the rapid urbanisation during the past three decades. The rate of urbanisation in China increased from 19.39 per cent in 1980 to 54.77 per cent in 2014. The urbanisation rate in Shenzhen and Beijing reached 100 per cent and 86.20 per cent, respectively. In the last three decades, many first-tier cities in China have achieved a high urbanisation rate. According to a World Bank report, the UK's urbanisation rate increased from 20 per cent in the 1750s to 100 per cent in the 1950s. France took 120 years to increase its rate of urbanisation from 25.5 per cent to 71.7 per cent. The US also took as long as 120 years to increase its rate of urbanisation from 25.7 per cent to 75.2 per cent.

In China, due to the aforementioned time constraints, reinvestment in any particular area is controlled if investment in this area has just been completed.

Meanwhile, due to the rapid urbanisation of China's cities, most investments have been made in areas surrounding cities, which increases potential ground rent in these areas. This in turn widens the rent gap. Once the rent gap is wide enough to permit reinvestment, gentrification may occur. This is the relationship between the rent gap and China's rapid urbanisation, as shown in Figure 12.5. Urbanisation in China occurs very quickly: most city areas receive several rounds of (re-)investment, which contribute to a continuous and rapid increase in potential ground rent. Therefore, any stagnation in investment rapidly creates a wide rent gap in the given city area.

Tourism gentrification in Shenzhen OCT and Beijing OCT

Tourism gentrification in the Shenzhen and Beijing OCT communities explains the rent gap in China. In the early 1980s, the Shenzhen OCT community was an urban–rural fringe located between two of Shenzhen's centres: Luohu district and Shekou district. At this time, Shenzhen OCT suffered from traffic disruption and poor infrastructure, so both actual and potential ground rent were very low. However, over the last three decades, Shenzhen OCT has evolved from an urban-rural fringe to a new centre of a global city through tourism and real-estate development. As a result, potential ground rent has increased continuously and rapidly over the last three decades. However, due to time constraints on

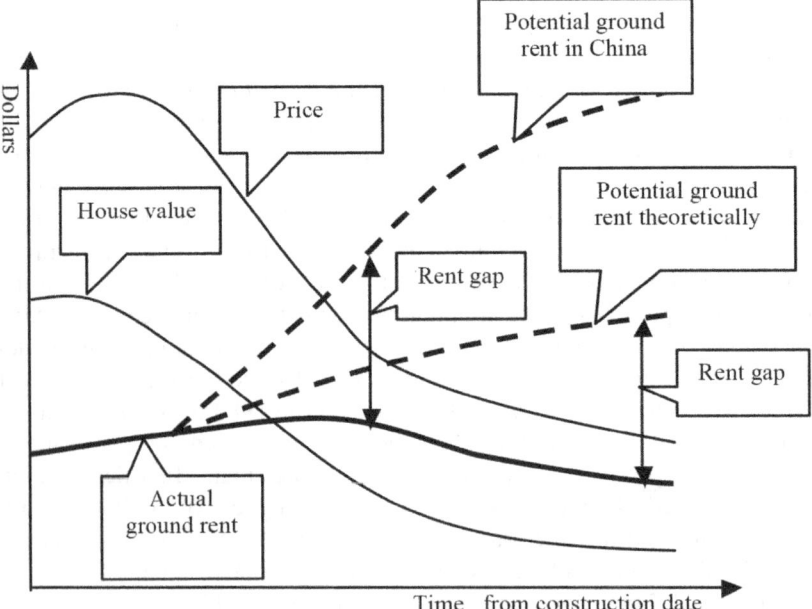

Figure 12.5 The evolution of the rent gap in China

Source: Created and adjusted by author based on Smith (2005: 63)

reinvestment, actual ground rent cannot increase at the same speed as potential ground rent. This has repeatedly created a rent gap and contributed to two consecutive phases of tourism gentrification.

Beijing OCT is located in the suburbs close to the centre of Beijing. Before 2000, it was a working-class residential area, close to rural communities. As this site lacked modern living facilities, convenient public transport and high-quality community service, both actual ground rent and potential ground rent were also very low. However, due to the rapid urbanisation and expansion of Beijing, Beijing OCT became the new centre of south-east Beijing, which led to a sharp increase in potential ground rent. Meanwhile, industrial development in this area led to a decrease in investment and a relatively low actual ground rent. This created a rent gap.

Conclusion and discussion

The rent-gap theory has been subject to continual debate over the last few decades. The geographical spread of tourism gentrification limits the generalisation of existing explanations of the phenomenon. In this study, tourism gentrification was considered in the context of China's rapid urbanisation, as illustrated by the case studies of the Shenzhen and Beijing OCT communities, to refine understanding of the role played by the rent gap in tourism gentrification in emerging countries such as China. Tourism gentrification is a generational or cohort-specific phenomenon that occurs not only in Western countries but also in China and other emerging countries. However, tourism gentrification in China may involve not only the super-gentrification of middle-class neighbourhoods as reported in Gotham's (2005) study in the US and illustrated by the case study of Shenzhen OCT, but also new-build gentrification arising in brownfield areas, as in the case of Beijing OCT. This indicates that tourism gentrification may occur in different ways in developed and emerging countries. The findings of this study also indicate that in China, the rent gap is the result of rapid urbanisation not of suburbanisation. Tourism gentrification in China is not a product of post-regression but a phase of rapid development and expansion. Any stagnation in investment in a given city area will rapidly create a wide rent gap which may lead to gentrification. Therefore, the definition and categorisation of gentrification, as well as the rent-gap explanation of tourism gentrification proposed by Gotham (2005), must be adjusted to fit the different contexts of China.

Additional issues require exploration in future studies. First, can the phenomenon of gentrification in China be considered a process of 'rapid gentrification?' The findings of this study suggest that the causes, process and socio-spatial consequences of gentrification in China differ from those of early cases. Second, the effects of location on tourism gentrification should be discussed. Tourism gentrification in Shenzhen OCT happened in a city centre while that in Beijing first occurred in a suburb. The differences between these models of tourism gentrification are clear. Third, it would be interesting to further examine the influence of time constraints on reinvestment on the rent gap.

Note

1 The state 5A-class is the highest level of tourist attraction in China. According to Standard of Rating for Quality of Tourist Attractions issued by General Administration of Quality Supervision, Inspection and Quarantine of the People's Republic of China (AQSIQ), there are five levels of tourist attraction from highest to lowest: 5A, 4A, 3A, 2A and 1A. In 2015, there were 213 5A-class tourist attractions in China.

References

Badcock, B. (2001) 'Thirty Years On: Gentrification and Class Changeover in Adelaide's Inner Suburbs, 1966–96', *Urban Studies*, vol. 38, no. 9, pp. 1559–1572.

Bao, J. (1994) 'A Study on the Distribution of Theme Parks', *Geographical Research*, vol. 13, no. 3, pp. 83–89 (in Chinese).

Bao, J. (1996) 'A Study on Market and Visitor's Behaviour of Theme Parks in Shenzhen City', *Architect*, no. 70, pp. 4–21 (in Chinese).

Beijing Municipal Statistics Bureau (2015) *Statistical Bulletin of the National Economic and Social Development of Beijing for 2014*, Beijing: Beijing Municipal Statistics Bureau (in Chinese).

Botterill, J. (1997) *The 'Fairest' of the Fairs: A History of Fairs, Amusement Parks, and Theme Parks*, British Columbia: Simon Fraser University.

Bourassa, S. C. (1993) 'The Rent Gap Debunked', *Urban Studies*, vol. 30, no. 10, pp. 1731–1744.

Bunting, T. E. (1987) 'Invisible Upgrading in Inner Cities: Homeowners' Reinvestment Behaviour in Central Kitchener', *Canadian Geographer*, vol. 31 no. 3, pp. 209–222.

Butler, T. (2002) 'Thinking Global But Acting Local: The Middle Classes in the City', *Sociological Research online 7*. Available at: http://www.socresonline.org.uk/7/3/butler.html [accessed 01 Feb 2017]

Butler, T. and Lees, L. (2006) 'Super-gentrification in Barnsbury, London: Globalization and Gentrifying Global Elites at the Neighbourhood Level', *Transactions of the Institute of British Geographers*, vol. 31 no. 4, pp. 467–487.

Clark, E. (1987) *The Rent Gap and Urban Change: Case Studies in Malmo 1860–1985*, Lund: Lund University Press.

Clark, E. (2005) 'The Order and Simplicity of Gentrification: A Political Debate', in R. Atkinson, R. and Bridge, G. (eds), *Gentrification in a Global Context: The New Urban Colonialism*, New York: Routledge, pp. 261–269.

Clark, E. and Gullberg, A. (1991) 'Long Swings, Rent Gaps and Structures of Building Provision – The Postwar Transformation of Stockholm's Inner City', *International Journal of Urban & Regional Research*, vol. 15, no. 4, pp. 492–504.

Davidson, M. (2007) 'Gentrification as Global Habitat: A Process of Class Formation or Corporate Creation?', *Transactions of the Institute of British Geographers*, vol. 32, no. 4, pp. 490–506.

Davidson, M. and Lees, L. (2005) 'New-build "Gentrification" and London's Riverside Renaissance', *Environment and Planning A*, vol. 37, no. 7, pp. 1165–1190.

Davidson, M. and Lees, L. (2010) 'New-build Gentrification: Its Histories, Trajectories, and Critical Geographies', *Population, Space and Place*, vol. 16, no. 5, pp. 395–411.

Engels, B. (1994) 'Capital Flows, Redlining and Gentrification: The Pattern of Mortgage Lending and Social Change in Glebe, Sydney, 1960–1984', *International Journal of Urban and Regional Research*, vol. 18, no. 4, pp. 628–657.

Glass, R. (1964) 'Introduction: Aspects of Change', in Center for Urban Studies (ed.), *London: Aspects of Change*, London: Macgibbon & Kee, pp. vii–xlii.

Gotham, K. F. (2005) 'Tourism Gentrification: The Case of New Orleans' Vieux Carré (French Quarter), *Urban Studies*, vol. 42, no. 7, pp. 1099–1121.

Gu, C., Wei, Y. D. and Cook, I. G. (2015) 'Planning Beijing: Socialist City, Transitional City, and Global City', *Urban Geography*, vol. 36, no. 5, pp. 905–926.

Hackworth, J. (2002) 'Postrecession Gentrification in New York City', *Urban Affairs Review*, vol. 37, no. 6, pp. 815–843.

Hackworth, J. and Smith, N. (2001) 'The Changing State of Gentrification', *Tijdschrift voor economische en sociale geografie*, vol. 92, no. 4, pp. 464–477.

Hannigan, J. (1998) *Fantasy City: Pleasure and Profit in the Postmodern Metropolis*, London: Routledge.

He, S. (2007) 'State-sponsored Gentrification Under Market Transition. The Case of Shanghai', *Urban Affairs Review*, vol. 43, no. 2, pp. 171–198.

He, S. (2010) 'New-build Gentrification in Central Shanghai: Demographic Changes and Socioeconomic Implications', *Population, Space and Place*, vol. 16, no. 5, pp. 345–361.

He, S. andWu, F. (2009) 'China's Emerging Neoliberal Urbanism: Perspectives from Urban Redevelopment', *Antipode*, vol. 41, no. 2, pp. 282–304.

Hines, J. D. (2007) 'The Persistent Frontier and the Rural Gentrification of the Rocky Mountain West', *Journal of the West*, vol. 46, no. 1, pp. 63–73.

Hines, J. D. (2011) 'The Post-industrial Regime of Production/Consumption and the Rural Gentrification of the New West Archipelago', *Antipode*, vol. 44, no. 1, pp. 74–97.

Hoyle, B.S. (1988) *Revitalising the Waterfront: International Dimensions of Dockland Redevelopment*, London, Belhaven.

Hubbard, P. (2008) 'Regulating the Social Impacts of Studentification: A Loughborough Case Study', *Environment and Planning A*, vol. 40, no. 2, pp. 323–341.

Lees, L. (2000) 'A Reappraisal of Gentrification: Towards a "Geography of Gentrification"' *Progress in Human Geography*, vol. 24, no. 3, pp. 389–408.

Lees, L. (2003) 'Super-Gentrification: The Case of Brooklyn Heights, New York City', *Urban Studies*, vol. 40, no. 12, pp. 2487–2509.

Lees, L., Slater, T. and Wyly, E. (2008) *Gentrification*, New York: Routledge.

Ley, D. (1986) 'Alternative Explanations for Inner-city Gentrification: A Canadian Assessment', *Annals of the Association of American Geographers*, vol. 76, no. 4, pp. 521–35.

Ley, D. (1996) *The New Middle Class and the Remaking of the Central City*, New York: Oxford Press.

Liang, Z. and Bao, J. (2015) 'Tourism Gentrification in Shenzhen, China: Causes and Socio-spatial Consequences', *Tourism Geographies*, vol. 17 no. 3, pp. 461–481.

Ma, L.J.C. and Bray, D. (2006) 'Social Space and Governance in Urban China: The Danwei System from Origins to Reform', *China Journal*, vol. 186, no. 55, pp. 472–474.

Mullins, P. (1991) 'Tourism Urbanization', *International Journal of Urban and Regional Research*, vol. 15, no. 3, pp. 326–342.

Mullins, P. (1994) 'Class Relations and Tourism Urbanization: The Regeneration of the Petite Bourgeoisie and the Emergence of a New Urban Form', *International Journal of Urban and Regional Research*, vol. 18, no. 4, pp. 591–608.

OCT Group, & Chaoyang District Government (2002) *The Joint Development Agreement on Beijing Overseas Chinese Town Theming Tourist Community*, Unpublished official document.

Phillips, M. (2004) 'Other Geographies of Gentrification', *Progress in Human Geography*, vol. 28, no. 1, pp. 5–30.

Phillips, M. (2010) 'Counterurbanisation and Rural Gentrification: An Exploration of the Terms', *Population, Space and Place*, vol.16, no. 6, pp. 539–558.

Robson, G. and Butler, T. (2001) 'Coming to Terms with London: Middle Class Communities in a Global City', *International Journal of Urban and Regional Research*, vol. 25, no. 1, pp. 70–86.

Sassen, S. (2001) *The Global City: New York, London, Tokyo*, (2nd edn), Princeton: Princeton University Press.

Shin, H. B. (2010) 'Urban Conservation and Revalorisation of Dilapidated Historic Quarters: The Case of Nanluoguxiang in Beijing', *Cities*, vol. 27, no. 1, pp. S43–S54.

Smith, D. P. (2002) 'Patterns and Processes of "Studentification"', *Regional Reviews*, vol. 12, no. 1, pp. 7–19.

Smith, D. P. (2005) '"Studentification": The Gentrification Factory?' in Atkinson R. and Bridge G. (eds), *Gentrification in a Global Context: The New Urban Colonialism*, London: Routledge, pp. 72–89.

Smith, D. P. and Holt, L. (2007) 'Studentification and "Apprentice" Gentrifiers Within Britain's Provincial Towns and Cities: Extending the Meaning of Gentrification', *Environment and Planning A*, vol. 39, no. 1, pp. 142–161.

Smith, N. (1979) 'Toward a Theory of Gentrification: A Back to the City Movement by Capital Not People', *Journal of the American Planning Association*, vol. 45, no. 3, pp. 538–548.

Smith, N. (1982) 'Gentrification and Uneven Development', *Economic Geography*, vol. 58, no. 2, pp. 139–155.

Smith, N. (2002) 'New Globalism, New Urbanism: Gentrification as Global Urban Strategy', *Antipode*, vol. 34, no. 3, pp. 427–450.

Smith, N. (2005) *The New Urban Frontier: Gentrification and the Revanchist City* (2nd edn), London: Routledge.

Smith, N. and Williams, P. (1986) *Gentrification of the City*, Boston: Allen & Unwin.

TEA, & AECOM (2015) *The Global Attractions Attendance Report for 2014*, Orlando: Themed Entertainment Association, 2015.

Wang, Y. P., Wang, Y. and Wu, J. (2009) 'Urbanization and Informal Development in China: Urban Villages in Shenzhen', *International Journal of Urban and Regional Research*, vol. 33, no. 4, pp. 957–973.

Wu, B. and Xu, X. (2010) 'Tourism-oriented Land Development (TOLD): A New Pattern of Tourism-Real Estate Development in China', *Tourism Tribune*, vol. 25, no. 8, pp. 34–38 (in Chinese).

Zhao, Y., Kou, M., Lu, S. and Li, D. (2009) 'The Characteristics and Causes of Urban Tourism Gentrification: A Case of Study in Nanjing', *Economic Geography*, vol. 29, no. 8, pp. 1391–1396 (in Chinese).

13 Super-gentrification and hyper-tourismification in Le Marais, Paris

Maria Gravari-Barbas

Introduction

The Le Marais district is one of Paris' most dynamic tourist areas. This is the result of a long and complex process in which four centuries of urban history have created a series of space appropriations by different populations who have gradually produced a space to meet their needs by differentially interpreting and using the urban forms of the past.

Tourists were one of the last populations to arrive in Le Marais, but today, tourism characteristics, visitor numbers and impact play a crucial role in the district's urban morphology. There are few Paris districts that enjoy a tourist imaginary (Gravari-Barbas and Graburn, 2016) as strong as that of Le Marais. This makes it a paradigmatic example of how tourism shapes historic urban centres, and an emblematic case of how tourism issues are managed by local stakeholders (politicians, public and private decision-makers and inhabitants).

Le Marais has also inspired the most academic works in France on residential gentrification (Clerval, 2008, 2010; Fijalkow and Oberti, 2001) and to a lesser extent, on commercial gentrification (Giraud, 2009; Mermet and Gravari-Barbas, 2013; Gravari-Barbas and Mermet, 2014; Mermet, 2016). However, despite the importance of tourism in this neighbourhood, not many studies have been conducted on the relationship between gentrification and tourismification (Verbeke, 1998; Dewailly, 2005) and on their potential synergies or antagonisms.

This chapter will link Le Marais' tourisimification to its heritagization and to commercial and residential gentrification. It will show that heritagization, gentrification and tourism are reciprocally supportive phenomena and form a dynamic relationship: although gentrification tends to encourage tourism in the first stages of 'rediscovery' through the reinvestment of heritagized areas by new 'gentry' (most often the new creative classes), it may also lead to a second stage that Lees (2003) calls super-gentrification, in which it becomes a factor of resistance to tourism. This is a very ambivalent relationship as tourists can also be super-gentrifiers and, as is the case in Le Marais, evict the first gentrifiers.

The relationship between super-gentrification (often related to advanced heritagization) and tourism is therefore complex. Super-gentrifiers eventually contest tourismification and are usually a critical element in tourism symptoms such as overcrowding, over-commercialization, souvenir shops, or street food stands. This chapter maintains that in the case of Le Marais, the continuous heritagization process and the steady interest from the French government over the past fifty years have contributed to super-gentrification. It also introduces the concept of hyper-tourismification which is a second stage of tourism development characterized by an extremely sophisticated and high-end tourism product. Contemporary Le Marais is characterized by a combination of super-gentrification and hyper-tourismification, two processes (now in their mature stages) that were implemented in the late 1960s and that are still ongoing today.

The chapter is organized in four sections: the first examines the origins and implications of the heritagization process in Le Marais; the second introduces an analysis of the gentrification patterns in the area; the third deals more specifically with commercial gentrification; and the fourth analyses tourismification patterns. The conclusion adopts a systematic approach to examine the inherent conflicts of heritagization, residential and commercial gentrification, and tourismification. In terms of methodology, this chapter draws on several studies conducted between 2009 and 2016 that combine cross methodologies [interviews with local stakeholders, *in situ* observation, and analysis of a tourist guide corpus, comments on social media (TripAdvisor) and notarial sales archives].

Le Marais: a case of burgeoning heritagization

Le Marais is a 'new' district built in the seventeenth century on the Left Bank of the River Seine (Auffray, 2001). Its name corresponds to an historical, not an administrative, identity and its area covers parts of Paris' 3rd and 4th arrondissements (Figure 13.1). Juliette Faure (1997) indicates that while the perimeter of Le Marais runs through seven Parisian neighbourhoods, only the Saint-Gervais neighbourhood is included in its entirety.[1] Le Marais lies on the marshland incorporated by the Charles V wall (1365–1420). In the seventeenth century it became the most fashionable area in Paris, where the aristocracy, who had fled the once old and overcrowded city centre, returned and built splendid *hotels particuliers* (private mansions).

From the eighteenth century, Le Marais had a certain decadence about it, which escalated with industrialization. The private mansions eventually fell into disrepair and in the mid-nineteenth century, small industrial businesses moved into the space, occupying the yards and gardens and contributing to the densification of the urban blocs. These changes transformed Le Marais into a working-class neighbourhood with an underlying sense of disaffection for the area and social disenfranchisement.

Le Marais' contemporary personality has been shaped by its inhabitants and in particular by the Ashkenazi Jews who moved to the neighbourhood between the late nineteenth century and the Second World War. Fleeing the poverty and persecution of Eastern Europe, they settled around rue des Rosiers in Pletzl, the old Jewish quarter (Brody, 1987).

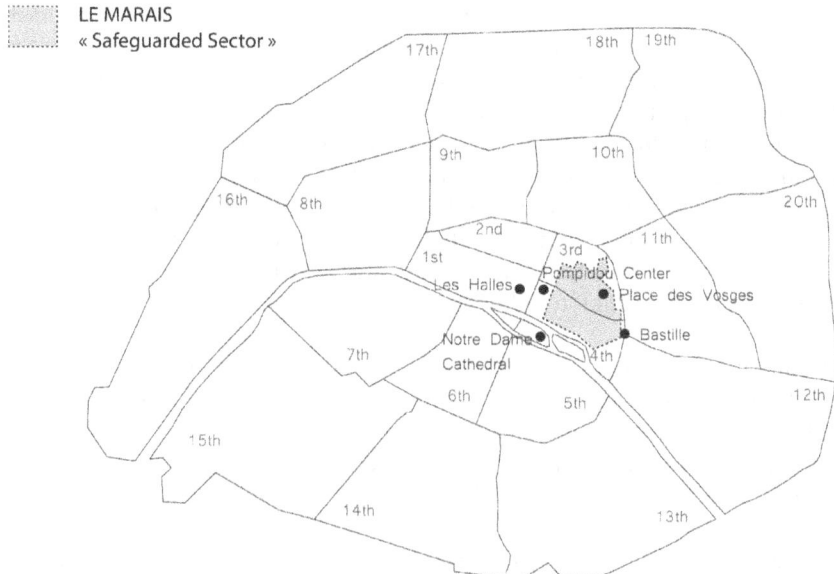

Figure 13.1a Paris arrondissements and the Le Marais area

Source: Author's own design

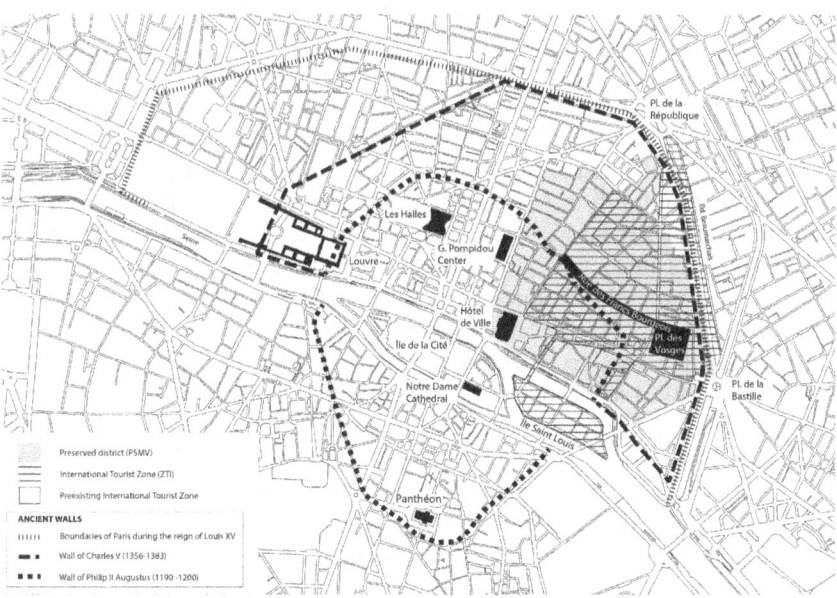

Figure 13.1b The Le Marais area

Source: Author's own design

The 'preservation' policy of Le Marais is not fundamentally different from other centrally located urban heritagized areas (Gravari-Barbas, 2014), but it has had a more spectacular effect because of the exceptional architecture and the substantial government funding the area has received for preservation and restoration purposes.

As early as the 1960s, Parisian elites were using every possible means to enhance the area's aesthetic, historical, and architectural value (Châtelain, 1967). The most notable actions were the mobilization of militant pro-Marais associations and specific events, especially the *Festival du Marais* founded in 1961, in a bid to raise Parisians' awareness of the area. Michel Raude, the festival's president and founder, said the following in his introductory speech:

> Despite the grime that covers it, despite the horrible buildings that clutter it up, despite the pastiches that disfigure it, Le Marais is a district apart, and one of the most beautiful ones at that. The sole aim of its theatre and music events is to raise its profile and make it more appreciated and therefore contribute to its preservation and renovation. (Neville and Raude, 2015: 141)

At the beginning of the 1960s, Le Marais was making a name for itself as the place to be for new creative, residential or commercial opportunities.

Le Marais was one of the first areas in France to be awarded 'safeguarded sector' (*secteur sauvegardé*) status following the 1962 'Malraux law'. This recognition is indicative of the renewed gaze towards this neighbourhood that in the 1930s was condemned to be demolished to make room for social housing. It is protected by the State Safeguarding and Enhancement Plan (*Plan de Sauvegarde et de Mise en Valeur*) (PSMV). The Le Marais PSMV, published on 16 April 1965 and covering 126 hectares[2] was eventually approved on 24 August 1996 and marks an important shift in approach towards the preservation of old urban fabrics. Between the 1939 modernist project of Hilt and Bodecher (in which old buildings were demolished and only a few isolated monuments were kept according to the principles of Le Corbusier's Voisin Plan) (Faure, 1997: 74) and the 1965 safeguarded sector, French society had slowly evolved from the notion of Historical Monument to the more global concept of Heritage (Choay, 1999; Tomas, 2004).

The district's restoration project was based on the 'Plan Turgot', an eighteenth-century map. On this map, Le Marais is shown as a low density district, much more so than in the mid-twentieth century. According to this, a massive de-densification of the population and buildings took place to preserve the district's heritage and in some cases, these selective demolitions (*curettages*) required the demolition of some nineteenth- and twentieth-century buildings in the yards and gardens of Le Marais mansions. These demolitions were typical of late twentieth-century restoration approaches and heritage representations.

It should be noted that, regardless of the urban doctrine that was gradually adopted (i.e. a complete or more 'selective' demolition of the insalubrious blocks) – the first PSMV did not take into consideration the neighbourhood's intangible heritage (workshops, local craftsmanship). Restoration projects simply ignored local residents and workers.

The safeguarded sector status definitively sealed the future of the area. Its application generated both a drastic drop in the number of residents and, in a quasi-consubstantial way, its tourism development. The tourismification of Le Marais is closely related to its systematic heritagization, a policy driven and pursued by the French government for several decades. In 1965, the PSMV contributed, probably for the first time in French urban history, to transform an "urban area into a commodity" (Scheppe, 2015: 19).

Le Marais is today one of the most highly heritagized areas of Paris. In addition to the PSMV, all of the existing heritage protection tools are implemented here: at the national level by classification of several buildings on the national historical monuments list, and at the international level as a UNESCO inscribed property on the World Heritage List (for the parts of Le Marais located along the banks of the River Seine). The 3rd and 4th arrondissements are, respectively, the third and first districts in terms of the number of listed historical monuments in the city of Paris with no fewer than 392 historical monuments and a density that is almost ten times higher than the Paris average.[3]

The 1965 PSMV, internationally hailed as one of the most successful examples of restoration, is analysed today much more critically. Although it avoided the worst-case scenario (the complete demolition of Le Marais), it did not manage to save the 'ecosystem' that had been gradually developing until the mid-twentieth century. According to Scheppe, this state-designed urban tool

> enabled an appropriation of the urban space by real estate and by invoking protection, forever changed the cityscape. This is how it managed to impose its objective, namely to remodel an iconic place into a pristine representation of its ideal, to create the blueprint for a second-class Marais that rose above the rubble left by the forced evictions and became a pretty sight. (Scheppe, 2015: 22)

In 2013, the PSMV was revised to promote a more adequate heritage policy (Guillaume, 2015) and to provide a better match for the 'historical product' created in the 1960s and contemporary market demand. While the 1965 PSMV mainly favoured pre-1800 historical periods, the revision aimed to more effectively take into account nineteenth- and twentieth-century architecture. On the one hand its objective was to identify and protect the most recent buildings of historic interest (e.g. the glass-covered workshops in the mansion gardens or yards), and on the other, to protect and enhance public space and activities related to leisure activities/events and to everyday neighbourhood life (shops and handicrafts, small-scale industries, tourism, etc.) (Mairie de Paris, 2009).[4]

The restoration, rehabilitation and reconstruction of Le Marais that continuously took place for half a century, ran parallel to the affirmation of its creative side. Several projects implemented in 2016[5] reflect Le Marais' highly sophisticated urban process (De Pieri, 2010) that corresponds to the arrival of big names in international "starchitecture" (Gravari-Barbas and Renard-Delautre, 2015), contemporary art projects, and major international commercial groups. This is the case for the project developed by architect Rem Koolhaas and his OMA Agency for the Galerie Lafayette foundation. Ironically, the project of this architect, [who, ironically, was behind the "fuck the context" concept (Koolhaas, 2011)], is nestled in an 1891 building that used to house the Department Store BHV's stocks. Another major former industrial building, the *Société des Cendres*, has been transformed by architect Pierre Audat and now houses the flagship store of Japanese brand Uniqlo. Today, the *Société des Cendres'* chimney is the only remaining Le Marais smokestack, a final vestige of its industrial past (Figure 13.2). Various other architectural projects from the 2010s symbolize the desire to produce contemporary architecture in this historical area. These projects that are sometimes designed to exploit particularly narrow spaces like the project by architectural firm Chartier-Corbasson that is nested on the edges of an urban block (Guillaume, 2015: 9).

The 2013 revised PSMV therefore enabled the preservation and functional transformation of buildings that could have been demolished in the 1965 PSMV. It marks the attractiveness of the few still existing industrial buildings that provide the most suitable locations for creative spaces. One example is Galerie Lafayette Foundation, which has created a new cultural centre to exhibit artistic productions sponsored by the Foundation and in situ installations in the reconverted building's basement workspaces.

Figure 13.2 The Smokestack of the *Société des Cendres*, one of the last vestiges of
Le Marais' industrial past

Source: Author, 2016

More than just a heritagized and carefully restored historic space, Le Marais is today at the forefront of diverse urban processes (artification, museumification, creativity, architectural iconicization) which co-exist free from apparent contradictions and antagonisms.

From gentrification to super-gentrification

The quality of the built environment is one of the main reasons for the arrival and installation of new populations in the area. The gentrifiers, for example, choose to live here because of the highly desirable and attractive heritage. Safeguarded sector status has sealed the fate of this territory. Since the 1965 PSMV the number of inhabitants has drastically decreased. From the 1960s, the heritagization of Le Marais created conditions that were suitable for the residential re-occupation of the neighbourhood. In 1962, an exhibition entitled '*Le Marais: golden age and renewal*' at the *Musée Carnavalet* (Guillaume, 2015: 7) documented the changes that had been developing.

The attractiveness of Le Marais reached a particular height in the 1980s (Giraud, 2009), a period in which the "rent gap" (Smith, 1987) was observed there. At the beginning of the decade, the urban fabric was far from being rehabilitated (particularly in some parts of the 3rd arrondissement and some streets of the 4th) with many commercial properties lying vacant and real estate prices that were cheaper than the Parisian average. This made the properties attractive and offered newcomers both symbolic capital and financial advantages. Gentrification began in parallel with the ongoing rehabilitation of both the historical buildings and the district's more modest urban fabric (Djirikian, 2004).

According to Sibalis (2004: 1743) "Le Marais had the highest gentrification rate of any neighbourhood in the capital in the 1975–82 period". Population decline was already evident in the 1950s and steadily continued its downward trend during the 1960s and 1970s. As the working class left, the middle class and white-collar workers moved in. Between the 1960s and the end of the twentieth century, Le Marais lost around 40 per cent of its inhabitants (Le Clere, 1985; INSEE, 2000).

Djirikian (2004) identified two periods of change that have taken place since the 1960s and the start of Le Marais gentrification process (Table 13.2). The first

Table 13.1 Number of residents in Le Marais

Year	Number of residents
1962	116,284
1968	110,281
1975	82,172
1990	68,903
1999	65,979

Source: INSEE (2000: 75/3)

Note: Population figures for the 3rd and 4th arrondissements.

Table 13.2 Comparison of the population evolutions in Le Marais (3rd and
4th arrondissements), 1st and 2nd arrondissements of Paris and the rest
of the Paris *intra-muros* districts (excluding arrondissements 1 to 4)

Arrondissements/ Periods	Le Marais (3rd and 4th arrondissements)	1st and 2nd arrondissements	5th to 20th arrondissements
1962–1982	−44.5%	−35.8%	−20.5%
1982–1999	−5.8%	−26.6%	−1.6%

Sources: INSEE (1962–1999); Djirikian (2004: 68)

was the early safeguarding operations between 1962 and 1985, a time when Le
Marais was a pioneering force of Parisian gentrification (44.5 per cent population
decrease). The second was the 1980–1990 period characterized by a moderate
decline (5.8 per cent between 1982 and 1999) (Djirikian, 2004: 68).

The first 'mass' departures of the Le Marais population were mainly caused
by the restoration operations and selective demolitions that significantly
de-densified the former urban fabric in order to preserve its historical buildings.
The restored *hôtels particuliers*, the large warehouses, were emptied of their
productive functions and then converted into lofts (Zukin, 1989), and the more
modest eighteenth- or nineteenth-century workshops or living accommodation
became spaces for new functions.

The establishment of a gay quarter with the arrival of gay boutiques, bars
and restaurants played an important role in the construction of Le Marais'
complex *gentryscape*: in the 1980s, it became one of the most important and
visible centres for gay cultural, leisure and commercial activities (Redoutey,
2004). The urban and real estate context of a still-affordable area influenced
the arrival of many gay stores. But as Giraud (2009) underlines, Le Marais is
not a classic case of *gaytrification* (Aldrich, 2004) as the "gay presence in the
Marais has been largely subsumed into a space where there is a great deal of
'competition' from other populations and commercial sectors" (Giraud, 2009)
and homosexual populations moved into the district once the gentrification
process was already under way. However, even if the gay neighbourhood was
limited to a few streets in the 4th arrondissement with just a couple of streets
and café terraces with clear gay identities, their presence contributed to the
creation of a special mix of functions, shops and services that still differentiate
Le Marais from the other central and more bourgeois districts such as the 6th
arrondissement.

Due to the restoration and recognition of its heritage, Le Marais also gained
a new urban and symbolic centrality comparable to the so-called creative neighbourhoods in New York, London, Rotterdam and Barcelona (see Table 13.3)
(Richards, 2005; Richards and Wilson, 2007; Gravari-Barbas, 2010) and its gentrification followed the same process that Soho underwent a few years previously
(Zukin, 1995). Here too, the urban fabric, which was now considered to provide
'character', experienced a depreciation cycle that made it more affordable and
conducive to new investment.

Table 13.3 Le Marais' attractiveness. Quotations from interviews with estate agencies, 2010

"The 4th arrondissement is very much appreciated because it's a fashionable district . . . it's trendy, very fashionable: Le Marais, Le Marais, Le Marais! I don't know what happened: I knew the district in the past and it's completely different!" (*Laforêt* estate agency).
"Its central position has a lot to do with it, as well as the fact that you can walk everywhere: it doesn't take very long to go from Ile Saint-Louis to Place des Vosges at Beaubourg. It's very easy to go from the Left Bank to the Right Bank. The district attracts foreigners as there's also the historical dimension. Some people feel that this is the place for them, in the same way that others feel about the 6th arrondissement and the Saint-Germain-des-Près district" (*Initiales* estate agency).

Source: IREST (2010)

These changes also had an impact on second homes. Between 1999 and 2006, the percentage of second or temporary homes in the 4th arrondissement increased from 9 per cent to 15.8 per cent. This phenomenon is linked to the proportion of properties purchased by foreign buyers (16.5 per cent according to the Paris Chamber of Notaries) and to the development of tourism residences (APUR, 2010: 4). The strongest growth in second or temporary homes is observed in the most touristic parts of Le Marais: the Islands (Ile Saint-Louis and Cité) and the most emblematic places such as Place des Vosges and rue des Rosiers.

One real estate agent cites several underlying reasons for this change:

> More and more apartments are being rented out on a short-term basis, and this is something that's happened over the last fifteen years. Before, there weren't as many furnished apartments available for short-term rentals. In tax terms, it's a good deal: 50% of rent and service charges are tax deductible under industrial and commercial income. However, for an unfurnished property, the tax allowance is around 20%, taxable under housing income. They're two different systems. The urban landscape is changing because it is adapting to the new population in Paris: foreigners. So, we rent apartments to foreigners.[6]

Obviously, this evolution creates a new social situation in Le Marais (Table 13.4).

These evolutions strengthen both the metropolitan and global centrality of the area: it is now *the* place to be for exclusive shops or global elites seeking a symbolic location. It has become a world-wide brand.

The term 'super-gentrification' was first used by Lees (2003) to describe the real estate market evolution in Brooklyn Heights, "the transformation of already gentrified, prosperous and solidly upper-middle-class neighbourhoods into much more exclusive and expensive enclaves". She localized this intensified re-gentrification in a few select areas of global cities such as London (Butler and Lees, 2006) and New York "that have become the focus of intense investment and conspicuous consumption by a new generation of super-rich 'financifiers' fed by fortunes from the global finance and corporate service industries" (Lees, 2003: 2487).

Table 13.4 Second homes in Le Marais. Excerpts from interviews with local associations, 2010

"You can't say that secondary residents are a burden, that would be ridiculous. The main problem with them is managing the co-ownership aspect because the people are less involved. They are also less involved in the preservation of the district because they're not really in their own homes. What's more, all the coming back and forth causes a problem for the permanent residents: passing people on the staircase with their suitcases, people who are just moving in and who aren't really interested in the property is not really a good thing . . . but they're not all like that, mind you" (Association *Vive le Marais*).

"It's a huge problem: first there's the empty apartments (with the lack of adequate housing, families are moving out of the area, especially once they have two children) and then there's the holiday homes. That's been a silent, surreptitious development . . . and once people realise what's happening it will be too late. The relationship with these residents is non-existent, which isn't good at all" (Association *Le 4ᵉ en action*)

"In the district there used to be local businesses: a book shop, grocers, cobblers . . . but now all the rich people have bought pieds-à-terre in the Islands (from €12000 to €15000 per square metre), so there are no free apartments and this has had an impact on local businesses . . ."

Source: IREST (2010)

This new stage of gentrification started, in the case studied by Lees, in the 1990s and does not follow a disinvestment stage. Rather, it corresponds to a new wave of gentrification and habitat renewal to even higher standards that have the potential to partially evict the early gentrifiers. According to Clerval (2008), it is not in line with the Marxist model of capital investment and disinvestment cycles in the urban space. Unlike gentrification that occurs in different phases and thus suggests that the process will finally reach a state of stabilization, super-gentrification emphasizes the ability of the process to constantly renew, which means that its end cannot be imagined (Clerval, 2008: 48).

Clerval used the term 'super-gentrification' to characterize some cases of the Parisian real-estate market. She stated that the analysis of gentrification patterns showed a cumulative process: even when gentrification is well advanced in a neighbourhood, as it is in the inner suburb of Saint-Antoine (near the Bastille, a few blocks away from Place de Vosges), for example, it is constantly trans-forming. Clerval concludes that a super-gentrification process has already taken place in some already-rehabilitated and gentrified places of the neighbourhood. Higher real estate prices for the 3rd and 4th arrondissements are proof of the steady demand for these central historical places in Paris (see Figure 13.3, Tables 13.5 and 13.6).

In 2015, the 4th arrondissement, with a price of €10,910 per square metre, was both the second most expensive area in *intra muros* Paris after the bourgeois 6th arrondissement and the area that still displayed a significant increase (5.4 per cent) in a context in which the Paris real estate market lost, between the sum-mer of 2012 and the end of 2015, 5.7 per cent of its value (approximately €500 per square metre).[7] This market is therefore favourable to super-gentrification:

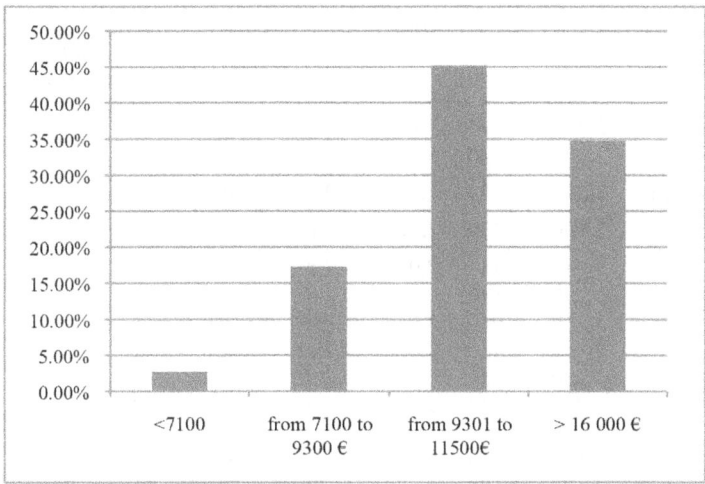

Figure 13.3 Median prices per square metre (apartments) in the 3rd and 4th arrondissements
(Le Marais) and evolution, 2008–2013

Source: Guinot and Pommier (2014)

upper social classes that purchased property in Paris in the 1990s can, in the
2010s, benefit from the 'double effect Kiss Cool' as real estate brokers use to
call it. The increase in over 150 per cent of the value of the property purchased
in the 1990s or the 2000s allows them to buy a more expensive property in a
market that has low purchase prices and low mortgage interest rates.

Table 13.5 Prices per square metre for apartments sold in Le Marais during the period
August 2015 to July 2016

District/neighbourhood	Median prices per square metre (31/12/2013)	Price evolution (5 years)
3rd arrondissement		
Arts et métiers	€9,550	31.7%
Enfants Rouges	€10,190	33.6%
Archives	€11,000	43.8%
Sainte-Avoye	€9,520	24.4%
Average for 3rd arrondissement	**€10,140**	**33.6%**
4th arrondissement		
Saint-Merri	Not provided	Not provided
Saint-Gervais	€10,970	24.4%
Arsenal	€10,770	30.1%
Notre Dame	€12,730	not provided
Average for 4th arrondissement	**€10,970**	**27.2%**

Source: EffiCity (2016)

Table 13.6 Median prices of apartments in Paris (euros) per square metre

Parisian arrondissement	1991 (€)	1996 (€)	1999 (€)	2007 (€)	2009 (€)	2013 (€)	Modification coefficient between 1991 and 2013
1st	4,600	2,590	2,990	7,800	8,050	10,400	2.26
2nd	2,750	2,020	2,940	6,610	6,720	9,610	3.49
3rd	**3,310**	**2,360**	**2,580**	**7,170**	**7,220**	**10,190**	**3.06**
4th	**3,990**	**2,900**	**3,220**	**7,920**	**8,260**	**10,930**	**2.74**
5th	4,210	3,010	3,440	7,960	8,220	10,420	2.49
6th	4,960	3,290	4,120	9,210	9,330	12,320	2.46
7th	4,850	3,310	3,930	8,650	8,860	11,910	2.33
8th	4,270	2,860	3,310	7,600	7,770	9,950	3.14
9th	2,800	2,080	3,220	6,220	6,320	8,760	3.16
10th	2,430	1,750	1,840	5,400	5,560	7,680	3.10
11th	2,640	1,970	2,180	5,860	6,030	8,190	2.75
12th	2,860	2,190	2,310	5,840	5,910	7,880	2.54
13th	3,050	2,190	2,180	5,770	5,750	7,750	2.54
14th	3,350	2,470	2,670	6,310	6,470	8,520	2.54
15th	3,480	2,560	2,770	6,440	6,400	8,640	2.48
16th	4,410	2,940	3,280	6,990	7,120	9,220	2.09
17th	3,050	2,230	2,430	6,110	6,290	8,390	2.75
18th	3,220	1,780	1,840	5,220	5,380	7,310	3.16
19th	2,470	1,760	1,790	4,910	4,490	6,730	2.73
20th	2,460	1,890	1,920	5,180	5,200	6,990	2.84

Source: BIEN database – Notaires de Paris Ile-de-France

The percentage of international transactions varies with apartment value. In 2011 in Paris, this was generally 5.9 per cent; for apartments worth over €4 million, 50 per cent of buyers were from abroad; and for those worth over €10 million, this percentage rises to 85 per cent. According to citylab (2012)[8] there is even a 'second generation' of foreign owners who have bought, sold, and bought again.

Commercial gentrification

Residential evolutions are embedded in commercial dynamics. When the PSMV introduced the selective demolition of the courtyards and gardens of the *hôtels particuliers*, this led to the departure of industrial and craft activities. However, it was essentially from the 1980s on, with a strong acceleration in the 1990s and 2000s, that the business structure of the district underwent drastic change (Djirikian, 2004).

A general upscaling of shops occurred alongside the appearance of new types of commerce that gradually substituted industrial and wholesale activities, respectively. These developments had a significant impact on the commercial landscape and, by extension, the urban one.

Rue des Rosiers is emblematic of these developments. In 1965, the food trade was the most popular followed by handicrafts or dressmaking (Faure, 1997). The 1990 inventory of shops shows that the prêt-à-porter trade had developed at the expense of the food trade and handicrafts. The latter have almost completely disappeared, although these developments have not completely modified the neighbourhood's business structure. However, in 2011, fashion boutiques (prêt-à-porter, shoes, jewellery, handbags, etc.), mainly represented by retail chains, became the dominant trade in the street and thus made rue des Rosiers *the* location for a brand. This development has been accompanied by the beautification of the street: partial pedestrianisation, recreation of the old paving and the creation of green areas. Rue des Rosiers is probably one of the streets in Le Marais that has been impacted upon the most by these commercial mutations (Mermet and Gravari-Barbas, 2013).

Between 1965 and 1990, the number of textile and clothing units fell from 590 to 400 (32 per cent decrease), jewellery/clocks from 218 to 123 (47 per cent decrease), and food from 445 to 246 (43 per cent decrease). At the same time, the number of boutiques selling art objects increased from 108 to 300 (177 per cent increase) and travel agencies from 23 to 77 (230 per cent increase) (Faure, 1997: 96).

This commercial gentrification is expressed by a so-called creative trade, in the broad sense of the term, which has gradually grown in size in the neighbourhood. Initially concentrated around Place des Vosges, art galleries have gradually spread to Haut-Marais (rue de Thorigny, Saint Anastase, des Coutures Saint-Gervais etc.).

This creative dynamic also resulted in the creation of new brands that are Le Marais brands born and bred, the success of which has contributed to their metropolitan, national, and often international distribution (e.g. Sandro opened his first boutique in Le Marais in 2004 and today has a thousand sale points). Similarly, Le Marais is also a Mecca for business innovation in the testing of new shop concepts, including a large number of concept, flagship and pop-up stores (Mermet and Gravari-Barbas, 2013).

As for the residential function, Le Marais heritage is an added value for brands, a symbolic capital that feeds their image. With its completely protected heritagized landscape, the district is certainly not a constraint for these brands, but a distinctive element and a competitive advantage. A paradoxical situation has arisen in which a modern business can be located in a boutique whose protected storefronts display the former functions (e.g. bakery or grocery) (Figure 13.4).

Two highly symbolic examples are the modified functions of two preserved and restored buildings, the Saint-Paul hammam and the *Société des Cendres* building (Figures 13.5 and 13.6).

In 1991, the Saint-Paul hammam, built in 1863 in rue des Rosiers and one of the most important places of memory for the Le Marais Jewish community, was sold to the clothing brand Chevignon, who opened a flagship store in the building. This inaugurated a long series of 'facadisms', in other words, the treatment of historic boutiques/places that retains the facade of the former function while

Figure 13.4 A clothing boutique installed in a former bakery with protected facades

Source: Author, 2015

the interiors are totally emptied to host the new activities. The hammam progressively hosted other functions including a café and the Jewish community's local radio station. However, in 2000,[9] a huge scandal erupted when McDonald's wanted to open a fast food restaurant there – an affair that turned political. The then mayor of the 4th arrondissement organized a protest that was rapidly backed by local inhabitants and businesses. In the end McDonald's proposal dropped out. In 2008, it was finally *COS (Collection of Style)*, the upscale brand of the H&M group, that took possession of the hammam. It is interesting to note that although the local inhabitants and shop owners reacted vehemently to the prospect of the fast food installation, they showed very mixed emotions for the other functions that have been hosted there previously. The destruction of tangible heritage (the hammam's interiors) and, to a larger extent, of the intangible heritage (the hammam's significance for the neighbourhood) did not generate any other reactions, just a bitter nostalgia for a rapidly disappearing past. It was only in 2015 that the new mayor of the 4th arrondissement, Christophe Girard, said that local residents and politicians should have fought to keep the hammam's function which would have allowed people "to stroll down rue des Rosiers in a peignoir after a hammam".[10]

It is a slightly different story for the Société des Cendres[11] that was built in the mid-nineteenth century. In 2014, twelve years after the cessation of its activities, it was taken over by the Japanese company Uniqlo.

Figure 13.5 The Saint-Paul Hammam, currently occupied by COS

Source: Author, 2015

A few years ago, developers would have probably demolished it. The ovens, the cast-iron wheels, the red brick smokestack. With the price per metre square in Le Marais, what a nice real estate transaction![12]

But times have changed. The Japanese company carefully preserved the vestiges of the past: "Above all we didn't want to change anything about the atmosphere of this place – that's what brought us here!" The industrial details were kept and

Figure 13.6 The *Société des Cendres*, currently occupied by Uniqlo
Source: Author, 2015

can be admired by shoppers. This solution satisfies local decision makers and the local conservation association *Vive le Marais*, whose president said in 2010 when SOFIMAR holdings (a major stakeholder of Société des Cendres) decided to protect the industrial elements:

> We still don't know what the site will be used for. An exhibition hall, a museum, a sports centre, a business incubator, a school . . . We wish the project the very best success with the condition that it conserves the facade and that it does not perturb an already fragilized urban balance by excessive visitation during weekends and the mono activity of a prêt-à-porter store.[13]

The new functions that progressively took place in Le Marais played a part in building its contemporary *culturescape* – the fusion between material (the district's location, sedimented urban fabric, and shops with storefronts recalling a former activity) and intangible (the artistic and intellectual productions) elements, creators (artists, designers) and urban populations that live, produce and consume there.

But this *culturescape* is also a *brandscape* (Klingmann 2007). In 2013, the department store BHV (*Bazar de l'Hôtel de Ville*), more well-known for its basement full of DIY and craft materials than for its fashion floors, started to be branded as 'BHV/Marais', obviously capitalizing on the profiles of its gentrifier clients.[14]

The area's evolution is the result of dynamics that, here more than anywhere else in the Parisian metropolis, are fighting for a location. This clearly becomes a marketing strategy for major commercial brands that vie to be established in this area, but the concentration of these exclusive commercial shops, located in buildings whose past has been expertly appropriated by the new brands, also creates a particularly attractive tourist destination (see Figure 13.7).

The new International Tourist Zones (ITZs)[15] allowing Sunday openings for one section of Le Marais asserted the symbiotic relationship between tourism and commerce. Indeed, unlike the former international tourist area introduced in 2007, limited to rue Francs-Bourgeois and Place de Vosges, nearly all of Le Marais became one of the twelve Parisian ITZs launched in 2015–2016. It is the largest ITZ in Paris (94 hectares) (see Table 13.7).

Table 13.7 The commercial structure of Le Marais

Le Marais	2003	2011	2014		Evolution 2003–2014		Networks	
	Number	Number	Number	%	Number	%	Number	%
Proximity to businesses and services	216	194	185	9.1	−31	−14.4	26	14.1
Bars and restaurants	323	325	332	16.4	9	2.8	23	6.9
Destination stores	956	1121	1160	57.3	204	21.3	400	34.5
Other businesses and services	337	329	348	17.2	11	3.3	128	36.8
Commercial businesses and services	**1832**	**1969**	**2025**	**100.0**	**193**	**10.5**	**577**	**28.8**
Vacant	192	176	145	5.9	−47	−24.5	0	
Other services and functions (basement)	363	290	279	11.4	−84	−23.1	11	3.9
Total	**2387**	**2435**	**2449**	**100.0**	**62**	**2.6**	**588**	**24.0**

Source: APUR, Les Zones Touristiques internationales à Paris, Diagnostic initial sur les commerces, Dec. 2015, p. 16, according to the BDcom Database 2014

Figure 13.7 A Sunday in rue Francs Bourgeois, Le Marais. The 'Boulangerie' sign, at the
right, no longer indicates a Bakery, but a fashion store. The street is practically
pedestrianized during Sundays in order to allow people to freely stroll and shop

Source: Author, 2016

A complex and diversified tourism development

This "shopping Mecca" (Chen, 2005) is also a tourism mecca. Le Marais has been
visited by tourists almost since the origins of tourism in Paris in the nineteenth
century, mainly for its *hôtels particuliers* (the Archives, hotel de Rohan-Soubise,
Musée Carnavalet, Bibliothèque historique de la Ville de Paris, hotel Le Peletier
de Souzy, Maison de Victor Hugo at Place des Vosges). The Baedecker Guide
(1924) presents the area as follows: "The Marais area has been till the 18th century
a distinguished neighborhood; it still possesses hôtels particuliers, but they are
rundown and occupied by commercial and industrial activities". The area's past is
still depicted in contemporary guides. As Pinçon and Pinçon-Charlot (2014: 86)
put it: "Guides tell groups of visitors, in various languages, of the long tormented
past of what was a popular quasi-ghetto, currently fading under the pressure of
trade and the pursuit of profit".

During the "creative destruction" (Harvey, 1989) of the industrial Le Marais by
its hypermodern counterpart, tourism initially played a limited role (Chapuis *et al.*,
2012). However, even if it was not at the origin of these transformations, it has
greatly benefited from the *culturescape* that progressively emerged and in which
tourists have themselves been incorporated both as consumers and producers.

Table 13.8 Number of visitors at the most important attractions of Le Marais or in the Marais vicinity

Notre-Dame Cathedral	14,300,000 (estimation)
Centre Pompidou	3,450,000
Notre-Dame Towers	517,424
Picasso Museum	206,195
Archaeological crypt	197,697
Maison européenne de la photographie	190,916
Victor Hugo's House	174,524
Musée d'art et d'histoire du Judaïsme	122,868
Musée Cognacq-Jay	51,845
Musée de la poupée	29,122

Source: OTCP (2015)

The inauguration of the Centre Pompidou in 1977 (Baudrillard, 1977) and of the nearby new commercial hub of Les Halles in 1979, have placed Le Marais into a hyper-touristic (Duhamel and Knafou, 2007) triangle, the angles of which are formed by Pompidou and two other major cultural, heritage and tourism icons – Place des Vosges and Notre-Dame Cathedral. The triangulation of Le Marais by three of the capital's most visited sites has crucially contributed to affirming its touristic centrality. The strategic location of Le Marais near the quays of the Seine, the plateau Beaubourg, Les Halles and Notre-Dame Cathedral, has created a new urban and tourism centrality in an area that was still considered to be marginal at the end of the 1960s. In its perimeter or vicinity are a large number of cultural and tourist attractions, some of which are the most iconic, including Notre-Dame Cathedral, which is the most visited site in Paris with estimated visitor numbers of 14.3 million in 2015 and which is located in the 4th arrondissement, just on Le Marais' doorstep.

Tourism visitation is not homogeneous: it is mainly concentrated in the centre of Le Marais, towards Le Marais' gay quarter and the few remaining traces of the Jewish Marais (around rue des Rosiers); most of the bars and art galleries are located in this area. The south of Le Marais (south of the axis Rivoli-Saint-Antoine) and the northern part experience relatively limited tourist visitor numbers (see Table 13.8).

Le Marais is perceived as one of the most touristic and attractive areas in Paris, with Paris travel tourism guides describing it as one of the must-see neighbourhoods. Table 13.9 presents some excerpts from different guides.

The number of comments on TripAdvisor also shows the importance of this neighbourhood to a mainly international tourist population. There are 23,333 comments for Notre-Dame Cathedral and 9,383 for the River Seine. There are also several comments on the main attractions of Le Marais, such as Place des Vosges or Musée Carnavalet, not to mention other major sites that are located close by such as the Islands (Ile Saint-Louis, Ile de la Cité) or Centre Pompidou.

Table 13.9 Different representations of Le Marais in the Paris guides

---

"Going back in time, this arrondissement unites modern art and the flamboyant gothic, the Centre Pompidou and Notre-Dame . . . and through the middle of it all flows the Seine".

(Routard, 2008)

"In a maze of twisting and turning streets, the aristocratic residences of Saint Paul are a beautiful sight at night-time and . . . through a secret passage, Place des Vosges".

(Routard, 2008)

"Rue Beautreillis, provincial charm in the middle of Paris".

(Le Routard des Amoureux, 2009)

"The outskirts of Beaubourg, the symbol of modern conviviality . . . is where you'll find the nice local restaurants"

(Guides Voir, 2003)

"Le Marais, an open-air museum"

(Routard, 2009, Un Grand Week-end à Paris 2007, Encyclopédie du Voyage 2006)

"At the corner of rue des Chantes and rue des Ursins is a distinguished, old house . . . Today it seems to blend into the background; . . . it looks very medieval, it's obviously a reproduction, but it is a real architectural masterpiece".

(Paris Secret et insolite, 19996).

Notes: Corpus of the studied guides: *PARIS*, Guides Bleues, Hachette, 2002; *PARIS*, Encyclopédies de voyage, Gallimard, 2007; *PARIS*, Le guide du Routard, Hachette, 2008; *Le Routard des Amoureux à Paris*, Hachette, 2009; *PARIS*, Guide Vert Michelin, 2007; *Un grand Week-end à Paris*, Hachette, 2006; *Paris braché*, Lonely Planet, 2009; *Paris des amoureux*, First, 2008; *Paris Secret*, Guides Gallimard, 198; *Paris secret et insolite*, Parigramme, 1996; *PARIS*, Guides Voir, Hachette, 2002.

More noticeable is that Le Marais is probably one of the very few urban areas in Paris recognized as a destination per se (there are 5,049 TripAdvisor comments on Le Marais as a neighbourhood in general) (see Table 13.10).

The comments analysis shows the important touristic dimension of the area. Le Marais represents *the* quintessential Paris, *the* place-to-be for heritage, fancy restaurants, premium museums, high-end art galleries, night-life or boutique shops (see Table 13.11).

The tourismification of the area is also most likely linked to the rise in available furnished apartments and in particular the website Airbnb, which has a rather large presence in the area. Le Marais represents 6 per cent of Airbnb rentals in Paris but only 1 per cent of the surface area of Paris (Hajdenberg, 2015).

The study by Cousin, Jacquot, Chareyon (in Hajdenberg, 2015) reports that 36 per cent of the owners of Airbnb rentals are multi-owners (Fouquet and Nussbaum, 2014). Internet sites[16] or specialized investors assist speculators in identifying the most rentable apartments for this type of property use by showing them the most attractive surfaces, costs or the most selected locations. Given the high profit from tourism rentals, even very high rents are not able to compete with revenues from daily rentals.

Table 13.10 Number of comments about Le Marais and Le Marais attractions on the TripAdvisor website

Sites	Rate	Number of quotes
Notre-Dame Cathedral	4.5	23,333
River Seine	4.5	9,383
Le Marais	4.5	5,049
Notre-Dame Towers	4.5	2,328
Ile de la Cite	4.5	2,293
Île Saint-Louis	4.5	2,225
Centre Pompidou	4	1,646
Place des Vosges	4.5	1,544
Musée Carnavalet	4	590

Source: TripAdvisor (2016)

Tourism therefore plays a role similar to the super-gentrifiers. It attacks the former landlords, most of them first-time gentrifiers. The Paris municipality tries to regulate this phenomenon, which causes not only the eviction of the population by driving up residential rents, but contributes to the acceleration of the cycle that started with the first gentrification. To paraphrase Scheppe (2015: 26), every year, permanent residents are being replaced by visitors who are just passing through, a change that will lead to the disappearance of all infrastructures. Luxury brands with a global presence have replaced essential everyday businesses (e.g. cobblers, grocers, stationers), producing a turnover that is directly achieved through tourism.

To conclude, the tourism phenomenon in Le Marais is closely linked with urban and cultural brand-spaces. Le Marais' hyper-tourismification can be expressed as a series of crossover symptoms.

Table 13.11 Selected comments from TripAdvisor

"This is a great area to walk around, stopping to enjoy the architecture, the old and new boutiques, and the Jewish quarter."
"This is one of the nicest neighbourhoods in Paris. I loved the feel while walking around. So many nice shops, lots of historical places, and just the right balance of everything. I followed one of the walking tours I downloaded online so we knew the stories and significance of the places. Next to the St. Germain area, I would also like to stay here in the future."
"Colourful area, with bars, boutiques, delis, little public parks, galleries and fantastic craft and art shops. For a tasty lunch on a budget, look no further than the numerous falafel joints."
"Amazing shops. The vintage scene is LARGE here, you can really get caught up in all the amazing shops. The cafes are quirky with a homely feel. The streets are tiny, the buildings are amazing."

Source: TripAdvisor (2016)

Tourism centrality

Over recent decades, Le Marais has gained the undisputed status of a central touristic destination in Paris. Along with Montmartre, it is the only urban area in Paris perceived as a sight in its own right (as opposed to isolated monuments such as the Eiffel Tower or the Louvre) and this can be seen in the TripAdvisor comments. However, tourism is not the exclusive function of the area and Le Marais is not perceived as a tourism enclave or a tourism bubble (Judd, 1999). Souvenir shops are not the dominant trade, as is the case in the vicinity of the Louvre or Montmartre.

Tourism embedded in everyday life

Hyper-tourismification is characterized by the embedding of tourism in all expressions of everyday life. Museums, galleries, cafés, restaurants, libraries, administrations and services all cater for diversified but co-existing populations that use, in different ways, the urban space and its components. Some of these publics are intrinsically touristic, whereas others belong to the more unclear categories of repeaters (or even multi-repeaters), temporary residents or second-home owners.

Globalization of real estate

The opening of the permanent home, second-home and tourist rental real estate sector to international markets has contributed to the cosmopolitanism of the area. Le Marais is one of the most desirable areas for global markets and this is closely linked to tourism rentals. For example, the *New York Times* advertised an apartment for sale in Le Marais not only with the advantages of living in this area ("The shops and restaurants of the Marais are just outside the door, and the apartment is within walking distance of the Louvre, Notre Dame and the Pompidou Centre"), but also with the investment potential through tourism rentals: "The current owner lives in Southern California and rents out the apartment when he's not in Paris, charging about $175 per night".[17] There is a clear continuity between the permanent residential market and tourism rentals. In some cases, tourism rentals function as super-gentrifiers.

A global culturescape

Today, Le Marais has an astonishing collection of art galleries and museums that are installed either in the restored *hôtels particuliers* or the former glass-roofed industrial buildings. Their close co-existence creates a particularly dense *culturescape* composed of both content (art works) and casing (historical seventeenth- to twentieth-century architecture). The quality of the architecture, the large international demand and the cost per square metre tend to produce a particularly artified urban context that is one of the district's major factors in its tourism attractiveness.

Hyper-aestheticization

This is linked to the hyper-aestheticization of the urban environment. Pedestrian streets, urban furniture, shop windows, architectural details, cafés and restaurants, modern and historical architecture, and parks and gardens all produce a perfectly estheticized urban setting offered for real or symbolic contemplation or consumption (Zukin, 1998).

Exclusive brandscape

Le Marais is *the* district for exclusive fashion brands, both independent creators and major chains. The concentration of these commercial offers creates an exclusive shopping destination that is concentrated in a rather limited space. Hyper-tourismification is closely related to the intensified commercial gentrification (Mermet, 2016) that tends to replace everyday shops with more sophisticated ones.

Creativity, design and architectural iconicity

Le Marais successfully combines a well-preserved historical setting with creative businesses, design and, more recently, contemporary architecture. What is happening in Le Marais can only be compared to other central historical areas of international metropolises such as Soho and the Meatpacking district in New York; a first stage of redevelopment that is characterized by the restoration of historical buildings and followed by the spectacular arrival of international names in contemporary art and architecture.

Tourism governance

The former Mayor of Le Marais, Dominique Bertinotti, understood perfectly that an internationally attractive tourism district implies specific governance, as it involves not only permanent residents, but also second-home owners, temporary residents and tourists. In a brief collaboration between researchers and the mayor,[18] a new *modus vivendi* was implemented that tried to bring together as best as possible these different populations that traditionally avoid each other. And indeed, in Le Marais, there are no exacerbated conflicts common in more recent urban destinations such as Barcelona, Venice or Berlin (Gravari-Barbas and Jacquot, 2016).

In hyper-tourismified areas, tourism is embedded in all the expressions of everyday life (see Figure 13.8). These areas therefore become testing grounds that lead to an alternative understanding of a new co-existence that takes place between all contemporary forms of mobilities that go far beyond tourism.

Figure 13.8 Urban landscape layers in Le Marais. On the right, one of the most important
architectural vestiges of the seventeenth century, the *Hôtel particulier*
Lamoignon that today houses the Public Library of the City of Paris. On the
wall the last marketing campaign by Camper (one of the brand's shops is
located close). In the forefront, the *culturescape* of Le Marais, indicated by
the superposition of all the signs indicating the museums and cultural centres
of the district

Source: Author, 2016

Conclusion

With the analysis of Le Marais' recent history, symbolically starting with the promulgation of the PSMV in 1965, it becomes clear that tourism was introduced after the start of the heritagization process. Although tourism did not play a major role in the beginnings of heritagization, it clearly drew the most benefit from the systematic restoration process undertaken in the area. It also benefited from the commercial gentrification process that endowed the area with the most exclusive shops or art galleries. Le Marais is today a heritage, culture, art *and* shopping destination, and the recent promulgation of the ITZ confirms the role of Le Marais as a shopping Mecca.

Tourismification has also benefited from residential gentrification for at least two reasons: the first is related to the occupants of the gentrified apartments that have become part of the site's attractiveness. Le Marais visitors are also attracted by the 'beautiful people' that live there. Gentrifiers visit other gentrifiers' places. The second is the gentrified apartments, many of which have been transformed into furnished rentals. Tourismification accompanied gentrification that was itself encouraged by heritagization.

The globalization that took place in the last decades of the twentieth century and the beginning of the twenty-first century further opened Le Marais real estate to global markets and intensified the gentrification phenomena, partially transforming gentrification into super-gentrification. It also had an impact on tourism in the area by creating one of the most exclusive urban destinations in the world; it has become a hyper-touristified area.

Hyper-touristification is characterized by: the centrality of the area with the metropolitan destination; the embedding of tourism into everyday life; globalization of the real estate tourism markets and the embedding of residential and tourism rentals; a global *culturescape* and *brandscape*; creativity, design and contemporary art; architectural iconicity; hyper-aestheticization; and the existence of a tourism governance. Tourism *permeates* Le Marais: it plays a part in neighbourhood planning issues, development and the nature of its commercial function (most of its shops cater mainly or also for tourists and visitors) and with the tourists' and locals' place imaginary.

In this sense, Le Marais is an iconic example of how the tourism phenomenon embeds an urban area. The maturity of the tourism development is visible in its governance: local decision makers are aware that tourism is not only a cash windfall but an increasingly constitutive component of Le Marais. Thus, in this central Parisian neighbourhood, a new governance is taking place that involves permanent and secondary residents (a growing proportion of whom come to Le Marais for similar reasons as the tourists), tourists (for whom residents form part of the attractiveness of the place) and local decision-makers who integrate, or even use tourism, as a part of cultural differentiation in their neighbourhood.

Contrary to several analyses conducted on touristified historic neighbourhoods, the situation in Le Marais is not symptomatic of a museumification.

Far from being a bubble or a tourist enclave, Le Marais seems to be a space of intense local–global exchanges between different populations and categories of users who are reciprocally attracted. They co-produce contemporary Le Marais, as can be understood by the evolution of the preservation procedures that are adapted to the needs and expectations of the local and global populations.

To paraphrase Mary McCarthy (1963), (*Tourist Venice is Venice*), "Tourist Le Marais *is* Le Marais", a space in which different operators – permanent or secondary residents, regional, national and international visitors, economic operators working at various scales, tourists and local decision makers – co-produce the contemporary Le Marais, as much by their conflicts as by their creative co-existence.

Notes

1 The others are Arts et Métiers, Enfants Rouges, Archives, Saint-Avoye, Saint-Merri, and Arsenal.
2 Safeguarded sector status applies from Place de la Bastille to Place de l'Hôtel de Ville and from the Seine to the Temple district; in total it covers an area of 126 hectares.
3 MERIMEE database, French Ministry of Culture.
4 Mairie de Paris, Brochure *Révision du Plan de Sauvegarde et de Mise en Valeur du Marais*, Mairie de Paris, September 2009.
5 "The Chartier-Corbasson agency divided a thin layer of land – 1 rue Turenne – into plots, running alongside an adjoining gable wall that slightly projected onto a building used for social housing, the façade of which was rather different . . . made of anodised aluminium panels tinted bronze and fixed onto a metallic structure". (Guillaume, 2015: 9)
     The architect Pierre Audat visually enhanced the 65-m high industrial chimney of the old Cendres workshop that was converted into a shop; he removed the base and made the red of the bricks more prominent.
6 Interview with *District* estate agency, IREST, 2010.
7 Data Coldwell Banker Paris, 2015.
8 www.citylab.com/housing/2012/09/housing-crisis-paris-blame-foreigners/3386/.
9 www.lexpress.fr/informations/un-mcdo-au-hammam_638644.html.
10 Taken from a description card of the interview with the Mayor in the exhibition "Le Marais en Heritages", 2015–2016, Musée Carnavalet, Paris.
11 The Société des Cendres used "ash washers" to treat jewellers' waste in order to recover the gold, silver or even platinum that was found in all sorts of dust.
12 www.lemonde.fr/economie/article/2013/12/03/uniqlo-transforme-la-derniere-usine-du-marais-en-temple-du-vetement_3524446_3234.html#yerLRrAYCMthLCMJ.99.
13 http://vivrelemarais.typepad.fr/blog/2010/06/nouveau-d%C3%A9part-pour-la-soci%C3%A9t%C3%A9-des-cendres.html.
14 "BHV is committed to this historic quarter and truly reflects the Le Marais spirit . . . its new name BHV/Marais represents a way of life, a life style . . . Le Marais is a real shopping destination for Parisians and tourists. Through this positioning, BHV/Marais is showing that BHV/Marais is the department store for creative urbanites, the one that will enable them to fully express their personality through all their areas of interest." www.e-marketing.fr/Thematique/Retail-1002/Breves/Le-BHV-devient-BHV-Marais-53333.htm#MO8Wc6veHVVAOIK2.
15 An international tourism zone is characterized by its "international influence created through the supply of world-class commercial, cultural, heritage and leisure offers". It must also be "served by transport infrastructure of a national or international scale" and

"have a great influx of tourists who live outside of France". Finally, "a significant flow of purchases must be made by tourists living outside of France; this is assessed by the amount of purchases or their share in the total turnover for the area" (APUR, 2015).

16 Such as Airdna.co.
17 www.nytimes.com/2009/04/29/greathomesanddestinations/29gh-sale.html?_r=0.
18 This paper draws on, among other sources, a request that the then mayor made to IREST/Paris 1 Panthéon-Sorbonne students to consider new ways of interaction between tourists and residents. The idea behind this project was to go beyond the usual conflict approaches of this "co-habitation" to create new social and cultural added values from tourism.

References

Aldrich, R. (2004) 'Homosexuality and the City: An Historical Overview', *Urban Studies*, vol. 41, no. 9, pp. 1719–1737.

APUR. (2001) *4e arrondissement, État des lieux, Éléments pour un diagnostic urbain, Paris en ses quartiers.*

APUR. (2010) *Paris 4e, Eléments de diagnostic. Préparation du PLH de Paris.* Available at : http://labs.paris.fr/commun/pdf/fichesDLH/4e.pdf. [accessed 29 Jan 2017].

APUR. (2015) *Les Zones Touristiques internationales à Paris, Diagnostic initial sur les commerces*, December.

Auffray, M-F. (2001) *Le Marais: la légende des pierres*, Tours: Hervas.

Baedecker, K. (1924) *Paris et ses Environs*, Leipzig : Karl Baedeker.

Baudrillard, J. (1977) *L'effet Beaubourg, implosion et dissuasion*, Paris: édition Galilée.

Bercé, F. (2000) *Des monuments historiques au patrimoine, du XVIIIe siècle à nos jours, ou 'Les égarements du cœur et de l'esprit'*, Paris: Flammarion.

Brody, J. (1987) 'Le quartier de la rue des Rosiers ou l'histoire d'un cheminement', in Gutwirth J. and Petonnet C. (eds), *Chemins de la ville: enquêtes ethnologiques*, Laboratoire d'anthropologie urbaine : CTHS, pp. 85–102. Available at: https://halshs. archives-ouvertes.fr/halshs-00076692/document [accessed 29 Jan 2017].

Butler, T. and Lees, L. (2006) 'Super-gentrification in Barnsbury, London: Globalization and Gentrifying Global Elites at the Neighbourhood Level', *Institute of British Geographers. Transactions*, vol. 31, no. 4, pp. 467–468.

Chapuis A., Fagnoni E., Gravari-Barbas, M., Jacquot, S. and Mermet, A-C. (2012) 'Dynamiques urbaines et mobilités de loisirs à Paris : pratiques, cohabitations et stratégies de production de l'espace urbain dans le quartier du Marais', in *Les quartiers historiques: Pressions, enjeux, actions*, Québec: Presses Universitaires de Laval, pp. 27–49.

Châtelain, P. (1967) 'Quartiers historiques et centre-ville: l'exemple du quartier du Marais', *Urban Core and Inner City: Proceedings of the International Study Week, Amsterdam,* 11–17 September 1966, Leiden: E. J. Brill, pp. 340–355.

Chen, A. (2005) 'A Shopping Mecca in the Marais', *The New York Times*, 2 January 2005.

Choay, F. (1999) *L'allégorie du patrimoine*, Paris: Seuil.

Clerval, A. (2008) *La gentrification à Paris intra-muros: dynamiques spatiales, rapports sociaux et politiques publiques*, PhD thesis, Paris 1 Panthéon Sorbonne University.

Clerval, A. (2010) 'Les dynamiques spatiales de la gentrification à Paris, une carte de synthèse', *Cybergeo*, no. 505. Available at: http://cybergeo.revues.org/23231?lang= en&hc_location=ufi. [accessed 29 Jan 2017].

De Pieri, F. (2010) 'Conservation in the Age of Gentrification: Historic Cities from the 1960s', *Contemporary European History*, vol. 19, no. 4, pp. 375–385, Cambridge University Press. Available at: http://porto.polito.it/2373899/ [accessed 29 Jan 2017].

Dewailly, J-M. (2005) 'Mise en tourisme et touristification', in Amirou, P.B., Dewailly, J-M., Malezieux, J. (eds), *Tourisme et soucis de l'autre. En hommage à Georges Cazes*, pp. 29–33.

Djirikian, A. (2004) *La gentrification dans le Marais, quarante ans d'évolution de la population et des logements*, Master Thesis in geography, Paris 1 Panthéon Sorbonne University.

Duhamel, P. and Knafou, R. (2007) 'Le tourisme dans la centralité parisienne', in Saint Julien and Le Goix (eds), *La métropole parisienne, Centralités, inégalités, proximités*, Paris: Belin, Mappemonde, pp. 39–62.

EffiCity. (2016) 'Prix immobilier au m2 Marais / Musée Picasso / Place des Vosges (Paris)'. Available at: www.efficity.com/prix-immobilier-m2/q_marais-musee-picasso-place-des-vosges_paris_75003/#16.00/48.8595/2.3630 [accessed 29 Jan 2017].

Faure, J. (1997) *Le Marais, organisation du cadre bâti*, Paris: L'Harmattan.

Fijalkow, Y. and Oberti, M. (2001) 'Urbanisme, embourgeoisement et mixité sociale à Paris', *Mouvements* vol. 1/2001, no. 13, pp. 9–21.

Fouquet, H. and Nussbaum, A. (2014) 'Paris Airbnb Cops Want to Know If Your Rental Is Legal', available at: http://www.bloomberg.com/news/articles/2014-08-07/paris-Airbnb-cops-want-to-know-if-you-re-rental-is-legal.

Giraud, C. (2009) 'Les commerces gays et le processus de gentrification, l'exemple du quartier du Marais à Paris depuis le début des années 1980', *Métropoles*, no. 5, pp. 79–115. Available at : http://metropoles.revues.org/3858 [accessed 01 Feb 2017].

Gravari-Barbas, M. (2010) 'Ville touristique, ville créative. Le touriste, co-opérateur de la créativité urbaine?', *Urbanisme*, no. 373, pp. 68–73.

Gravari-Barbas, M. (2014) 'Patrimoine, culture, tourisme et transformation urbaine; le *Lower East Side Tenements Museum*, NY', in Djament-Tran G. and San Marco P. (eds), *La métropolisation de la culture et du patrimoine*, Paris: Editions Le Manuscrit, pp. 143–181.

Gravari-Barbas, M. and Graburn, N. (eds) (2016) *Tourism Imaginaries at the Disciplinary Crossroads: Places, Practices, Media*, New York: Routledge.

Gravari-Barbas, M. and Jacquot, S. (2016) 'No Conflict? Discourses and Management of Tourism-related Tensions in Paris', in Novy J. and Colomb C. (eds), *Resistance and Protest in the Tourist City, Contemporary Geographies of Leisure*, London: Routledge/ Taylor & Francis, pp 31–51.

Gravari-Barbas, M. and Mermet, A.-M. (2014) 'Commerce et Patrimoine', in Gasnier A. and Lemarchand, N. (eds), *Le commerce dans tous ses états. Espaces marchands et enjeux de société*, Rennes: Presses Universitaires de Rennes, pp. 81–91.

Gravari-Barbas, M. and Renard-Delautre, C. (eds) (2015) *Starchitecture(s): Celebrity Architects and Urban Space*, Paris: L'Harmattan.

Guillaume, V. (2015) 'Le Marais en héritage(s). Cinquante ans de sauvegarde, depuis la loi Malraux', in Musée Carnavalet, *Le Marais en héritages*, Paris : Paris Musées, pp. 7–15.

Guinot, D. and Pommier, A.-H. (2014) 'Immobilier : les prix à Paris quartier par quartier", *Le Figaro*, 30 March 2015. Available at: http://immobilier.lefigaro.fr/article/immo bilier-les-prix-a-paris-quartier-par-quartier_5e0fc62e-b4f9-11e3-92ad-7803e0daa19d/ [accessed 29 Jan 2017].

Hajdenberg, M. (2015) 'Comment Airbnb squatte la France', *Mediapart*, 31 July 2015.

Harvey, D. (1989) 'From Managerialism to Entrepreneurialism: The Transformation in Urban Governance in Late Capitalism', *Geografiska Annaler*, Series B, Human Geography, The Roots of Geographical Change: 1973 to the Present, vol. 71, no. 1, pp. 3–17.

INSEE (Institut National de la Statistique et des Etudes Economiques) (2000) *Populations légales: recensement de la population de 1999: France*. Paris: INSEE.

IREST (2010) *Le diagnostic touristique du 4ème arrondissement de Paris*, report by the students of the Master Développement et Aménagement Touristique des Territoires, under the supervision of A. Chapuis and A.-C. Mermet, Paris 1 Panthéon-Sorbonne University.

Judd, D. (1999) 'Constructing the Tourist Bubble', in Judd, R.J. and Fainstain, S. (eds), *The Tourist City*, Yale University Press.

Klingmann, A. (2007) *Brandscapes: Architecture in the Experience Economy*, Cambridge. MA: MIT Press.

Koolhaas, R. (2011) 'Bigness ou le problème de la grande dimension', in *Junkspace*, Paris: Edition Manuel Payot, 2011, pp. 29–42.

Le Clere, M. (ed.) (1985) *Paris de la préhistoire à nos jours*, Saint-Jean-d'Angély: Bordessoules.

Lees, L. (2003) 'Super-gentrification, the Case of Brooklyn Heights, New York City', *Urban Studies*, vol. 40, no. 12, pp. 2487–2509.

McCarthy, M. (1963) *Venice Observed*, Harvest book.

Mermet, A.-C. (2016) 'The Global Retail Capital and the City: Towards an Intensification of Gentrification', *Urban Geography*, pp. 1–24.

Mermet, A.-C. and Gravari-Barbas, M. (2013) 'Commerce et Patrimoine, l'exemple du Marais à Paris', *Figures nouvelles, figures anciennes du commerce en ville*, *Annales de la Recherche Urbaine*, vol. 108, no.1, pp. 56–67.

Neville, S-L. and Raude, M. (2015) 'Entretien avec Michel Raude', Musée Carnavalet, *Le Marais en héritage(s). Cinquante ans de sauvegarde, depuis la loi Malraux*, pp. 140–144.

OTCP (2015) *Tourism in Paris, Key figures*, Available at: http://asp.zone-secure.net/v2/index.jsp?id=1203/1515/65204&lng=en [accessed 01 Feb 2017].

Pinçon, M. and Pinçon-Charlot, M. (2014) 'Les enjeux de la mixité', *Sociologie de Paris*, Paris: La Découverte, Repères.

Préfecture de Paris, DULE, Bureau de l'habitat. (2007) *Le parc de logements à Paris*.

Redoutey, E. (2004) 'Le Marais, un quartier gay?', *Urbanisme*, no. 337, pp. 20–23.

Richards, G. (2005) 'Creativity: A New Strategic Resource for Tourism', in Swarbrooke, S. and Onderwater (eds), *Tourism, Creativity and Development: ATLAS Reflections 2005*, Association for Tourism and Leisure Education, Arnhem, pp. 11–22.

Richards, G. and Wilson, J. (2007) *Tourism, Creativity and Development*, New York: Routledge.

Scheppe, W. (2015) 'L'invention de l'image de la ville. Notes sur les métamorphoses du Marais', in Musée Carnavalet, *Le Marais en héritages*, pp. 19–27.

Sibalis, M. (2004) 'Urban Space and Homosexuality: The Example of the Marais, Paris' "gay ghetto"', *Urban Studies*, vol. 41, no. 9, pp. 1739–1758.

Smith, N. (1987) 'Gentrification and the Rent Gap', *Annals of the Association of American Geographers*, vol. 77, no. 3, pp. 462–465.

Tomas, F. (2004) 'Les temporalités du patrimoine et de l'aménagement urbain', *Géocarrefour*, vol. 79, no. 3, pp. 197–212.

Verbeke, M. (1998) 'Tourismification of Historical Cities', *Annals of Tourism Research*, vol. 25, no. 3, pp. 739–742.

Zukin, S. (1995) *The Culture of Cities*, Oxford: Blackwell.

Zukin, S. (1998) 'Urban Lifestyles: Diversity and Standardisation in Spaces of Consumption', Urban Studies, vol. 35, no. 5/6, pp. 825–839.

Zukin, S. (1989) *Loft Living: Culture and Capital in Urban Change*, New Brunswick: Rutgers University Press.

Index

Abe, D. 257
Abram, Ruth 153–154, 155, 156, 158–159, 161, 168
accommodation: Cartagena de Indias 76, 85–88, 92, 94, 97, 98, 99n4; Colonial City 37, 39, 43, 48; design of Airbnb accommodation 197–198; Lisbon 258, 259–260, 263–265, 268, 271; Reykjavík 52–53, 56–69; *see also* apartments; hotels; housing; rent gap; rent increases; rental accommodation
activism 109, 143; *see also* protests; resistance
aestheticization 8, 16, 59, 321, 323
African immigrants 156, 158, 161, 169, 170
agency 234, 235–236, 248
Aiello, G. 226
Airbnb 7, 15–16, 54–56, 111, 181–183, 209; Barcelona 124–125; Berlin 183–201; Colonial City 37, 39, 43, 48; facilities and design of accommodation 197–198; interaction between hosts and guests 195–197; Lisbon 256, 264, 265; Le Marais 318; Marseille 10; monetary dimension 193–195; motives of users 192–193; New York 166; personality profiles of users 191–192; Reykjavík 52, 56–69
Alemanno, Gianni 138, 148n10
"allochronism" 234
Alto del Cabo 27
Amsterdam 4–5
apartments: Airbnb in Berlin 184–185, 189, 194–195, 198, 199–200; Cartagena de Indias 83, 93; Colonial City 37, 38; conversion to Airbnbs 7, 55; Lisbon 258, 262, 264–265, 266, 268, 270, 271; Le Marais 308, 310, 318, 320, 323; New York 164; Reykjavík 62, 64–65, 67, 68; speculation 128n1; touristification 260

architecture 8, 11; Cartagena de Indias 76, 80, 81, 83–84; Colonial City 30, 48; domestic or vernacular 30, 233, 234–235, 237, 242; Le Marais 302, 303–304, 320–321; New York Tenement Museum 164, 171; Rethemnos 242, 244; Reykjavík 59; Rome 138; self-gentrification 236; South Street Seaport 214–215, 216, 217, 221
Archontakis, Dimitrios 243
Argentina 78–79
art 8, 134, 135–136; Le Marais 303, 318, 321; Rome 138, 139, 141, 142–146, 147; South Street Seaport 224
art galleries 4, 8; Berlin 188, 197; Cartagena de Indias 88; Lisbon 266; Le Marais 304, 311, 317, 318, 320, 323; New York 153, 163, 171–172; Raval 118; Reykjavík 59; Rome 143–144, 146
artification 8, 16, 134–137; Le Marais 305, 320–321; Rome 140, 142–146, 147
artisanship 240, 243, 244, 246
Ascher, F. 1
Atkinson, R. 78
attractions: Bangkok 239; Beijing 12, 288, 291; Cartagena de Indias 90; Chinese rating system 296n1; Colonial City 34, 35; Lisbon 262; Le Marais 317–318, 319; Shenzhen 281, 284; *see also* theme parks
Audat, Pierre 304, 324n5
authenticity 5, 14, 182, 234, 235; Airbnb 198; Berlin 188; Cartagena de Indias 86, 90, 97; Lisbon 259, 266; marketing strategies 209; New York 155, 164; South Street Seaport 218; staged 110

Back, K-J. 7
"back to the city" movements 2, 278
Balaguer, Juan 31, 32

Baltimore 4, 8, 13, 214
Bangkok 17, 237–241, 245–248
Bao, J. 277, 280, 285
Barata-Salgueiro, Teresa 255–275
Barcelona 9, 10, 14, 17, 107–133, 138;
　Airbnb 7, 15, 54, 55, 62, 124–125, 127;
　conflicts 321; hotel beds 265; Le Marais
　compared with 306; public space 258
Barrera, Alberto 83
Barrio Chino 9, 10, 107, 113, 126
bars 4, 279; Berlin 188; Cartagena de
　Indias 86, 88, 90, 91, 92, 96; Lisbon
　266; Le Marais 315, 317, 319; New
　York 171–172; opening hours 200;
　Raval 118; South Street Seaport 219
Baudry, Sarah Lilia 134–152
Bauman, Z. 109
beautification 8, 136, 234; Bangkok 239,
　246; Cartagena de Indias 76; Colonial
　City 32; Le Marais 311
Beijing 2, 12, 234, 277, 280, 287–291,
　292, 294–295
Belleville 136
belonging 218, 225–226, 227
Berlin 8, 16, 17, 255; Airbnb 7, 16, 55, 63,
　181, 183–201; complaints about tourism
　259; conflicts 321; entertainment 13
Bertinotti, Dominique 321
Betancur, J. 79
BHV 315, 324n14
Bianchi, R. 111
Bilbao 8, 11–12, 136
Bloomberg, Michael 173n4, 208, 210
Bock, K. 5
Booking.com 256, 265, 266
Boston 9–10, 214
Botsman, Rachel 181–182
Botterill, J. 285
Bouchier, M. 137
Boyer, Christine 215
branding 219–220, 262
brandscape 315, 321, 323
Bridge, G. 12, 78
Bridgetown 26
business travellers 194
Butler, T. 210

cafés: Berlin 188; Cartagena de Indias 76,
　88, 90, 91, 97; "explorer-tourists" 183;
　Lisbon 256, 258; Le Marais 319, 320;
　Reykjavík 59, 66
Camagüey 26
capital 11, 17, 173; accumulation 1; circle
　of investment 78; global flows of 98,
　271; Lisbon 259, 260

capitalism 12, 13, 75, 78, 218–219, 255
Caribbean 25–27, 47
Cartagena de Indias 10, 13, 15, 26, 75–78,
　79, 80–98
Catholic Church 29
CCTV 42
Cerdà, Ildefons 112, 113
Chapuis, A. 4–5
Charmes, E. 136
Chile 75, 78–79
China 12, 13, 236, 276–298
Chinese immigrants 156, 158, 159,
　161–162, 163, 166, 168–169, 170
Cienfuegos 26
"City-labelling" 135
Clay, P. 258
Clerval, A. 308
Clos, Joan 128n2
Cócola Gant, A. 7, 258, 265
Cohen, R. 111
collaborative consumption 181–182, 190
"collage tourism" 209
Colonial City, Santo Domingo 15, 25,
　27–49
commercial gentrification 4, 8; Cartagena
　de Indias 88–91; Latin America 79; Le
　Marais 299, 300, 310–316, 321, 323; *see
　also* shops and retail
commercialization 26
commodification 11, 153, 166, 209, 260
community 225–226, 280
competition 8, 14, 209; access to resources
　112; China 293; entrepreneurial city
　222; Reykjavík 66, 68
competitiveness: Barcelona 107; Colonial
　City 36, 40; Lisbon 256; Rome 138
conflict 9, 112, 321; Lisbon 259, 271;
　Rome 140; second-home ownership 54
conservation: Cartagena de Indias 79,
　80, 81–82, 84; circle of investment
　78; Colonial City 34, 35; Greece 237;
　Rethemnos 238, 243; South Street
　Seaport 218; Thailand 237; *see also*
　preservation
conspicuous consumption 7, 183, 307
consumption: aesthetic 134; Barcelona
　109, 111–112; Beijing 288, 291;
　collaborative 181–182, 190;
　conspicuous 7, 183, 307; cosmopolitan
　112, 127; Lisbon 256; Shenzhen
　281, 286; South Street Seaport 224;
　spatial division of 107, 255; tourism
　gentrification 279
"consumptionscapes" 6–7
convents 85–86

Coro 26
cosmopolitanism 112, 257; Beijing 288;
    Raval 108, 109, 118–119, 122, 127;
    Rome 147; Shenzhen 281
creative class 8–9, 136, 147, 189
Crete 236, 237–245, 247–248
"crypto-colonialism" 238, 239, 246, 248
Cuba 26, 27
cultural capital 143, 183; Greece 248;
    Lisbon 258; Rethemnos 244; Rome 135,
    146–147; self-gentrification 236
cultural tourism 9–11, 16–17, 75;
    Caribbean cities 25; Cartagena de
    Indias 82; Colonial City 34, 35; *see also*
    heritage; museums
"cultural turn" 135–137
culture: artification 136–137; Beijing 288,
    290, 291; commodification of 209;
    Greece 239; Lisbon 258; local 3; New
    York 166, 168; Raval 118; Rome 135,
    138, 139–140, 148n6; Shenzhen 281,
    284–285; temporary cultural events 223;
    Thailand 238, 239, 240, 246; tourism as
    a cultural intermediary 13–14
culturescape 315, 316, 320–321, 322, 323
Cunin, E. 81
currencies 46–47

data collection: Airbnb in Berlin 190–191;
    Airbnb in Reykjavik 56–58; Barcelona
    108; Cartagena de Indias 76–77; Lisbon
    256; Le Marais 300; Rome 135; South
    Street Seaport 208–209
Davidson, M. 266, 268
demolitions 302, 306, 310
deregulation 75, 78
Detroit 16
Diaz de Paniagua, R. 79, 80, 84, 99n12
Dickinson, G. 226
"displaced time" 234
displacement 27, 53, 278, 280, 293; Airbnb
    68, 69; Beijing 291; Colonial City 32,
    42; "induced-displacement" 2; Lisbon
    260, 266, 269–270, 271; Shenzhen 281,
    285, 286; *see also* evictions
Dissanayake, E. 134
Djirikian, A. 305–306
Dolkart, Andrew 157
domestic space 236
Dominican immigrants 168, 169, 170
Dyckhoff, T. 188

economic crisis: Berlin 189, 195;
    Cartagena de Indias 86; collaborative
    consumption 182; Greece 241; Iceland

52, 59, 60; Santo Domingo 31; South
    Street Seaport 219; Spain 120, 121,
    122, 127
economic development 8, 9, 233; Beijing
    287, 289, 290, 291; Cartagena de Indias
    81; entrepreneurial cities 222; Greece
    242; Shenzhen 282–283, 284, 285;
    South Street Seaport 212
Edensor, T. 110
educational institutions 93–94
elites 5–8, 11; Cartagena de Indias 83,
    85, 88; China 12; crypto-colonial
    248; Le Marais 302, 307; Rome 146;
    South Street Seaport 12, 227–228;
    transnational 17, 221, 278
Elliott, A. 111
embourgeoisement 243
employment 269, 289, 290
empty homes 43–45, 263, 270, 308
Eng, David 171
entertainment 13; Beijing 291; experiential
    economy 209; festival marketplaces
    218; Lisbon 258; New York 163–164;
    Shenzhen 286; South Street Seaport
    207, 217; *see also* nightlife
entrepreneurship 14, 209, 222, 293
environmental degradation 30
Esso Plan (1967) 33, 34
ethnicity: Colonial City 25; demographic
    changes in New York 164–166; New
    York Tenement Museum 154, 156–157,
    158, 168, 170; *see also* immigrants
ethno-tourism 159
events 98, 223, 224, 256, 262, 302; *see
    also* festivals
evictions 234; Berlin 187; Colonial City
    31–32, 34, 48; Lisbon 262, 267, 270; Le
    Marais 303, 319; *see also* displacement
exoticism 244
experiential economy 209, 217
experiential turn 5
"explorer-tourists" 7, 181, 183
expulsions *see* evictions

Fabian, Johannes 234
Fainstein, S.S. 11, 16, 112
fashion 8; Beijing 291; Cartagena de
    Indias 88–89; Colonial City 47; Lisbon
    267; Le Marais 311, 312, 316, 321;
    South Street Seaport 219, 224
Faure, Juliette 300
Fernández, M. 107
festival marketplaces 13, 208, 214,
    216–219, 222, 223, 228
festivals 136, 302

Fincher, Ruth 110
Fontaine, P. 68
forced displacement *see* displacement
fragmentation 40, 47, 263, 265
France 293; *see also* Paris
Franco, Francisco 33, 113
Franklin, A. 108
Füller, H. 54, 188, 259
funding 25, 32–33

gastronomy 8
gated communities 36, 46, 281
gay culture 306
gender 239
gentrification 1–3, 49, 134; Airbnb 181,
   200; artification 137; Berlin 181, 183,
   185, 188–189; Caribbean cities 25–27;
   Cartagena de Indias 75–76, 79, 80,
   82–91; China 276–277, 280–295;
   Colonial City 28, 30, 33, 37, 40–47, 48;
   cosmopolitan lifestyles 112; creative
   class 136; definitions of 3, 259, 278;
   domestic architecture 233; evolution
   of 277–278; "generalized" 276, 278;
   indirect 53; international elites 5–8; Latin
   America 78–80, 98; Lisbon 256, 257,
   258–259, 260–268, 269, 270–271; local
   tourism strategies 8–14; Le Marais 299,
   300, 305–316, 323; as multidimensional
   process 226; multiple actors 198; nature
   of tourism gentrification 278–280; neo-
   liberalization 75; New York 153–154,
   161–162, 164, 166, 168, 171–173,
   210; Raval 108, 122, 127; research
   agenda 15–16; Reykjavík 52, 65–68;
   Rome 146; second-home ownership
   54; self-gentrification 17, 235–237,
   243–244, 246, 248; sharing economy
   200; short-term rentals 52–53, 55, 56;
   South Street Seaport 208, 227; tourism
   as gentrifying process 5, 255; *see also*
   super-gentrification
German immigrants 155–156, 157, 166,
   170, 173n3
ghettos 158–159
Girard, A. 209
Girard, Christophe 312
Giraud, C. 306
Giuliani, Rudolph 13
Glass, Ruth 1, 278
globalization 3, 48, 111, 276, 323;
   capitalist 13; Cartagena de Indias 79,
   88; China 280; flows of 1; institutional
   connections 278–279; tourism practices
   and policies 209

González-Pérez, Jesús 25–51
González, S. 107
Goss, J. 216, 217, 218–219
Gotham, Kevin 3–4, 53, 75, 226, 259,
   276–277, 278–280, 295
governance 1, 198–201, 321, 323; *see also*
   local government; policies; regulation
Gravari-Barbas, Maria 1–21, 217,
   299–326
Greece 236, 237–245, 247–248
Grunenberg, H. 190
grunge authenticity 164
Guanaybo 27
Guggenheim museum 11–12, 136
guided tours 90, 155, 172–173
Guimarães, Pedro 255–275
Guinand, Sandra 1–21, 207–232

Haarich, S. 11
Hackworth, Jason 53, 154
Hadid, Zaha 139
Hall, M. 111
Hall, Peter 208, 215
Hannam, K. 110
Hannigan, J. 126
Harvey, David 107, 226, 255
Havana 26, 27, 81
Heinich, N. 136
Heinrichs, H. 190
Helsinki 6
heritage 9–11, 15, 27, 75, 233; Bangkok
   238; Caribbean 47; Cartagena de Indias
   79; Colonial City 25, 29–30, 34, 36,
   37, 40, 42, 48; domestic architecture
   234–235; international institutions 78;
   Latin America 79; Le Marais 302, 303,
   311, 312, 318; "musealization" of 40;
   New York Tenement Museum 153, 155,
   158–159; quality of 78; Rethemnos 245,
   247; Rome 137, 138, 139–140, 146;
   self-gentrification 236; South Street
   Seaport 217
heritagization 8, 15; artification 136; Le
   Marais 4, 299, 300–305, 311, 323
Herzfeld, Michael 3, 233–252
historic city centres: Bangkok 237;
   Barcelona 117, 119, 120, 122, 124;
   Caribbean cities 26–27; Cartagena de
   Indias 76–78, 81–84, 86–87, 93–95, 98,
   99n5; investment in 257; Latin America
   75, 78, 79; Lisbon 256, 258, 260–262,
   263–265, 266, 269; Le Marais 299–324;
   Rome 138, 139
historiography 233, 238–239, 240
Holm, A. 189

hostels: Cartagena de Indias 86–88, 92, 98; Lisbon 258, 260, 263–264
hotels: Barcelona 122, 128n2, 265; Beijing 291; Berlin 186–187, 194; Caribbean cities 25; Cartagena de Indias 76, 79, 82, 85–88, 90, 92–94, 96, 98; Colonial City 36, 37, 38–40, 47; Crete 242; Latin America 75; Lisbon 263–264, 266, 267–268, 269; New Orleans 53; New York 172; Reykjavík 64; Shenzhen 286; South Street Seaport 213, 227; tourism gentrification 279
housing: Airbnb gentrification 7; Beijing 288, 289, 291; Berlin 184–185, 189, 194–195, 198, 199; Cartagena de Indias 84, 92, 93, 94; China 276; Colonial City 28, 29–32, 33–36, 41, 42–46, 48; competition for 112; impact of short-term rentals 55, 56, 199; "Latino gentrification" 79; Lisbon 256, 262; Le Marais 302; New York 163, 164–165, 174n5; policies 1; Raval 108, 116, 117–118, 122–126, 127; Rethemnos 241, 243; Reykjavík 52, 59–61; right to 109; Shenzhen 281, 283, 284–285, 286; South Street Seaport 211, 212–213, 227; *see also* accommodation; real estate; rent gap; rent increases; second-home ownership
Howard Hughes 207–208, 210, 219, 221, 223, 226–228
human rights 32
Hurley, Andrew 215
Huxtable, Ada Louis 212
hyper-aestheticization 321, 323
hyper-tourismification 299, 300, 319, 320, 321, 323

identity: American 163, 170; art 137; class 279; elites 6–7; European 237–238, 248; indigenous 235; Lisbon 269; loss of 218; Lower East Side 173; South Street Seaport 221, 223
ideology 237
IMF *see* International Monetary Fund
immigrants: Beijing 291, 292; Berlin 196; Greece 247; Lisbon 269; New York Tenement Museum 17, 153–154, 155–156, 157–162, 168–169, 170, 172–173; New York's history 163–164; Raval 107–108, 119–120, 121, 128n3; Shenzhen 281; *see also* migration
individualism 236, 290
"induced-displacement" 2
inequalities 112, 121–122

Inter-American Development Bank 25, 78
interim interventions 223–224
International Development Bank 78
international institutions 78
International Monetary Fund (IMF) 31
International Tourist Zones (ITZs) 315, 323, 324n15
internet 110–111, 182, 199; *see also* Airbnb; social media
interstitial tourism 144
investment: Cartagena de Indias 79, 82–85, 86, 98; China 276, 282–284, 285–286, 288, 291, 293–294; circle of 78; Colonial City 42, 46, 48; entrepreneurial city 222; increase in urban tourism 257; Latin America 78; Lisbon 256–257, 262, 266, 268, 271; neo-liberalization 75; New York 166–168, 169, 173; Raval 115, 122, 123–124; super-gentrification 307; touristification 260
Inzulza-Contardo, J. 79
Irish immigrants 155–156, 162, 166
Italian immigrants 154, 155–156, 157, 158, 162, 168, 170
ITZs *see* International Tourist Zones

Jacobs, Jane 110
Jaramillo Giraldo, L. 99n5
Jewish immigrants: Le Marais 300, 311, 312, 317; New York Tenement Museum 154, 155–156, 157–158, 162, 168, 170
Johannesburg 136
Jones, G. 27
Judd, D.R. 11, 16

Kagermeier, Andreas 16, 54, 181–206
Kaplan, R.A. 54
Kessner, Thomas 156
Kingston 27
Kirkup, M. 260
Koolhaas, Rem 304
Kuala Lumpur 11

land-use planning: Cartagena de Indias 92, 94; Colonial City 33, 36, 48; *see also* planning
landlords 66–68, 243–244, 319
Latin America 26–27, 40, 42, 75, 78–80, 98, 161
Laughlin, Tim 171–172
Le Galès, Patrick 222
Leahy, Nicholas 171
Lees, L. 12, 78, 99n9, 210, 268, 299, 307–308
Liang, Zeng-Xian 12, 13, 276–298

Liedtke, A. 192
lifestyles 13, 111, 112, 126, 279;
    artification 137; Beijing 288, 291;
    elites 6–7; Lisbon 269; multiple 209;
    Shenzhen 281, 282
Lindsay, Mayor 212
Lisbon 7, 13, 15, 138, 255–275
liveability 109, 112, 118, 127
Liverpool 8
local actors 3–4, 198, 235–236, 257
local government 16, 78, 222, 263,
    282–283, 287, 290, 291–293; *see also*
    governance
local residents 68, 237–238; Bangkok 238,
    246–247, 248; Beijing 288, 290, 291,
    292; Cartagena de Indias 85, 90, 92,
    95–97, 98; Lisbon 256, 259, 265–266,
    268–271; Le Marais 302, 319, 321,
    323–324, 325n18; New York 163; Raval
    114–115, 122, 127; retail gentrification
    183; Rethemnos 238, 241–245, 248;
    Reykjavík 66–67, 69; Rome 146–147;
    self-gentrification 236–237, 248;
    South Street Seaport 216, 220; tourism
    gentrification 279; *see also* low-income
    households
localization 3, 278–279
London 2, 278; Airbnb 54, 55–56, 63,
    188; hotel room costs 187; Le Marais
    compared with 306, 321; super-
    gentrification 307; Tate Modern 136
Lonely Planet 141
low-income households: Berlin 189;
    Cartagena de Indias 80, 85; China 279,
    284, 291; Colonial City 42–44, 45; New
    York 174n5; *see also* local residents;
    working class
Lower East Side (New York) 11, 17,
    153–173
Lower Manhattan Development
    Corporation (LMDC) 154, 160, 162,
    166–168, 173n2, 173n4
Lucas, L. 209
luxury leisure developments 11–12

MacCannell, D. 7
Madrid 16, 63
Maitland, R. 110, 112
Manhattan 153, 162–168, 169; Lower
    Manhattan Development Corporation
    154, 160, 162, 166–168, 173n2, 173n4;
    South Street Seaport 9, 12, 207–209,
    210–228; *see also* New York
Le Marais 4, 7, 13, 15, 299–326

Marcuse, P. 260
marginalization 213
Marino, Ignazio 148n10
marketing 11, 12, 134; experiential
    economy 209; Lisbon 256, 263; Lower
    Manhattan museums 160; Le Marais
    315; South Street Seaport 222, 225–226
Marseille 10
Martinotti, G. 112, 255
masculinity 239
Massey, D. 110
Master Plan for the Colonial City (MPCC)
    25, 33–36, 40, 48, 49n4
Mathiesen, A. 60
Matthey, Laurent 222
McCarthy, Mary 324
McNeill, D. 109
Mejía, M. 32
Mele, Chris 164
memorialization 162
Mendes, Luis 255–275
Mermet, Anne-Cécile 52–74
metropolization 1
Mexico 75, 78–79
Michel, B. 54, 188, 259
middle class 3, 17, 53, 260; Barcelona 119,
    122, 123, 127; Beijing 287, 288, 291,
    292; Cartagena de Indias 81, 83, 85;
    China 236, 279; Colonial City 32, 36;
    Greece 247; international postmodern
    112; "Latino gentrification" 79; Lisbon
    258, 270, 271; London 278; Le Marais
    305; Shenzhen 281, 284, 286; super-
    gentrification 276, 307; theme parks
    285; tourism gentrification 279; *see also*
    social class
migration 111; Berlin 189; Raval 10,
    116–117, 118, 119–120, 121–122,
    127, 128n4; Rethemnos 241; *see also*
    immigrants; mobility
Millspaugh, Martin 8
Mitchell, A. 260
mixed-use neighbourhoods 2; Cartagena
    de Indias 92, 93, 94; Colonial City 29,
    36; Latin America 79
mobility 111, 209; changes in 2, 257;
    lifestyle 126; transnational 6, 17, 110,
    117; *see also* migration
"monumental time" 233, 234
monuments: Bangkok 245; Cartagena de
    Indias 76, 77, 81; Le Marais 302, 303;
    Rethemnos 242; Treaty of Montevideo
    99n16
Morel, E. 32

Moreras, J. 120
mortgages 60, 68
MPCC *see* Master Plan for the Colonial City
"multi-unit hosts" 55, 65
municipal departments 95
Múñoz, Francese 109
"musealization" 40, 48
museums 8, 16–17, 137; Bilbao 11–12;
    Cartagena de Indias 76; Colonial City
    35; Manhattan 160, 162, 163–164; Le
    Marais 305, 318, 320, 322; New York
    Tenement Museum 10–11, 16–17,
    153–177; Raval 113; Reykjavík 59;
    Rome 139, 141, 142–144, 146, 147;
    South Street Seaport 212, 213, 215–216,
    219, 227

Nadler, M.L. 54
Nanjin 259–260
Naukkarinen, O. 134–135
neoliberalism 1, 3, 11, 27, 48, 75, 134,
    276; Airbnb 55; Barcelona 107,
    109; Caribbean cities 26; China 280;
    Colonial City 36; inequalities 112;
    Latin America 78; Lisbon 268; South
    Street Seaport 209; tourism growth
    strategy 255
New Orleans 3, 53, 75, 226, 276–277
New York 13, 17; Airbnb 7, 63; Le Marais
    compared with 306, 321; One World
    Tower 11; South Street Seaport 9, 12,
    207–209, 210–228; super-gentrification
    307; Tenement Museum 10–11, 16–17,
    153–177
"niche spaces" 134
nightlife 183, 188, 291, 318; *see also*
    entertainment
9/11 terrorist attacks (2001) 159, 162–163,
    164, 219
Nom 248
Non, A. 236
nostalgia 14, 217, 279; Beijing 288;
    China 281; Le Marais 312; New York
    164, 173

OCTs *see* Overseas Chinese Towns
"off-the-beaten-track" tourism 5, 8,
    54, 126, 134, 182; artification 137;
    collaborative consumption 181;
    marketing strategies 209; post-tourism
    14; Rome 141
Olympic Games (Barcelona, 1992) 10,
    108, 112, 138
Organization of American States 78

Overseas Chinese Towns (OCTs) 12,
    280–295
overseas development cooperation 32–33

Panama 79, 235
Paniagua, R. 79, 80, 84, 99n12
Papandreou, Andreas 240
Pappalepore, I. 183
Paris 17, 138; Airbnb 16, 54, 55, 188; art
    and culture 136, 137; hotel room costs
    187; Le Marais 4, 7, 13, 15, 299–326
Paris, C. 54
Pavel, F. 258, 264–265
pedestrian areas: Colonial City 35, 38,
    41; Le Marais 311, 316; South Street
    Seaport 212
personality profiles of Airbnb users
    191–192
Philips, W-J. 7
Piano, Renzo 139
Pinçon-Charlot, M. 316
Pinçon, M. 316
Piñeros, Sairi T. 75–103
place: global sense of 110; loss of 218;
    "place luck" 16; place-making 224
planned evictions *see* evictions
planning: Cartagena de Indias 76, 92, 94,
    99n6; Colonial City 25, 27, 31, 33–36,
    40–42, 48; Lisbon 256, 262, 263, 268;
    Le Marais 302, 303; Raval 113–116;
    Rome 138; shifts in planning practices
    222; South Street Seaport 210, 211–212,
    213; *see also* policies
Plaza, B. 11
plazas 96–97, 211
policies: Beijing 287; Cartagena de Indias
    98; China 276; Colonial City 28, 29,
    31–47; Iceland 60; Latin America
    78–79; Lisbon 256–257, 262, 268; Le
    Marais 303; New York 208, 210; Rome
    135, 138, 140, 146; Shenzhen 282,
    283–284; *see also* planning; regulation
politics 239–240, 243, 248
pollution 81, 82
"poly-topical" living 5–6
Pom Mahakan 237–241, 245–248
pop-up events 223, 224
population growth 1; Beijing 289;
    Cartagena de Indias 81; Colonial City
    36; Lisbon 260; New York 227, 228n7;
    Raval 121; Shenzhen 281
ports: Cartagena de Indias 76, 99n3;
    Colonial City 35, 41; Manhattan 210;
    see also waterfronts

Posso, L. 80
post-industrial societies 285, 287, 290
post-materialism 182
post-tourism 14, 208, 209
Potsdamer Platz 13
poverty: Bangkok 247; Colonial City 27,
    30, 48; Greece 247; Rethemnos 240,
    241, 242, 243–244, 245
preservation 279; China 277; Lower East
    Side 173; Le Marais 302, 304, 324;
    South Street Seaport 212, 213, 218, 219;
    see also conservation
prices: Airbnb 193–195; hotels 187; real
    estate 65–66, 115, 124, 128n1, 163, 189,
    195, 199, 260, 270, 286, 291, 308–310;
    see also rent gap; rent increases
privatization 27, 75; Latin America 78;
    Santo Domingo 40; South Street
    Seaport 227
property rights 276, 293
"prosumers" 182
protests: Airbnb 188; Barcelona 109, 113;
    Berlin 188–189; Le Marais 312; New
    York Tenement Museum 154; *see also*
    activism; resistance
public space: Beijing 290; competition in
    access 112; Lisbon 258, 263, 269, 270;
    South Street Seaport 213, 215, 216,
    217–218, 223
Puebla 27
Puerto Rican immigrants 168, 170
Puerto Rico 27

Quaglieri Domínguez, Alan 10, 15,
    107–133

Rabinowitz, Richard 156
Raggi, Virginia 148n10
Raude, Michel 302
Raval 9, 10, 107–108, 112–127
real estate: Airbnb impact on 16, 54,
    65–68, 199; Baltimore 8; Berlin 189,
    195; Bilbao 12; Caribbean cities 25;
    Cartagena de Indias 79–80, 83, 85;
    China 277, 279–280, 286; Colonial City
    25, 46, 48; Greece 248; institutional
    connections 75; Latin America 78;
    Lisbon 256, 259–260, 263, 266, 270,
    271; Le Marais 303, 307, 308–310,
    320, 323; New York 154, 166,
    173, 227; Raval 10, 108, 114–115,
    122–126, 127, 128n1; Rethemnos 242;
    Reykjavík 52–53, 57–58, 59–61, 65–68;
    speculation 1, 30, 32, 43–44, 48, 55, 56,

79–80, 85, 128n1, 227; World Heritage
    Sites 26; *see also* accommodation;
    housing
regeneration *see* urban regeneration
regulation: Airbnb market 55, 67–68,
    184–185, 198–199; Cartagena de Indias
    81, 84, 92, 99n6; Lisbon 257, 263, 267;
    Le Marais 302; *see also* governance;
    planning; policies
rehabilitation 75, 78; Barcelona 109,
    115, 123–124, 126–127; Cartagena de
    Indias 84; Colonial City 34; Lisbon 255,
    256–257, 258, 261–262, 263, 267, 268,
    270; Le Marais 303, 305; South Street
    Seaport 212, 216; *see also* restoration;
    revitalization; urban regeneration
rent gap 78, 257, 276; Cartagena de Indias
    98; China 277, 291–295; Le Marais 305;
    Reykjavík 65–68
rent increases: Beijing 295; Berlin 187,
    188, 189, 195, 198; Lisbon 259, 268,
    270; Le Marais 319; Shenzhen 285,
    294–295; South Street Seaport 218
rental accommodation 54–56; Barcelona
    122, 124–125, 127; Colonial City 37,
    39, 48; Lisbon 258, 259–260, 264–265,
    271; Le Marais 307, 318, 320, 323;
    Reykjavík 52–53, 56–69; *see also*
    Airbnb; short-term rentals
representation 154, 155, 161, 172–173
resistance: Bangkok 238, 240, 246–247;
    Colonial City 32; Greece 239; New
    York Tenement Museum 159–160, 161–
    162; super-gentrification 299; Thailand
    239; *see also* activism; protests
restaurants 4, 8, 13, 75, 279; Beijing 291;
    Berlin 188; Cartagena de Indias 76, 83,
    85, 86, 88, 90–91, 92, 94, 96; Colonial
    City 29, 37, 47; "explorer-tourists" 183;
    Lisbon 256, 258, 265–266, 267–268,
    269; Lower East Side 166; Le Marais
    315, 318, 320; opening hours 200; Raval
    118; Reykjavík 59; sharing economy
    200; South Street Seaport 216, 219,
    224–225
restoration 279; Cartagena de Indias 79,
    81–82, 83–85, 86, 88; Le Marais 302,
    303, 306, 321; Rethemnos 244; Rome
    138, 139; South Street Seaport 212; *see
    also* rehabilitation; revitalization; urban
    regeneration
retail *see* shops and retail
Rethemnos 17, 237–245, 247–248
revanchism 211

revitalization 8, 27, 279; Barcelona 108, 120; Colonial City 25, 34, 48; New York 153, 162–163; Rome 140; South Street Seaport 219; *see also* rehabilitation; restoration; urban regeneration
Reykjavík 16, 52, 56–69
Riis, Jacob 159
Rinaudo, C. 81
Rio de Janeiro 75
Roberts, P. 257
Rockefeller, David 210, 212
Rodrigues, W. 258
Rofe, M. 6
Rogers, Roo 181–182
Rojas, E. 88
Rome 16, 135, 137–147
Roseman, Barry 168
Rotterdam 306
Rouse, James 8, 207, 208, 214, 216, 218, 222, 224, 228
Rubino, S. 75
Russo, A.P. 110
Rutelli, Francesco 138, 139

"safeguarded sector" status 302, 303, 305
Saint George 26
Salvador Bahia 79
Sampson, Elissa 153–177
Samudio, A. 81
San Francisco 7, 54, 55, 63, 69n5
San Juan 26, 27, 81
Sanders, J. 224
"sanitization" 213, 216
Santa Cruz de Mompox 26, 49n2
Santo Domingo 8, 13, 15, 25, 26, 27–49, 79
Santurce 27
Scarnato, Alessandro 10, 15, 107–133
Scarpaci, J. 90
Schaller, S. 223
Scheppe, W. 303, 319
schools 93
second-home ownership 5, 17, 54; Cartagena de Indias 10, 79, 83; Colonial City 44, 48; Le Marais 307, 321; Shenzhen 281
segregation: Barcelona 117; Colonial City 25, 27, 33, 36, 40, 46; forced displacement 27
self-gentrification 17, 235–237, 243–244, 246, 248
self-managed social centres 141
service sector 136

Shanghai 12, 248
Shapiro, R. 136
sharing economy 7, 16, 17, 54, 55, 181–206; *see also* Airbnb
Shenzhen 280–287, 294–295
Shin, H.B. 277
SHoP Architects 219, 221
shops and retail 4, 13, 75, 279; Beijing 290, 291; Berlin 188; Cartagena de Indias 76, 83, 86, 88–89, 94, 95–96; Colonial City 29, 30, 35, 37, 47, 49n3; "explorer-tourists" 183; Lisbon 256, 258, 260, 263, 266–268, 269–270, 271; Le Marais 310–316, 318, 319, 321, 323; New Orleans 53; New York 163–164, 166, 171–172; Raval 118; retail gentrification 183; sharing economy 200; Shenzhen 286; South Street Seaport 207, 217, 218, 219; *see also* commercial gentrification
short-term rentals 52–53, 54–56; Barcelona 124–125, 127; Berlin 183–201; Lisbon 258, 260, 264–265; Le Marais 307; Reykjavík 56–69; *see also* Airbnb; rental accommodation
Sibalis, M. 305
sightseeing tours 90
Sklair, L. 13
Skowronnek, A. 194
Smith, N. 53, 98, 164, 257, 276
social capital 258
social class 2, 276, 279; China 277, 290; Colonial City 40; *embourgeoisement* 243; image of community 226; Lisbon 259; super-gentrification 210; *see also* low-income households; middle class; working class
social media 5, 14, 182, 199, 224; *see also* TripAdvisor
social problems: Cartagena de Indias 83, 92; Raval 107, 108, 113, 115, 126
social spaces 96–97
"social time" 233, 234
SoHo 4
Sorkin, Michael 215
South Street Seaport 9, 12, 207–209, 210–228
souvenirs: Colonial City 47; Lisbon 266, 267, 269; Le Marais 320; New York Tenement Museum 172; super-gentrification 300
Spain 33, 80; *see also* Barcelona
squares 96–97, 113–114, 138
St. John's 27

Stampfl, Nora 182
Stanford, Norma 212
Stanford, Peter 212, 213
"starchitecture" 8, 11, 139, 303
"state-sponsored gentrification" 276
Steel, Robert K. 173n4
Stock, M. 5–6, 209
Stors, Natalie 16, 54, 181–206
storytelling 222
street art 142–143, 144–145, 147, 163
street sellers 42, 82, 83, 245
streets: Cartagena de Indias 96; Colonial
    City 29, 30, 35, 37; Raval 115
suburbanization 277, 278, 295
suburbs: Barcelona 117, 118; Cartagena
    de Indias 80; China 295; gentrification
    in 278
super-gentrification 278, 299–300; Airbnb
    7; China 276–277, 280, 286, 295; Le
    Marais 7, 300, 307–310, 320, 323; New
    York 12, 208, 210, 227; North America
    279; *see also* gentrification
surveillance 27, 42
sustainability 182, 209, 226
symbolic capital 11, 107, 311

T-Spoon 140
Talinn 2
Tate Modern 136
tax incentives 79, 80; Cartagena de Indias
    84, 86, 98; Le Marais 307; New York
    169, 174n5
tempo 233
tenant rights 161–162
Tenement Museum (New York) 16–17,
    153–177
Testaccio – Ostiense 146–147
Thailand 237–241, 244, 245–248, 249n11
theme parks 13, 98, 279; Beijing 288,
    290–291; Shenzhen 281, 283, 284,
    285–286
Theodossopoulos, D. 235
Thoem, J. 55
Thompson, Benjamin 207, 214
time 233–234
Times Square 13
Toronto 13
"tourism playgrounds" 4
touristification 27; Airbnb 196; Barcelona
    10, 108; Berlin 187–188; circle of
    investment 78; Colonial City 30, 33,
    36–37, 40; hyper-tourismification 299,
    300, 319, 320, 321, 323; Lisbon 255,
    257–260, 269, 270–271; Le Marais 299,
    303, 318, 323; Rome 135, 139–140, 147

tours 90, 155, 172–173
traffic: Cartagena de Indias 84, 92;
    Colonial City 35, 39–40, 41;
    Lisbon 262
Trinidad 26, 49n1
TripAdvisor 90, 317–319, 320
Trujillo, President 31
Tulinius, K. 60

UNESCO 75, 78; Caribbean cities 26;
    Cartagena de Indias 15, 76, 77, 83–84;
    Colonial City 25, 33; Le Marais 303;
    Raval 116; Rome 137
Uniqlo 304, 312–314
United Kingdom 293; *see also* London
United States 13, 277, 293; *see also*
    New York
urban regeneration 1, 8, 15, 134;
    Caribbean cities 25; Colonial City 25,
    28, 31–47; "cultural turn" 135–137;
    Lisbon 255, 256–258, 259–260,
    263, 268, 270; Rome 139, 140, 146;
    South Street Seaport 208; *see also*
    rehabilitation; restoration; revitalization
urban sprawl 1, 28, 59
"urbanalization" 109
urbanization 280, 284, 285, 287,
    291–294, 295
Urry, J. 111

Vallat, C. 138
Valley de los Ingenios 26, 49n1
values 182, 222, 226, 228, 241
Varley, A. 27
Veblen, T. 7
Veltroni, Walter 138
Venice 321
Vertovec, S. 111
Vieux Carré 3, 53, 75, 226, 276–277,
    280, 293
visitor numbers: Beijing 291; Berlin 184,
    186–187; Colonial City 36, 37, 40;
    Iceland 58; Lisbon 263; Le Marais 317;
    New York Tenement Museum 159, 169,
    173n1; Shenzhen 285, 286; South Street
    Seaport 207
Vivant, E. 136

wastelands 16, 134–135, 136, 137, 139,
    140–141, 147
waterfronts 2, 13, 278; Baltimore 8;
    Colonial City 40; Reykjavík 58–59;
    South Street Seaport 9, 12, 207–209,
    210–228
While, A. 109

Willemstad 26
working class: Beijing 288, 289–290; Cartagena de Indias 81; Colonial City 37, 48; Greece 248; London 278; Le Marais 300; new tourism experiences in working-class neighbourhoods 134; Raval 116, 117, 127; Rome 138, 146; Shenzhen 281, 283, 285, 286; tourism gentrification 279; *see also* low-income households; social class
World Bank 31, 78

World Heritage Sites 75; capital investment 78; Caribbean cities 26; Cartagena de Indias 15, 76, 77, 83–84; Colonial City 25, 33; Le Marais 303; Raval 116; Rome 137

Zea, Gloria 83
Zhao, Y. 259–260
zoning 35, 36, 46, 48
Zukin, Sharon 164
Zuleta Jaramillo, L.A. 99n5